OPTIMIZATION

OPTIMIZATION

*Symposium of the Institute of Mathematics and Its
Applications*

University of Keele, England, 1968

Edited by

R. FLETCHER
A.E.R.E. Harwell

1969

ACADEMIC PRESS · LONDON AND NEW YORK

ACADEMIC PRESS INC. (LONDON) LTD
Berkeley Square House
Berkeley Square,
London, W1X 6BA

U.S. Edition published by
ACADEMIC PRESS INC.
111 Fifth Avenue,
New York, New York 10003

SBN: 12-260650-7

Library of Congress Catalog Card Number: 79-92408

Printed in Great Britain by
HARRISON AND SONS LIMITED
BY APPOINTMENT TO HER MAJESTY THE QUEEN,
PRINTERS, LONDON, HAYES (MIDDX.) AND HIGH WYCOMBE

Contributors

J. ABADIE, *Electricité de France and Institut de Statistique de l'Université de Paris, Paris, France*

V. DE ANGELIS, *Department of Mathematics, University of Birmingham, Birmingham, England.*

Y. BARD, *IBM Ltd, New York, U.S.A.*

E. M. L. BEALE, *Scientific Control Systems Ltd., London, England.*

B. BERNHOLTZ, *Department of Industrial Engineering, University of Toronto, Toronto, Ontario, Canada.*

J. CARPENTIER, *Electricité de France and Institut de Statistique de l'Université de Paris, Paris, France*

W. C. DAVIDON, *Haverford College, Haverford, Pennsylvania, U.S.A.*

D. DAVIES, *I.C.I. Ltd., Wilmslow, England.*

R. FLETCHER, *Mathematics Branch, Atomic Energy Research Establishment, Harwell, England.*

J. L. GREENSTADT, *IBM Ltd., New York, U.S.A.*

D. GOLDFARB, *Courant Institute of Mathematical Sciences, New York University, New York, U.S.A.*

P. HUARD, *Conseiller Scientique, E.D.F. Service, I.M.A., Paris, France.*

L. P. HYVARINEN, *IBM European Systems Research Institute, Geneva, Switzerland.*

T. O. M. KRONSJO, *University of Birmingham, Birmingham, England.*

A. P. MCCANN, *Computing Laboratory, Leeds University, Leeds, England.*

G. P. MCCORMICK, *Research Analysis Corporation, McLean, Virginia, U.S.A.*

W. MURRAY, *National Physical Laboratory, Teddington, England.*

B. A. MURTAGH, *Imperial College, The University, London, England.*

J. D. PEARSON, *Research Analysis Corporation, McLean, Virginia, U.S.A.*

M. J. D. POWELL, *Mathematics Branch, Atomic Energy Research Establishment, Harwell, England.*

J. D. ROODE, *Atomic Energy Board, Pelindaba, Pretoria, S. Africa.*

v

R. W. H. SARGENT, *Imperial College, The University, London, England.*

L. E. SCHWARTZ, *Center for Naval Analyses, The University of Rochester. Arlington, Virginia, U S.A.*

B. M. E. DE SILVA, *Mechanical Engineering Laboratory, English Electric Company, Leicester, England.*

W. SPENDLEY, *ICI Ltd, Billingham, England.*

W. H. SWANN, *ICI Ltd., Wilmslow, England.*

G. S. TRACZ, *Department of Industrial Engineering, University of Toronto, Toronto, Ontario, Canada.*

A. W. TUCKER, *Princeton University, Princeton, New Jersey, U.S.A.*

G. M. WIENBERG, *IBM European Systems Research Instute, Geneva, Switzerland.*

P. WOLFE, *IBM Ltd., Yorktown Heights, U.S.A.*

Preface

The Keele Conference

In 1967, Philip Wolfe, mathematical programmer from IBM Yorktown Heights, reported on the differences between the study and use of optimization techniques in the U.S. and U.K.: "There has been a considerable difference in the approaches taken in British and American research in optimization problems (by which I mean here problems of the mathematical programming type, in reasonable numbers of variables—say, one to ten thousand—rather than problems in the calculus of variations). American work has usually been closely related to operations research problems, having many variables for which non-negativity is an important constraint, and whose other constraints are often nearly linear. British work has concentrated on problems arising from engineering and from the chemical industry, having relatively fewer variables, for which non-negativity is often unimportant, but which are involved in highly non-linear objective functions and constraints; and most of the work has been concerned with maximizing a function without constraints. (This division is of course a vast oversimplification, and does violence to much work on both sides of the Atlantic; but it is true that most of the recent literature on procedures for unconstrained optimization has appeared in the *Computer Journal*.) The different kinds of problems have led to the development of quite different computational techniques, which I guess are each appropriate for their problems; but as British problems get bigger and America's get kinkier, we are going to have to merge our techniques."

The Keele conference on Optimization and Nonlinear programming was organized to facilitate this sort of interaction between the "hill climbers" and the "mathematical programmers".

Don Davies agreed to supervise the organization of the conference and Harry Greenwood offered hospitality on behalf of Keele University and undertook a lot of local organization. The conference was sponsored jointly by the British Computer Society, who agreed to deal with publicity and promotion, and the Institute of Mathematics and its Applications who dealt with the secretarial work and who really organized the whole thing. Bill Sherman put in a great deal of his time in this respect. Norman Clarke, secretary of the I.M.A. found a publisher for the proceedings and I agreed to edit on behalf of the I.M.A. Thus anyone who buys this book is contributing to the finances of the I.M.A.: an organization which fulfills a great need among the numerical analysts of the United Kingdom.

The programme was formulated by a committee consisting of those people mentioned in the last paragraph and Mike Box, Martin Beale, Mike Powell and Walter Murray. Contributed papers were circulated amongst members of the committee and discussed at a meeting in December 1967. There was too much material, and it was decided that no papers on integer programming should be accepted. Even so a number of interesting papers had regrettably to be rejected. Also many authors took advantage of discussions at the conference to revise their papers, though, in the interests of publishing the papers as quickly as possible, another round of refereeing was dispensed with.

The conference was held from 25th March to 28th March 1968. The arrangement of the programme mirrored to some extent the different backgrounds of the participants. Papers on unconstrained optimization, constrained optimization by hill climbing methods and constrained optimization by Simplex-like methods were presented. An applications session was also held. The papers are published here in the order in which they were read. At the last moment it transpired that Professor Zoutendijk would be unable to be present which was a great disappointment to us all. However, his colleague Dr Roode provided an excellent substitute. Another speaker who does not appear in these pages is Professor Vajda who spoke most amusingly at the conference dinner.

The second half of this preface is based on the notes used by Philip Wolfe for his lecture at the Conference in which he reviewed progress in each separate field and relating them to each other.

The final session of the conference was a panel discussion chaired and lead by Phil Wolfe with a number of eminent personalities. With comments interjected from the floor, an interesting discussion developed on many important topics in the subject. The whole was tape recorded, edited, and it appears as the final chapter of this book. Amongst many other things, you can learn what introductory texts are available in the subject and where computer programs can be obtained. The tape recorder was also put to good use in recording the discussions which followed many of the papers. These are given here following the paper to which they refer. On some occasions the quality was not very good and it has not been possible to make out the speaker's name. In such cases he has been ranked with the immortals as ANON.

As I write now over a year later, it is clear that the conference marked a turning point in the development of nonlinear programming. It was obvious that we already know a great deal about conjugate gradient and Davidon rank 2 methods for unconstrained optimization, as evidenced by McCormick and Pearson's talk. However, the conference marks the beginning of the sudden growth of interest in rank 1 type methods for this problem. Davidon published a paper on this subject in the *Computer Journal* just before the conference and followed this up in his paper at the conference. Murtagh and Sargent describe the approach in their paper at the conference. Other authors also discovered the method independently at about the same time. The advantages in doing away with the need for a linear search and in the flexibility in the choice of corrections are obvious.

Another feature of the conference is that it marks a transition from penalty function methods to more direct methods for solving nonlinear programs by hill-climbing methods. My joint paper with Tony McCann emphasizes that there is a limited amount to what can be done with penalty functions, however sophisticated the approach. New and more direct approaches are being described, evident in the paper of Abadie and Carpentier describing the GRG method which is the best code in Colville's nonlinear programming study, and there is also the method described in Murtagh and Sargent's paper.

The methods which attempt to reduce nonlinear programming to a succession of linear programs are also at the cross-roads. The classical cutting plane and decomposition methods seem to be falling into disuse, perhaps because little success has been obtained with them. Separable programming is a technique which has worked well in practice and is described by Vanda de Angelis in her paper. The range of problems solvable by this technique is of course limited. Martin Beale describes the situation in his review paper on the subject.

Acknowledgements

I would like to thank all those people, far too many to mention by name, who have helped in the preparation of this book. Also may I thank the authors of the papers, who have spent much time in preparing them and in correcting proofs. Finally may I thank Norman Clarke of the Institute of Mathematics and its Applications for his enthusiasm for the publication of these proceedings.

I would like to dedicate the book to my wife Mary, who listens with sympathy to all my hard luck stories, and to my daughter Jane, of whose existence I first knew on returning home from the Keele conference.

R FLETCHER

A.E.R.E. Harwell
June 1969

Introduction

Philip Wolfe's Review of Nonlinear Programming

This is a precis of Philip Wolfe's review on the basis of some notes which he kindly provided. The sentiments are meant to be those expressed by the speaker: I hope I do not put too many words into his mouth. The problem under consideration is to

$$\text{maximize} \quad f(x) = f(x_1, x_2 \ldots, x_n)$$
$$\text{subject to} \quad g_i(x) \leqslant 0 \quad i = 1, 2 \ldots, m.$$

This is the general nonlinear programming problem, allowing that the constraint set might be empty (unconstrained optimization) or that strict equality may be asked for in the constraints. It is not however practicable to look at algorithms which will solve this general problem, rather it is better to break it down into a whole spectrum of lesser problems. If this is done it will be seen that the size of problem which can be handled varies with the complexity of the problem. This is illustrated in Fig. 1. Size here is taken to mean the number of equations plus variables, and the shaded area represents the size of problem which can be solved at each level of complexity. Linear programming problems (f and g_i all linear in x) of size 1000 and more can be solved, and if they have some simplifying structure then this figure goes up even beyond 10^6. Yet if we permit f to be quadratic then the size comes down by a factor of 10 from the linear case.

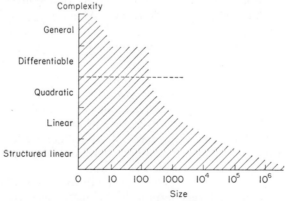

FIG. 1. Max $f(x) = f(x_1, x_2, \ldots, x_n)$ subject to $g_i(x) \leqslant 0$, $i = 1, \ldots, m$.

In many problems all that can be asked of the object function is that its derivatives be calculated. In these cases the upper limit on problem size is around 100. However on occasions the only piece of information which can be given about the object function is its value for any value of the variables. This might be the result of some physical experiment for instance, or possibly of a numerical calculation for which even the operation of differentiation defies analysis. The size of such problems which can be tackled is small indeed.

The dotted line on the graph defines the boundary between the relatively small but complex problems tackled by the "hill climbers" and the large simple problems tackled by the mathematical programmers. In the past, very little dialogue has taken place between the two camps. It is hoped that this conference will remove this dichotomy and that each shall listen to and perhaps take advantage of ideas expressed by the others.

Much early work was done assuming conditions of convexity and/or concavity on the object and constraint functions. Often these conditions were just so as to be able to assume a global maximum to the problem, and in these cases convergence to some local maximum could be relied upon when the functions did not satisfy the convexity/concavity conditions. Of course it is not often the case that these conditions are satisfied in general nonlinear programming problems. Some methods, however, may not even converge to a local maximum if the conditions are not satisfied. For example in the cutting plane method a cut may remove the maximum from the point set being considered. Only with these sort of methods need one worry if the convexity/concavity conditions are not satisfied.

Much important unconstrained optimization is based on the gradient ∇f of the object function, and methods such as steepest descents, Newton's method and the pseudo-Newton methods are mentioned in the review paper on that subject. It is as well to emphasize that there is a lot of work yet to do in this field. For steepest descents we know that convergence can be proved and also that the rate of convergence (Akaike) is

$$\frac{f(x_{k+1})}{f(x_k)} \approx c \left(\frac{\Lambda - \lambda}{\Lambda + \lambda} \right)^2$$

where Λ and λ are upper and lower bounds on the eigenvalues of the Hessian. Similarly the order of convergence is known to be 1. However for many of the other methods, especially the newer variable metric methods, we know very little indeed about convergence properties. To these should be added questions about the relevance of the "quadratic termination" properties which are often proved, and perhaps questions about what happens when the functions are near quadratics.

When linear constraints are added to an unconstrained problem, methods using the gradient are still very important. In these cases the problem can be considered as optimization of a function on the linear manifold defined by the intersection of some of the constraints. Thus we can project everything into this manifold, getting the gradient projection method of Rosen, and Goldfarb's version of the variable metric method. Alternatively, new variables can be defined which are the normals to some of the constraints and the gradient transformed into this basis. This leads to the Reduced Gradient method. Convergence proofs for these methods, however, are more hazardous. For instance, it is possible to construct a function for which convergence occurs to a point which is not a local maximum or even a stationary point. Again work on convergence proofs is lacking here: thinking geometrically about these problems is dangerous.

The great theorem of mathematical programming concerns duality and generalized Lagrange multipliers. Although duality presents interesting results about the structure of the problem, it has been of little use in suggesting methods for nonlinear programming. Lagrange multipliers have been used, notably by Arrow *et al.* (as described in Chapter 12), however, the technique has never really caught on for solving practical problems.

Mathematical programmers like to have techniques which reduce nonlinear problems to a sequence of linear problems which can be solved. An example of this is the Frank–Wolfe algorithm for quadratic programming about which nice convergence results can be proved (Canon and Cullum). However, the convergence is in fact first order, and the technique is now superseded by Beale's method.

For general nonlinear problems there are two ways in which one can perform a linearization. One is to linearize the functions by

$$f(x) \approx f(x^k) + \nabla f(x^k) \cdot (x - x^k)$$

and solve the linear program obtained. The point obtained is used to construct further linearizations giving rise to a larger linear program which is again solved, and so on. This is the cutting plane method; each linearization of a constraint cuts off a part of the feasible region in which the solution cannot lie. The computation is carried out using the dual simplex method which readily permits the addition of extra equations as described. The method only converges at first order however and is fraught with numerical difficulties. Another method based on this type of linearization is the method of Approximation Programming described briefly in Chapter 3.

An alternative approach is to linearize the variables, that is

$$x \approx \sum_k s_k x^k, \qquad s^k \geqslant 0, \qquad \sum_k s_k = 1$$

so that functions can be replaced by the linearizations

$$f(x) \approx \sum_k s_k f(x^k).$$

This is the basis of the decomposition method, which is dealt with in some detail in Chapter 3. The particular importance of decomposition is in separable programming in which it is assumed the nonlinear functions are

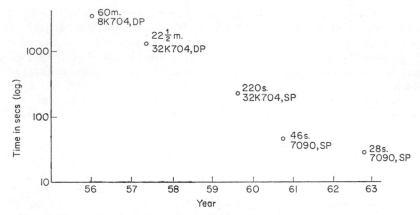

FIG. 2. Computing times for a linear programming problem with 67 equations.

separable as functions of a single variable, that is

$$f(x) \equiv \sum_{j} f_j(x_j)$$

which are then linearized as described. Separable programming is described further in both Chapters 3 and 4 of this book.

So much then for the typical approaches to the solution of nonlinear programming problems. Although we possibly do not yet have the ultimate weapon for solving these problems, let us take heart from our rate of progress in linear programming. Take for instance the time to solve one particular LP problem with 67 equations (Fig. 2). This varied from 60 min. on an 8K IBM 704 with double precision in 1956, to 28 sec. on an IBM 7090 with single precision in 1963. These figures are almost certainly down by another factor of 10 in recent years. The improvement is by no means all due to faster computers;

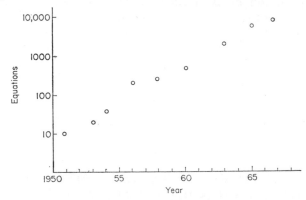

FIG. 3. Linear programming capacities.

better methods and better programming have all played their part. Similarly the size of problem handled by an LP code has increased steadily (Fig. 3). Who knows what developments await in the field of nonlinear programming?

In conclusion some remarks on the proliferation of papers in this subject might be relevant. Too many authors are publishing algorithms without relating them to other work. What is needed are far more comparisons between algorithms so that we can constantly keep in mind that our aim is to produce the best method for solving any particular problem. To this extent the comparative study of Colville has been extremely valuable, showing up many of the *ad hoc* methods in their true light and focusing attention on a small sub set of methods in which we should become really interested. The study is restricted to only a small number of problems, however, and much more information on these lines is wanted.

Contents

1. A Review of Methods for Unconstrained Optimization

R. Fletcher*

Computing Laboratory, University of Leeds, Leeds, England

1. Introduction

The problem to be considered is that of finding a local minimum of a function $F(\mathbf{x})$ of n variables $\mathbf{x}' = (x_1, x_2, \ldots, x_n)$. The gradient of F will be denoted by $\mathbf{g}(\mathbf{x})$ or $\mathbf{g}$ with ith element $\partial F / \partial x_i$, and the matrix of second derivatives by $\mathbf{G}$ with i, jth element $\partial^2 F / \partial x_i \, \partial x_j$. Some continuity condition on F is assumed, at least that of differentiability, so that the condition $\mathbf{g} = \mathbf{0}$ at a minimum can be used. Methods are all iterative and $\mathbf{x}_1, \mathbf{x}_2, \ldots$ will be used to denote successive approximations to the minimum. Similarly $\mathbf{g}(\mathbf{x}_i)$ will be written $\mathbf{g}_i$ and the notation $\Delta \mathbf{x}_i = \mathbf{x}_{i+1} - \mathbf{x}_i$ will be used.

Nearly all methods are based on the iteration

$$\mathbf{x}_{i+1} = \mathbf{x}_i + \alpha_i \, \mathbf{s}_i$$

where $\mathbf{s}_i$ is determined as a correction to $\mathbf{x}_i$ such that $\mathbf{x}_{i+1}$ be the minimum for certain simple functions. However, it may be that F_{i+1} would be worse than F_i if this correction were applied in general, hence the parameter α_i is introduced and chosen so that $F_{i+1} \leqslant F_i$. In fact α_i is usually chosen to minimize $F(\mathbf{x}_i + \alpha_i \, \mathbf{s}_i)$. One may imagine $\mathbf{x}_i$ as a point in n dimensional space (Fig. 1) and $\mathbf{s}_i$ as a "direction of search" emanating from $\mathbf{x}_i$, with $\mathbf{x}_{i+1}$ as the optimum point along this direction. A fairly foolproof algorithm for the solution of this subproblem (the "linear search") can be written: an excellent description of this has been given by Box *et al.* [1] which, together with the paper by Powell [2], provides good general reading on the subject. Adopting the iteration based on a linear search means that a method can be specified merely by showing how the direction of search $\mathbf{s}$ is calculated.

It is beyond the scope of this paper to cite actual figures for the comparisons between methods; my sources of information are the papers themselves,

---

* Present address: Theoretical Physics Division AERE, Harwell, England.

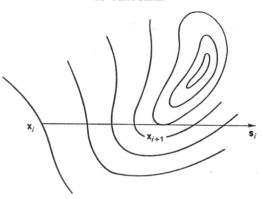

FIG. 1. The search along a line for a minimum.

reviews by Box [3] and Fletcher [4] and also experience in practical applications to a wide range of scientific problems arising in a university environment. To some extent, however, this paper is a personal assessment of the present state of the art.

2. Gradient Methods

One of the oldest methods is that of Steepest Descents in which **s** is chosen simply as the downhill gradient vector $-\mathbf{g}$. It is based on the fact that, local to the current approximation, the direction $-\mathbf{g}$ is that along which $F(\mathbf{x})$ decreases most rapidly. In practice it usually improves F rapidly on the first few iterations and then gives rise to oscillatory progress and becomes unsatisfactory. The reason lies in the failure of the theory to represent adequately functions with minima. The only functions for which the steepest descent property holds along the whole direction of search and which still have a minimum are those with spherical contours, and this does not adequately represent the minimum of a general function.

It seems reasonable then that a more broadly based assumption must be made about the nature of F, and this can be done by assuming that it is represented adequately by a quadratic function

$$Q(\mathbf{x}) = \text{constant} + \mathbf{a}'\mathbf{x} + \tfrac{1}{2}\mathbf{x}'\mathbf{G}\mathbf{x}$$

where **a** is a constant vector and **G** a constant matrix of second partial derivatives. $Q(\mathbf{x})$ has a minimum if and only if **G** is positive definite, and subsequently the term "quadratic function" will imply that a minimum is present. Following a suggestion of Philip Wolfe, I state that a method based on the properties of such a function has property Q. This supercedes the term "quadratic convergence" which is often confused with the term "second-order convergence".

In the Taylor series method (so called because it can be derived by truncating the series $g(x + \Delta x) = (x) + G\Delta x + O(\Delta x^2)$: it is also often called the Newton method, but this can cause confusion with that method for solving nonlinear equations) the properties of quadratic functions are used directly to generate a direction of search $= -G^{-1}g$. As under these circumstances the method converges in one iteration it will be said to have property $Q1$. For nonquadratic functions the process is applied iteratively, evaluating g and G at each iteration. The convergence of this type of method is very rapid in the region of any minimum when $F(x)$ is adequately represented by the first two terms of its Taylor series. However it has various disadvantages; it requires the matrix of second derivatives to be provided and calculated at each iteration, and it requires the solution of a set of linear equations to determine each direction of search. More significantly the matrix G may be locally singular, in which case the iteration breaks down, and also s may be orthogonal to g, in which case no further progress is made.

A method could be made to avoid this last failing if it was ensured that directions of search were "downhill", that is if a reduction in F could be guaranteed along s. Such a condition would be necessary to prove convergence of a method for general functions. One way to do this is to generate s from $s = -Hg$ where H is a positive definite matrix. The Taylor series method can be modified as suggested by Greenstadt [5] by finding the eigensolution of G at each iteration, replacing any negative eigenvalues by their modulus and any zero eigenvalues by unity. This method is even more disadvantageous however on the grounds of convenience and housekeeping time.

To use the properties of quadratic functions without requiring G to be calculated means that the restriction of $Q1$ must be removed, and replaced by say Qn. This gives some extra flexibility which can be used to ensure that the directions of search are downhill. One way of doing this is to invoke the properties of so-called "conjugate" directions. If the vectors $s_1, s_2, ..., s_n$ have the property

$$s_i' G s_j = 0 \quad (i \neq j); \qquad s_i' G s_j \neq 0 \quad (i = j)$$

with regard to a positive definite matrix G, then they are said to be "conjugate". Methods which generate conjugate directions of search when applied to quadratic functions have property Qn, that is they minimize a quadratic function with matrix of second derivatives G in at most n iterations (see [6]).

There are two ways in which a method can be made to generate conjugate directions.

2.1. *Parallel subspace method*

If $\mathbf{x}$ and $\mathbf{y}$ are the minima of quadratic functions in two parallel subspaces, then the minimum in the subspace of one higher dimension lies along the line joining $\mathbf{x}$ and $\mathbf{y}$. This is illustrated in Fig. 2(a) for subspaces of dimension one and two respectively. This theorem can be used recursively to generate a set of conjugate directions. Given a set of linearly independent vectors $\mathbf{p}_1, \mathbf{p}_2, ..., \mathbf{p}_n$, then from an initial approximation $\mathbf{x}_1$ the first search is made along $\mathbf{s}_1 = \mathbf{p}_1$ to give $\mathbf{x}_2$ (Fig. 2(b)). This point is displaced by adding to it a proportion of $\mathbf{p}_2$, giving a point $\mathbf{z}_2$, and the minimum in the parallel subspace is now found by minimizing along $\mathbf{s}_1$ again to give $\mathbf{y}_2$. The line $\mathbf{y}_2 - \mathbf{x}_2$ is taken as $\mathbf{s}_2$ and the minimum along $\mathbf{s}_2$ is $\mathbf{x}_3$. It can be shown that $\mathbf{s}_1$ and $\mathbf{s}_2$ are then conjugate.

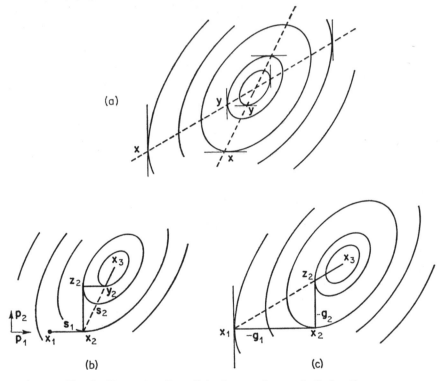

Fig. 2. Properties of parallel subspaces for quadratic functions.

When the subspace is of dimension $i-1$, then conjugate directions $\mathbf{s}_1, ..., \mathbf{s}_{i-1}$ have been accumulated such that $\mathbf{x}_i$ has been found from $\mathbf{x}_1$ as the minimum along these directions. Again $\mathbf{x}_i$ is displaced along $\mathbf{p}_i$ to give $\mathbf{z}_i$; the minimum in the parallel subspace is found by minimizing along $\mathbf{s}_1, ..., \mathbf{s}_{i-1}$

giving y_i; s_i is taken as $y_i - x_i$ and x_{i+1} is found by minimizing along s_i. Again it can be shown that s_i is conjugate to $s_1, ..., s_{i-1}$. Extrapolating to $i = n$ shows that a set of conjugate directions is generated.

2.2. *Projection method*

This makes use of the property of a quadratic function that $G\Delta x = \Delta g$. A set of linearly independent vectors $p_1, ..., p_n$ is taken and s_i is obtained from p_i by projecting out the components $\Delta g_1, ..., \Delta g_{i-1}$. That is

$$s_i = p_i - \beta_1 \Delta g_1 - \beta_2 \Delta g_2 - ... - \beta_{i-1} \Delta g_{i-1},$$

where the β_j are chosen so that $s_i' \Delta g_j = 0$. Now for a quadratic $\Delta g_j = G\Delta x_j$ and $\Delta x_j = \alpha_j s_j$, so we have $s_i' G s_j = 0$ showing that the directions of search are conjugate.

With gradient methods it is natural to choose $p_1, ..., p_n$ (in both sections 2.1 and 2.2) as downhill gradient vectors $-g_1, ..., -g_n$ so that the downhill property follows automatically. In this case further simplifications also follow. In the Partan method, Shah *et al.* [7] showed that x_{i+1} lies along the line joining z_i with x_{i-1}: see Fig. 2(c) for $i = 2$. Hence the only points which need to be determined by linear searches starting from x_1 are $x_2, z_2, x_3, z_3, ...,$ x_n, z_n, x_{n+1} and the method has property $Q2n-1$. Similarly the method of Conjugate Gradients [8] makes use of the fact that $\beta_1, ..., \beta_{i-2}$ are zero in Section 2.2 enabling s_i to be calculated simply as the sum of two vectors: various rearrangements of this formula are possible depending upon exactly which properties of the quadratic are used. The method has property Qn and is the simplest of this type of method. It produces the same set of conjugate directions as the Partan method. Another method due to Powell [8] also uses the properties of parallel subspaces and is the same as the Partan method if $n = 2$ (Fig. 2(c)); it differs for $n > 2$, although it still has property $Q2n-1$. It can be shown that the last n search directions are conjugate, although not the same set as generated by the previous methods.

All these methods find the minimum of a quadratic function in a given number of searches. For nonquadratic functions the methods can be applied iteratively merely by repeating the complete cycle until convergence occurs, although other possibilities are mentioned in the source papers. The performance of these methods on nonquadratic problems is much better than Steepest Descents and the minimum is usually found successfully. Each iteration is simple to apply and housekeeping time is only of $O(n)$ or $O(n^2)$.

Little is known about the relative performance of these methods mostly due to the fact that another method given by Davidon [9] has proved considerably superior to them all. In this method (see also [10]) a positive definite approximation H to G^{-1} is updated at each iteration, and is used to generate directions of search $s = -Hg$.

If $\mathbf{H}$ is initially chosen as the unit matrix, then for a quadratic function the method generates exactly the same sequence of points $\mathbf{x}_1, ..., \mathbf{x}_{n+1}$ as do the Conjugate Gradient and Partan methods, and so property Qn is ensured. To show this consider the method of Conjugate Gradients which can be written theoretically as

$$\mathbf{s}_i = -\mathbf{P}_i\,\mathbf{g}_i$$

where $\mathbf{P}_i$ is a projection matrix with the property

$$\left.\begin{array}{l}\mathbf{P}_i\Delta\mathbf{g}_j = 0 \\[2mm] \mathbf{P}_i\mathbf{q}\ \ = \mathbf{q}\end{array}\right\}\ j = 1, 2, ..., i-1 \text{ and } \mathbf{q} \perp \Delta\mathbf{g}_j.$$

$\mathbf{P}_i$ could be updated by

$$\mathbf{P}_{i+1} = \mathbf{P}_i - \frac{\mathbf{P}_i\,\Delta\mathbf{g}_i\,\Delta\mathbf{g}_i'\,\mathbf{P}_i}{\Delta\mathbf{g}_i'\,\mathbf{P}_i\,\Delta\mathbf{g}_i}$$

thus ensuring $\mathbf{P}_{i+1}\,\Delta\mathbf{g}_i = 0$. It can be shown that addition of the term $(\Delta\mathbf{x}_i\,\Delta\mathbf{x}_i')/(\Delta\mathbf{x}_i'\Delta\mathbf{g}_i)$ to this does not affect the generation of directions of search: and this is precisely the updating formula used by Davidon (with $\mathbf{H}_i$ replacing $\mathbf{P}_i$). The only difference in the quadratic case is that $\mathbf{P}_{n+1}$ is zero, whilst $\mathbf{H}_{n+1}$ is $\mathbf{G}^{-1}$ (except in cases where less than n iterations are required to find the minimum).

Because Davidon's method generates exactly the same conjugate directions as the Partan and Conjugate Gradients methods, and because it is much superior to them for nonquadratic functions, it should be clear that we cannot associate fast convergence solely with the property of quadratic convergence. However the formula for the updated matrix $\mathbf{H}_{i+1}$ is such that it possesses the property $\mathbf{H}_{i+1}\,\Delta\mathbf{g}_i = \Delta\mathbf{x}_i$. This feature is the analogue of the property $\mathbf{G}^{-1}\Delta\mathbf{g}_i = \Delta\mathbf{x}_i$ and is valid to first order for any function if $\mathbf{G}^{-1}$ exists. It may well be this sort of feature which accounts for the more rapid convergence of this method away from the minimum. Methods which are closely associated with the Taylor series method (such as this one and Greenstadt's method) are known as pseudo-Taylor series (or pseudo-Newton) methods.

To summarize, the following properties are of value in methods for minimization.

(1) Simplicity—low housekeeping	Avoid evaluating second derivatives and solving equations at each iteration.
(2) Fast convergence near minimum	Property Q.
(3) Stability	Downhill property (to some extent).
(4) Fast convergence remote from minimum	Property Q to some extent. Possibly pseudo-Taylor series property.

Davidon's method fits all these requirements, although troublesome cases still arise. Slow convergence can be expected when $\mathbf{G}$ is singular at the minimum, for example $F(\mathbf{x}) = x_1^4 + x_2^2$, but this trouble would be common to all methods with property Q. Sometimes $\mathbf{H}$ can tend to become singular and then the minimum is found only in a subspace of the complete space. This behaviour can be caused by not paying attention to the scaling of the variables.

3. Direct Search Methods

If those methods for solving problems in which only the function can be evaluated are now considered, a very similar situation is found. The most obvious method is that of Alternating Directions in which n searches are made along successive coordinate directions and this whole cycle is repeated until convergence. The method turns out to be highly oscillatory and usually fails to converge.

Workers such as Hooke and Jeeves [11] and Rosenbrock [12] noticed that the line $\mathbf{x}_{n+1} - \mathbf{x}_1$ joining the first and last points of a cycle of alternating directions often proved to be a fruitful direction of search. Davies, Swann and Campey [13] developed Rosenbrock's method to include provision for a linear search. After a cycle of alternating directions, the vectors $\mathbf{x}_{n+1} - \mathbf{x}_1, \mathbf{x}_{n+1} - \mathbf{x}_2, \ldots, \mathbf{x}_{n+1} - \mathbf{x}_n$ are set up and orthogonalized by the Gram–Schmidt process; these then become the search directions for another cycle of alternating directions. Although the method does not have property Q, it can be shown that in the limit the directions tend to lie along the axes of the quadratic function, and that these axes (eigenvectors of $\mathbf{G}$) are a special case of conjugate directions. As has been said earlier, the effect of property Q is most noticeable near the minimum when the method is applied to nonquadratic functions. At this time the DSC method also has a good approximation to directions which are locally $\mathbf{G}$-conjugate and so behaves on nonquadratic functions very much like a method with property Q.

Another method which works well in some cases is given by Nelder and Mead [14] in which the function is first evaluated at a basis set or "simplex" of $n+1$ points and the set is altered systematically—dropping some points and adding others—until the region of the minimum is reached. Its precise location is then found by interpolating a quadratic function at suitably chosen points. It has been found however that the method does not work so well if the number of variables becomes large.

There are also methods which set out to construct conjugate directions explicitly. Smith's method [15] follows the parallel subspace approach above, using coordinate directions for the vectors $\mathbf{p}_1, \ldots, \mathbf{p}_n$; the method thus has property $Q\frac{1}{2}n(n+1)$. Powell [16] noticed that the direction $\mathbf{p}_n$ was only used once in each cycle and substituted an alternative cycle based on displacing

x_i to give z_i (see Section 2.1) by a sequence of minimizations along $p_i, p_{i+1}, \ldots,$ p_n so that every direction is used as often in the cycle. Also the directions $s_1, \ldots, s_n$ thus obtained are carried forward to be the $p_1, \ldots, p_n$ for the next cycle. Together with a device for preventing linear dependence, a method with much better performance than that of Smith was arrived at. It has property Qn^2 and is probably marginally more powerful than the DSC method.

TABLE I

Rate of convergence of Stewart's Method

Test function	n	Powell's Method		Stewart's Method			
				as published		more simple linear search	
Parabolic Valley	2	37	153[a]	23	152[a]	44	208[a]
Helical Valley	3	56	180	20	130	24	138
Powell's '62 function	4	77	229	18	158	28	179
Chebyquad	2	11	36	5	25	4	19
	4	28	82	9	55	8	53
	6	103	281	13	126	19	167
	8	192	520	—	—	24	264

[a] Numbers in each entry are (left column) "linear searches", and (right column) "function evaluations". The accuracy required is that F shall be reduced to within 10^{-k} of the minimum, where k varies from problem to problem but is in the range 11 ± 2.

The most recent development is that of Stewart [17] who modified Davidon's method to allow difference approximations to derivatives. The interval over which the differences are taken is determined automatically to balance truncation and round-off error, and tests show that three to six figure accuracy can be obtained for g on a twelve digit machine even with the crudest difference formula. The performance of this method as against other methods with property Q seems to parallel that for gradient methods. Some tests have been done by Lill [18] in which a different linear search was used (this should have marginal effect) to determine the reliability of Stewart's figures. These results, shown in Table 1, reinforce the conclusion that this method may prove to be the best alternative currently available. One interesting result is that many fewer iterations are required than in other methods but more function evaluations are required per iteration. This is because of the n evaluations required for the differences and means that only a small proportion of the total evaluations are used in the linear search. The implication is that it is worth using a fairly sophisticated linear search, possibly using extra evaluations, to ensure that a good minimum is found at each iteration.

4. Sums of Squares

Here the special case is considered in which $F(x)$ is the sum of squares of m nonlinear functions $f_i(x)$, collectively $f(x)$ or f, so that $F = f'f$. This arises frequently and the problem can be solved by minimizing F with one of the above methods, in which case the relationship $g = 2J'f$ may be required, where J is the matrix of first derivatives of f (Jacobian) with $J_{ij} = \partial f_i/\partial x_j$. This is usually the safest line of approach but often much more rapid convergence can be obtained by taking into account the special nature of F.

One way to do this is to assume that f is linear and can be expressed as $Jx-b$ where J is a constant matrix and b a constant vector. This implies that F is quadratic with $G = 2J'J$. In this case the Taylor series method of an earlier section $s = -G^{-1}g$ becomes the Generalized Least Squares method with $s = -(J'J)^{-1}J'f$. The method assumes that the derivatives of f are available; it is much more reliable however as $J'J$ is at least positive semi-definite so that the downhill property usually applies. The time required to solve equations may be less important too because this is $O(n^3)$ and the time to evaluate say g is $O(mn)$. Hence when $m \gg n$ the former computation is relatively less time consuming. The major snag is still that an $m \times n$ matrix J must still be provided. Powell [19] has published a version of the method in which both $(J'J)^{-1}$ and J are approximated by differences and updated at each iteration.

When $m = n$ the problem becomes that of solving a system of nonlinear equations $f = 0$. In this case the Generalized Least Squares method reduces to Newton's method with $s = -J'f$. Both these methods can also be derived by truncation of the appropriate Taylor series; in the latter case $f(x+\Delta x) = f(x)+J\Delta x+O(\Delta x^2)$. Newton's method also requires derivatives, and a method due to Barnes [20] shows how difference approximations to J can be set up and updated at each iteration. Both methods still require the solution of equations, however, and this can be avoided by updating an approximation to J^{-1}. This is done by Broyden [21] and essentially also in the Secant method of Wolfe [22].

In a recent paper Fletcher [23] shows that all these methods (from the Generalized Least Squares method onwards) can be unified as special cases of a Generalized Inverse method, because both $(J'J)^{-1}J'$ and J^{-1} are generalized inverses of J when J is of rank n. This method also shows how to proceed when J is not of rank n—a case when the earlier methods break down. The new method is proved to have the downhill property and it also shows how a difference approximation to the generalized inverse can be set up and updated in a reasonably simple way.

Unfortunately in practice the situation is not as rosy as it might appear, especially when $m = n$. Highly oscillatory, and hence very slow, convergence can occur which render the methods valueless on certain problems. In some cases it is found that the search direction s makes an angle of nearly 90° with the downhill gradient vector and this is associated with J being nearly singular. However I have also known of oscillatory behaviour in the region of a solution where J^{-1} is well defined. Yet in other cases very rapid convergence is obtained, much faster than by using a general minimization method. It seems that more work needs to be done in this area of the subject to determine what factors are important.

One method which attempts to cover cases where J is ill-conditioned is due to Levenberg [24] in which a positive parameter λ and a diagonal matrix of positive elements D are used to calculate $s = -(J'J+\lambda D)^{-1}g$; there is a similar method by Marquardt [25] in which the diagonal elements are uniquely determined by the scaling of the variables. This has the effect that the search direction obtained interpolates in some way between the steepest descent direction and that of the Generalized Least Squares method. The precise determination of the parameter λ is one obvious problem which arises. However I know of no comprehensive comparison between these methods and the Generalized Least Squares method, nor of any work to adapt the methods for those problems where derivatives are not available.

Finally, in the case when $m = n$ and equations are being solved, and especially when J is ill-conditioned, it is worth looking for rearrangements of the equations in the form

$$\mathbf{x}_{i+1} = \mathscr{F}(\mathbf{x}_i)$$

to be solved by a simple iterative approach. Another method worthy of mention was suggested by Freudenstein and Roth [26] and also by Deist and Sefor [27] in which a system of equations $\phi(\mathbf{x}, \mathbf{a}) = 0$ is constructed with the property that $\phi(\mathbf{x}, \beta) = \mathbf{f}(\mathbf{x})$ and also that the solution $\mathbf{x}(\alpha)$ to $\phi(\mathbf{x}, \alpha) = 0$ is known. The vector $\mathbf{a}$ is then varied between α and β in small increments and the solution $\mathbf{x}(\mathbf{a})$ is traced. At each stage $\mathbf{x}$ can be determined either as the solution of a differential equation or by solving equations in the usual way. These methods sometimes enable convergence to an exact solution to take place when other methods which minimize the sum of squares do not.

References

1. M. J. Box, D. Davies and W. H. Swann (1969). "Non-linear Optimization Techniques" I.C.I. Monograph No. 5, Oliver and Boyd, London.
2. M. J. D. Powell (1966). Finding minima of functions of several variables. *In* "Numerical Analysis" (J. Walsh, ed.) Academic Press, London.
3. M. J. Box (1966). A comparison of several current optimization methods and the use of transformations in constrained problems. *Comput. J.* **9**, 67.
4. R. Fletcher (1965). Function minimization without evaluating derivatives —a review. *Comput. J.* **8**, 33.
5. J. Greenstadt (1967). On the relative efficiencies of gradient methods. *Maths Comput.* **21**, 360.
6. R. Fletcher and C. M. Reeves (1964). Function minimization by conjugate gradients. *Comput J.* **7**, 149.
7. B. V. Shah, R. J. Buehler and O. Kempthorne (1964). Some algorithms for minimizing a function of several variables. *J. Soc. ind. appl. Math.* **12**, 74.
8. M. J. D. Powell (1962). An iterative method for finding stationary values of a function of several variables. *Comput. J.* **5**, 147.
9. W. C. Davidon (1959). "Variable Metric Method for Minimization" Argonne National Laboratory, ANL-5990 Rev.
10. R. Fletcher and M. J. D. Powell (1963). A rapidly convergent descent method for minimization. *Comput. J.* **6**, 163.
11. R. Hooke and T. A. Jeeves (1961). "Direct search" solution of numerical and statistical problems. *J. Ass. Comput. Mach.* **8**, 212.
12. H. H. Rosenbrock (1960). An automatic method for finding the greatest or least value of a function. *Comput. J.* **3**, 175.
13. W. H. Swann (1964). "Report on the development of a new direct search method of optimization". I.C.I. Ltd. Central Instr. Lab. Res. Note 64/3.
14. J. A. Nelder and R. Mead (1965). A simplex method for function minimization. *Comput. J.* **7**, 308.
15. C. S. Smith (1962). "The automatic computation of maximum likelihood estimates". N.C.B. Sc. Dept. Report S.C. 846/MR/40.
16. M. J. D. Powell (1964). An efficient method of finding the minimum of a function of several variables without calculating derivatives. *Comput. J.* **7**, 155.
17. G. W. Stewart III (1967). A modification of Davidon's minimization method to accept difference approximations to derivatives. *J. Ass. Comput. Mach.*, **14**, 72.
18. S. A. Lill (1968). private communication.
19. M. J. D. Powell (1965). A method of minimizing a sum of squares of non-linear functions without calculating derivatives. *Comput. J.* **7**, 303.
20. J. G. P. Barnes (1965). An algorithm for solving non-linear equations based on the secant method. *Comput. J.* **8**, 66.
21. C. G. Broyden (1965). A class of methods for solving non-linear equations. *Maths Comput.* **19**, 577.
22. P. Wolfe (1959). The secant method for simultaneous non-linear equations. *Communs Ass. Comput. Mach.* **2**, 12.
23. R. Fletcher (1968). "Generalized inverse methods for the best least squares solution of systems of non-linear equations. *Comput. J.* **10**, 392.
24. K. A. Levenberg (1944). A method for the solution of certain non-linear problems in least squares. *Q. appl. Math.* **2**, 164.

25. D. W. Marquardt (1963). An algorithm for least squares estimation of nonlinear parameters. *J. Soc. ind. appl. Math.* **11**, 431.
26. F. Freudenstein and B. Roth (1963). Numerical solutions of systems of nonlinear equations. *J. Ass. Comput. Mach.* **10**, 550.
27. F. H. Deist and L. Sefor (1967). Solution of systems of non-linear equations by parameter variation. *Comput. J.* **10**, 78.

Discussion

BROYDEN. (University of Essex). The downhill property of the Davidon algorithm is true in theory but doubtful in practice and Dr Fletcher states that in some cases the matrix **H** becomes singular. I have known cases where **H** becomes decidedly nonpositive definite and negative values of α have to be taken to reduce the function. I would also like to question one of the sacred cows; that we minimize the function along a given direction. In a similar algorithm of my own for minimizing a sum of squares, I found that if I insisted on minimizing at each search, then convergence was inhibited. In fact I simply chose the parameter α as one, which I could do because the method was a pseudo-Newton method.

DAVIDON. The cumulative effects of rounding-error can certainly affect the **H** matrix and one should become suspicious if a large number of steps are required by the algorithm. It takes some time to determine whether a matrix is becoming positive definite. I agree with the second point, and have suggested a modification of the original algorithm in which precisely this feature is used.

FLETCHER. The reason for minimizing along the line in many algorithms is that it provides an important orthogonality condition in the proofs of property Q. However if it could be shown in the general case that more rapid convergence was usually obtained without this feature, then I would go along with this.

BARD. In the question of whether to minimize along the line, or to choose α equal to one, I consider either extreme as unnecessary except where there is a theoretical requirement to go to the minimum such as in the Davidon-Fletcher-Powell method. The first extreme can be too expensive, and the other too dangerous in that the function value may not be reduced. In our experience we have found that a choice which reduces the function value but does not pay too much attention to finding the actual minimum, proves satisfactory in terms of the number of evaluations of the function which are required.

DAVIDON. In my recent modification of the Variable Metric method, in which α is chosen as one, wild behaviour is impossible because each iteration is always started from the best point obtained on the previous iteration. It would seem that in these circumstances one can obtain the aesthetic appeal of property Q without the computational disadvantage of having to minimize along a line.

BARD. In this case there is an excessive number of iterations required for the solution, because of the number of false steps which would be taken. It is still my opinion that the optimum strategy is somewhere between the two extremes, regardless of the simplicity of the $\alpha = 1$ method.

GLASS. (English Electric Company). From practical experience in engineering problems, I find that if we get deep into a valley then we take very small steps giving oscillation. If we are satisfied with going part way to the minimum, then we are surveying the valley from a fairly constant level, and we can contour around valley. In practical problems the computer time required to estimate the exact minimum along a line is excessive.

2. Variance Algorithms for Minimization*

WILLIAM C. DAVIDON

Haverford College, Haverford, Pennsylvania, U.S.A.

1. Introduction

Cauchy's method of steepest descent has been used for over a century to search for minima of real functions over manifolds [1]. While simple and stable, the method often converges slowly, because the direction of steepest descent and the direction to the minimum may be nearly perpendicular. The direction of steepest descent depends not only on the function being minimized, but also on the metric properties of the manifold. The distinguishing feature of the algorithms considered here is that they iteratively adjust the metric in an effort to make the direction of steepest descent point toward a minimum. Equivalently, the metric is adjusted to make each contour surface for the function equidistant from the minimum.

We first summarize a few basic properties of manifolds on which different metrics, or no metric at all, may be defined. The rate of change of a function along a specified trajectory in a manifold is independent of any metric properties of the manifold. If the vector velocity is v and the vector gradient

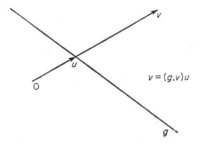

$$v = (g.v)u$$

FIG. 1. Graphic representation of velocity and gradient vectors.

* Revised version of a paper presented at the Conference on Optimization, University of Keele, Staffordshire, England, on 25–28th March 1968.

of the function is g, then the rate of change of the function is $g \cdot v$, and depends linearly on both g and v. In an N-dimensional manifold, the sets of all gradients g and all velocities v in the neighborhood of a point form dual N-dimensional vector spaces. We can add vectors in each space and multiply them by real numbers, and each velocity together with a gradient determines a real number, the corresponding rate of change of the function. However, in the absence of a metric, the magnitudes of velocities in different directions cannot be compared and no direction of steepest descent can be defined. (The velocity space is the space of contra-variant vectors, and the gradient space is the space of co-variant vectors.)

In Fig. 1, we represent each veclocity vector by an arrow, and each gradient vector g by the set of all velocity vectors v satisfying $g \cdot v = 1$. In Fig. 2, we illustrate the parallelogram rule for adding both velocity and gradient vectors. This rule for addition is defined only for pairs of vectors which are not parallel; continuity can be used to extend the definition to all pairs.

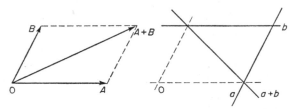

FIG. 2.   Graphic addition of velocity and gradient vectors.

Metrics can be represented graphically by the ellipsoid of all velocity vectors of unit length. Each metric determines a linear map V from gradient space to velocity space, for which the image Vg of the gradient g is in the direction of steepest descent and has a length equal to the magnitude of the gradient. A graphical construction for Vg given V and g is illustrated in Fig. 3. (In terms of projective geometry, Vg is the pole of g for the ellipse V.)

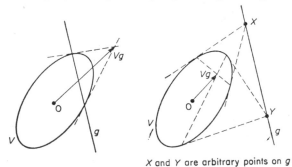

X and Y are arbitrary points on g

FIG. 3.   Graphic construction of velocity Vg from variance V and gradient g.

The map V from gradient space to velocity space determined by a metric has these properties:

Linearity: $V(g+h) = Vg + Vh$

$V(ng) = nVg$

Symmetry: $g.Vh = h.Vg$

Non-negative: $0 \leqslant g.Vg$

for all gradient vectors g and h and all real numbers n. We define a *variance* as any map from gradient to velocity space with these properties. (The variance matrix in statistics determines just this type of map.)

Each concave (upward) function with second derivatives determines a variance V at each point, with the defining property that $Vg' = v$ for all v, where g' is the rate of change in the gradient resulting from a motion with velocity v. If this variance were constant along the line from an arbitrary point x to the minimum, then the displacement of x from the minimum would be Vg, where g is the gradient at x. The algorithms considered here attempt to locate the minimum by an iterative process, subtracting displacements of the form Vg from successive estimates of the location of the minimum. In each iteration, the variance V is also changed to incorporate new information about the function and its gradient.

2. Variance Algorithms

We define a *variance algorithm* as one for searching for minima of functions, whose iterations have the following properties:

(1) Each begins with an estimate x for the location of the minimum and an estimate V for the variance of the function in the region between x and the minimum. (In addition, other information may be available from previous computations.)

(2) Values of the function and its gradient are computed at one or more places in the neighborhood of x.

(3) New estimates are made for x and V to begin the next iteration.

(4) When minimizing quadratic functions with unique minimum, x and V converge to the correct minimum and variance from almost all initial estimates.

This definition of variance algorithms is similar to Broyden's definition of quasi-Newton algorithms [2]. It includes those of his quasi-Newton algorithms for which his matrix H is symmetric, non-negative, and in the quadratic case converges to the variance of the function.

Figure 4 illustrates a change in a variance estimate from V to V^+ in one iteration of a variance algorithm. The change $V^+ - V$ is also a map from gradient space into velocity space, although it may not be a variance since it need not be non-negative. The range of this map (that is the set of all velocity vectors of the form $(V^+ - V)g$ for some gradient g) is a subspace of velocity space whose dimension is by definition the *rank* of the change. The N-dimensional gradient space can be decomposed into two subspaces A and B, where A is the set of all gradients a with $Va = V^+a$, and B is the set of all gradients b with $b.Va - b.V^+a = 0$ for all a in A. In other words, A is the set of all gradients on which V and V^+ produce the same effect, and B is the set of all vectors each of which is orthogonal (using either metric) to all vectors of A. The rank of the change is equal to the dimensionality of B, and the sum of the dimensionalities of A and B is equal to the dimensionality of the manifold. In Fig. 4, the rank of the change is one, and the subspace A of gradients for which $Va - V^+a$ consists of all gradients parallel to the common tangents of the two ellipses.

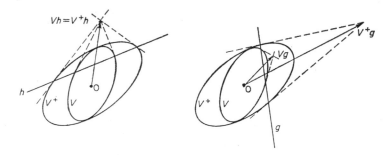

FIG. 4.   Rank one change in a two dimensional variance.

The rank by which the variance estimate is changed in each iteration provides one useful classification of variance algorithms; both rank one and rank two algorithms have been described in detail [3, 4, 5]. At the present time, proofs of convergence for some nonquadratic functions have been obtained by Goldfarb [6] only for rank one algorithms, and these are generally simpler both conceptually and computationally. However, rank one changes in the variance consistent with $V^+(g^+ - g) = x^+ - x$ can only be made under more restricted conditions.

In addition to rank, there are other useful classifications of variance algorithms applicable to all minimization algorithms. An algorithm is:

stable, if the sequence of function values for successive iterations is always non-increasing, whatever the nature of the function being minimized;

quadratically terminating, if in the absence of round-off, the algorithm terminates in no more than N iterations when minimizing a quadratic function over an N-dimensional manifold;

a linear search algorithm if each iteration includes a search for a minimum along a line.

Stability of algorithms appears to be highly desirable to ensure reasonable behavior particularly for nonquadratic functions. It ensures that the best approximation to the minimum is always preserved. The value of quadratic termination is more doubtful. Although most variance algorithms now in use possess this property, it seems to be more the result of historical accident and ease of analysis rather than systematic evaluation. The relative merits of linear search algorithms also need more exploration, for these searches are often time consuming yet largely irrelevant to subsequent computations.

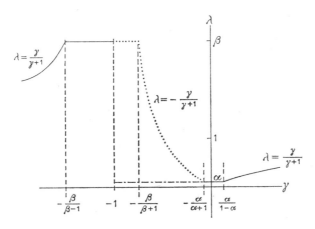

FIG. 5. Graph of $\lambda(\alpha)$. $\cdots$ 1967 algorithm, —–—present algorithm, - - - common to both.

3. Modifications in a Variance Algorithm

A rank one, stable, quadratically terminating algorithm without linear searches was proposed by the author in 1968 [4]. This was intended as a rudimentary algorithm, to which various improvements could be incorporated. As noted then in proof, changes in step 3 would eliminate situations in which successive iterations would improve neither the position nor variance estimates. One such change which has been found quite satisfactory is as follows:

Step 3. Define $\gamma = -g.r/\rho$.

    If $-1 < \gamma < \alpha/(1-\alpha)$, define $\lambda = \alpha$.

    If $-\beta/(\beta-1) < \gamma \leqslant -1$, define $\lambda = \beta$.

    If neither of these, define $\lambda = \gamma/(\gamma+1)$.

This changes the dependence of λ on γ, as shown in Fig. 5. In the earlier version, $\lambda(\gamma)$ was chosen to minimize the difference between $V^+(g^+-g)$ and x^+-x using the metric $(V^+)^{-1}$, while this revision minimizes the same difference but using the metric V^{-1}. In both cases, V^+ is restricted by the condition that the ratio of its determinant to that of V must be between α and β.

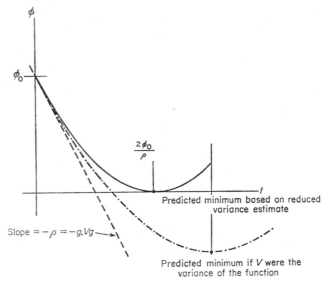

FIG. 6. Reduction in V estimated for non-negative functions.

Another change in the algorithm which has been found useful is to make use of the fact that in many cases, the function being minimized is known to be non-negative. In that case, we can reduce the variance estimate if the reduction in the function expected in an iteration is greater than the value of the function itself. This is illustrated in Fig. 6. The additional step for accomplishing this reduction in V is inserted prior to the steps of the algorithm as it was previously published [4] and is as follows:

Step 0.   If $g.Vg - 2\phi \leqslant \varepsilon$, go to step 1.

If not, define $r = Vg$

$$\rho = g.r$$

$$\lambda = 2\phi/\rho$$

$$V^+ = V + (\lambda - 1)rr/\rho.$$

Begin new iteration,

This algorithm fails to make full use of information obtained about the function and its gradient in each iteration, and it is to be expected that utilizing this information can accelerate convergence. It makes no use of the computation of function values other than to choose which point is closest to the minimum. Also it does not use the fact that the gradient computed within one iteration would be orthogonal to the steps taken in all preceding iterations if the function being minimized were quadratic, so that any lack of orthogonality could give additional information about the function.

One feature of this algorithm which is essential to Goldfarb's proofs of convergence for certain nonquadratic functions is that $V^+ - V$ is either non-negative or non-positive. (This is not the case for existing rank two algorithms.) As a consequence of this, if the initial variance estimate V is greater than variance of the function at the initial location and if the sequence of variances of the function at successive iterations is non-increasing then the sequence of variance estimates will also be non-increasing and converges to the variance of the function at the minimum. Or dually, the sequence of variance estimates will be non-decreasing if the dual conditions are fulfilled (that is interchanging increasing and decreasing throughout). This monotonicity in the variance estimates is not only essential for the present convergence proofs for certain nonquadratic functions, but also appears on the basis of some computational experience to result in better variance estimates. This is another subject which needs considerable further analysis and experience.

References

1. A. Cauchy (1847). *C.r. hebd. Seanc. Acad. Sci., Paris*, **25**, 536.
2. C. G. Broyden (1967). Quasi-Newton methods and their application to function minimisation. *Maths Comput.* **21**, 368.
3. W. C. Davidon (1959). "Variable metric method for minimization". AEC Research and Development Report ANL 5990 (Rev. TID 4500, 14th edition).
4. W. C. Davidon, (1968). Variance algorithm for minimization. *Comput. J.* **10**, 406.

5. R. Fletcher and M. J. D. Powell (1963). A rapidly convergent descent method for minimization, *Comput. J.* **6**, 163.
6. D. Goldfarb (1969). Sufficient conditions for the convergence of a variable metric algorithm. *In* "Optimization" (R. Fletcher, ed.) Chap. 18, this Volume, Academic Press, London.

Discussion

BROYDEN. The algorithm that Dr Davidon describes seems to be similar to a method which I suggested in *Mathematics of Computation* [2] last year. The algorithm was a single rank method and I wonder how this fits into the general picture, and whether perhaps Dr Goldfarb could analyse this in the way he analyses other methods in his talk. (This point is taken up by Davidon in this revised paper at the beginning of Section 2. *ed.*)

3. Nonlinear Optimization by Simplex-like Methods

E. M. L. BEALE

Scientific Control Systems Ltd., London, England

1. Introduction

1951 was an important year: in particular two papers were published that provided the main sources of the two streams of activity being discussed at this conference. Dantzig's work [1] is the fundamental reference on linear programming, this being the name given to problems requiring the maximization of a linear objective function of variables subject to linear constraints: Box and Wilson's [2] paper was the direct or indirect source of much of the present interest in hill-climbing methods for maximizing general functions. Work in both subjects can be traced back earlier; but as Dantzig [3] points out development of their practical aspects had to wait until computers became available to do the arithmetic involved in applications. It is therefore very natural that most of this work has been done in the last two decades.

Since 1951, linear programmers have widened their interests to include nonlinear objective functions and constraints, and acknowledged this by changing the name of their subject to mathematical programming. Hill-climbers meanwhile have widened their interests to include inequality constraints on the values of the variables. It is interesting to note, therefore, that although these two streams of activity have come close together, they have not merged.

This paper attempts to explain, in the context of general computing methods for optimization, why mathematical programming has developed as it has. Section 2 outlines the most relevant features of practical linear programming problems, and how these features are exploited in the usual method for solving these problems, known as the simplex method. Section 3 shows why a special version of the Simplex method, known as the product form of inverse matrix method, is much more efficient on medium sized or large linear programming problems. Many years ago Philip Wolfe said at a conference that everyone working on optimization problems should under-

21

stand this product form method—since then I have become increasingly convinced of the wisdom of this remark. The product form of inverse is crucial to the development of both linear programming itself and its extensions to more general mathematical programming; although this does not mean, of course, that it should be used in all optimization problems.

Section 4 discusses some ways in which nonlinear problems can be represented in linear programming terms. It also outlines some standard extensions of the simplex method to truly nonlinear problems. The techniques mentioned are separable programming, quadratic programming and the method of approximation programming. Finally, Section 5 discusses some general points about the relative merits of different methods for solving practical optimization problems.

It should perhaps be pointed out that this paper does not describe any new mathematical programming methods; it does not even describe old methods in much detail. More details can be found for example in Wolfe [4] and Beale [5, 6], and other methods, including some new ones, are discussed by Abadie [7] and in other papers in this volume.

Throughout this paper the term "mathematical programming" is used to mean "the process of obtaining numerical solutions to practical mathematical programming problems". This gives the subject a rather fragmented look, which may not please people interested in the more theoretical aspects of the subject; but I hope it will be forgiven as this paper was presented at a conference concerned with optimization methods.

2. Linear Programming and the Simplex Method

The main feature that distinguishes linear programming and its extensions from other optimization methods is the size of the problems tackled. A merely quantitative difference may not seem very fundamental, but the point is that it amounts to about two orders of magnitude. A typical hill-climbing problem may have five variables, although there are real problems with as few as two variables. Problems with twenty variables must be considered large, even though some with a hundred variables have been solved; corresponding numbers of variables for linear programming problems are all up by a factor of a hundred, although a problem with two thousand variables might not be considered large nowadays.

A large problem is not necessarily more difficult than a small problem, but it is important to take advantage of special features which make a large problem relatively easy. In a typical programming problem the variables represent the amount of time, in a particular time period, that a plant is operated in a particular way, or the amount of some material that is sent from one place to another. Most of the constraints are linear, not as a result of simplifying assumptions but because of the additive nature of accounting processes: material balance constraints represent the fact that the amount of some material at a given place at the beginning of any time period, plus the amount bought, shipped in or made, minus the amount sold, shipped out or used, equals the amount left at the end of the time period; and capacity constraints represent the fact that the sum of the amounts of time spent operating a plant in various ways must be less than or equal to the total time available. Other important constraints that are often taken for granted are that the variables themselves must be non-negative. This word "non-negative" is used rather than "positive", because variables can be zero. Indeed it is generally only tolerable to contemplate thousands of variables because we know that most of them will not occur in an optimum solution, that is they will take the value zero.

Linear programming problems are normally solved by the simplex method. As Zoutendijk [8] has emphasized, this can be regarded as a method of Feasible Directions. This means that we have a point representing a feasible trial solution, that is one that satisfies all the constraints without necessarily optimizing the objective function, and we seek a direction of motion from this point along which the trial solution remains feasible and the objective function improves. If we find such a direction, we move along it until either the objective function ceases to improve or alternatively the trial solution is about to become infeasible. If we cannot find such a direction, we have a local optimum solution to the problem; and for a linear programming problem this must also be a global optimum.

At this level of generality there is no difference between the simplex method and a hill-climbing method. To understand the special features of the simplex method we must study it algebraically. By introducing *slack variables*, representing the differences between the left and right hand sides of inequality constraints, we can represent any linear programming problem as one of maximizing a linear function of non-negative variables subject to equality constraints. (Any variable that does not have to be non-negative can be eliminated from the problem by using one equation to solve for it in terms of the remaining variables.) These equations can be written in the form

$$\sum_{j=0}^{n} a_{ij} x_j = b_i \quad (i = 0, 1, ..., m) \tag{1}$$

or as a matrix equation

$$A x = b \tag{2}$$

where the variable x_0 represents the value of the objective function and occurs with a unit coefficient in Eqn (1) and zero coefficients elsewhere, that is $a_{00} = 1$ and $a_{i0} = 0$ for $i \neq 0$.

The next step is to solve Eqn (1) for x_0 and m of the remaining x_j in terms of the remaining variables. If the selected variables are denoted by x_0, $X_1, ..., X_m$, the resulting equations can be written as

$$x_0 = \bar{a}_{00} + \Sigma \bar{a}_{0j}(-x_j)$$
$$X_i = \bar{a}_{i0} + \Sigma \bar{a}_{ij}(-x_j) \qquad (i = 1, ..., m) \qquad (3)$$

it being understood that the variables on the right-hand side do not include those on the left-hand side.

There is nothing original about using the constraints to solve for some variables in terms of the others. We may think of the variables on the left-hand side of Eqn (3) as dependent variables, and those on the right-hand side as independent variables. The array of coefficients in Eqn (3) is known as a *tableau*, and associated with the tableau we have a trial solution obtained by putting all the independent variables equal to zero. The values of the dependent variables are then the constant terms $\bar{a}_{i0}$. We assume that none of the $\bar{a}_{i0}$ is negative for $i > 0$, so that the trial solution is feasible. Ways of obtaining a tableau with a feasible trial solution are discussed in Dantzig [1] or any textbook on mathematical programming. We further assume that none of these $\bar{a}_{i0}$ is zero; the practical problems arising from zero valued $\bar{a}_{i0}$ are not fully discussed in the literature. They do not have a tidy solution, since they depend essentially on tolerances, that is rules for deciding when to regard a small number as zero; but they can be ignored by users of good linear programming codes.

We now have a feasible trial solution at which the only *active constraints* are that the independent variables must be non-negative; these are especially simple constraints to handle. We see that the coefficients $\bar{a}_{0j}$ of the independent variables in the expression for the objective function are minus the components of the gradient of the objective function regarded as a function of the independent variables. They are known as the *reduced costs*. If they are all non-negative, then x_0 is maximized at the trial solution. Otherwise we find some x_j, say x_q, for which $\bar{a}_{0q} < 0$. We can then improve the trial solution by increasing x_q as far as possible without making any dependent variable negative. This means putting $x_q = \theta$, where,

$$\theta = \min_{i, \bar{a}_{iq} > 0} \left(\frac{\bar{a}_{i0}}{\bar{a}_{iq}} \right) = \frac{\bar{a}_{p0}}{\bar{a}_{pq}}, \qquad \text{say.} \qquad (4)$$

We then have a new and improved trial solution in which $X_p = 0$ and $x_q > 0$. It is now more difficult to analyse the consequences of increasing another independent variable, since this might require a decrease in x_q in order to prevent X_p from becoming negative, or alternatively it might allow a further increase in x_q without X_p becoming negative. The difficulty is due to the fact that we have an active constraint that is not a simple lower bound on the value of one of the independent variables. The essential point of the simplex method is that we overcome this difficulty by changing the set of independent variables at each iteration. We use the equation for X_p to solve for x_q in terms of X_p and the other independent variables, and then treat X_p as an independent variable instead of x_q. This operation is known as *pivoting* on the element $\bar{a}_{pq}$. The problem is now back in the same form as Eqn (3), and the pivoting process is repeated as often as necessary to reach an optimal solution, that is one with $\bar{a}_{0j} \geqslant 0$ for all j.

Because the set of independent variables changes at each iteration, they are given a special name, that is *nonbasic variables*, the dependent variables are called *basic variables*. We also say that these variables are *in the basis*, although we shall see in the next section that it is more correct to say that the basis consists of the columns of coefficients of the basic variables in the initial formulation of the problem.

3. The Product Form of the Inverse Matrix Method

The simplex method as outlined above works well for small linear programming problems, but larger problems have another characteristic feature that must be exploited. This is the *sparseness* of the A-matrix, that is the fact that most of the coefficients a_{ij} in Eqn (1) are zero. In a practical problem most variables will occur in not more than half a dozen constraints; so that a problem with two-hundred rows will typically be about 3 per cent dense, that is 3 per cent of the elements will be nonzero, while a problem with twelve hundred rows will be only about 0·5 per cent dense. When the constraints are transformed into the solved form of Eqn (3) they are normally far denser, so in order to take proper advantage of sparseness we must work with the original matrix **A**. We therefore consider the matrix $\mathbf{B}^{-1}$, where the square matrix **B**, known as the *basis*, consists of the columns of A representing the coefficients of the basic variables $x_0, X_1, ..., X_m$. If we premultiply Eqn (2) by $\mathbf{B}^{-1}$, we have the equation

$$\mathbf{B}^{-1} \mathbf{A} \mathbf{x} = \boldsymbol{\beta}, \tag{5}$$

where $\boldsymbol{\beta} = \mathbf{B}^{-1} \mathbf{b}$; and if we now write Eqn (5) as a set of algebraic equations we see that it reduces to Eqn (3), except that all variables are on the left-hand sides. The current values of the basic variables, the $\bar{a}_{i0}$, are the components of β, and the tableau elements, the $\bar{a}_{ij}$, are the components of $\mathbf{B}^{-1}\mathbf{A}$. It is therefore possible to avoid computing all the $\bar{a}_{ij}$ at each iteration, and instead to work with the original matrix $\mathbf{A}$ and some representation of $\mathbf{B}^{-1}$. This leads us to the Inverse Matrix or Revised Simplex method.

The inverse matrix method would be very unattractive if we had to perform a matrix inversion to find $\mathbf{B}^{-1}$ after each iteration when the variable x_q replaces X_p in the set of basic variables. But this is not necessary. The new $\mathbf{B}^{-1}$ can be derived from the previous one by premultiplying by an *elementary transformation matrix* $\mathbf{T}_k$ which is a unit matrix except for the pth column (if rows and columns are numbered from zero). In the pth column the diagonal element is $1/\bar{a}_{pq}$ and the ith element for $i \neq p$ is $-\bar{a}_{iq}/\bar{a}_{pq}$.

The matrix $\mathbf{T}_k$ can be used to update an *explicit inverse* $\mathbf{B}^{-1}$; but in the *product form*, due to Dantzig and Orchard–Hays [9], $\mathbf{B}$ is represented entirely as a product of elementary matrices applied to an initial inverse consisting of a unit matrix. This unit inverse can be created by introducing *artificial variables* representing the differences between the left and right hand sides of each equation and making them all basic. In any feasible solution to our original problem, all artificial variables must take the value zero, so the first few matrices $\mathbf{T}_k$ will normally represent the effects of replacing artificial variables in the basis by genuine variables.

In addition to the matrix $\mathbf{A}$ and some representation of $\mathbf{B}^{-1}$, the inverse matrix method works with the right-hand side vector β and the ordered list of basic variables $x_0, X_1, \ldots, X_m$. Each iteration of the simplex method can be sub-divided into five steps as follows.

Step 1. Produce a pricing vector, that is a row vector π that can be multiplied into any column of $\mathbf{A}$ to determine the reduced cost of the corresponding variable. The point is that, assuming the current trial solution is feasible, we want to examine the coefficients $\bar{a}_{0j}$, that is the top row of the matrix $\mathbf{B}^{-1}\mathbf{A}$. This can be regarded as forming the matrix $\mathbf{c}\mathbf{B}^{-1}\mathbf{A}$ where $\mathbf{c}$ is the row vector $(1, 0, 0, \ldots, 0)$. Since

$$\mathbf{B}^{-1} = \mathbf{T}_r \mathbf{T}_{r-1} \ldots \mathbf{T}_3 \mathbf{T}_2 \mathbf{T}_1 \qquad (6)$$

we first form the row vector

$$\pi = \mathbf{c}\mathbf{B}^{-1} = \mathbf{c}\,\mathbf{T}_r \mathbf{T}_{r-1} \ldots \mathbf{T}_3 \mathbf{T}_2 \mathbf{T}_1 \qquad (7)$$

by forming the row vector $\mathbf{c}\mathbf{T}_r$, then postmultiplying by $\mathbf{T}_{r-1}$, and so on.

Step 2. Price out the columns of $\mathbf{A}$ by forming the inner products $\bar{a}_{0j} = \Sigma \pi_i a_{ij}$. If these are all non-negative, then the current trial solution is optimal;

otherwise pick out the most negative, say $\bar{a}_{0q}$. The variable x_q has then been selected to enter the basis.

Step 3.   Form the column vector

$$\alpha = T_r T_{r-1} \ldots T_3 T_2 T_1 \mathbf{a}_q$$

where $\mathbf{a}_q$ is a column whose components are a_{iq}, by forming the column vector $T_1 \mathbf{a}_q$, then premultiplying by T_2, and so on. The components of α are then $\bar{a}_{iq}$.

Step 4. Perform a ratio test between the elements of α and the elements of β to determine the variable X_p to leave the basis, using Eqn (4).

Step 5. Update the list of basic variables (substituting x_q for X_p), form the new elementary transformation matrix T_{r+1} to add to the representation of B^{-1}, and update β by premultiplying by T_{r+1}.

Since each iteration adds another matrix T_k, the representation of B^{-1} becomes more and more cumbersome, and the work involved in Steps 1 and 3 becomes more tedious. It is therefore expedient to stop from time to time and to enter a process known as *reinversion*. We then abandon all previous work except for the names of the variables that are now basic. We return to our unit inverse associated with the artificial basic variables and perform m special iterations to introduce the variables we really require in the basis, one at a time, instead of the artificials. There is a great deal of scope for choosing the order in which to take these variables, and the order in which to remove the artificials, so as to reduce the number of nonzero coefficients in the resulting product form representation of B^{-1}. And the main advantage of the product form over the explicit inverse form is that the inverse of a typical sparse matrix can be represented very much more compactly in this form. As an extreme example, consider a matrix B that is a diagonal matrix except for a full last row and column. Then B^{-1}, written out as an explicit matrix, is full. But if it is formed by pivoting on the diagonal elements in order, then all the elementary transformations other than the last have only two nonzero coefficients in their non-unit columns. In this way B^{-1} can be stored far more compactly in product form than it could as an explicit matrix. As the simplex method progresses, the product form representation will build up with the addition of a probably fairly full elementary transformation at each iteration; but for some time the product form of inverse will remain much more compact, and a further reinversion can be made before this ceases to be so. This will incidentally control the build-up of rounding-off error in the calculations.

4. Nonlinear Problems

We have already noted that a practical solvable optimization problem normally involves fewer than one hundred nonlinear variables, although more can be tolerated in a nearly linear problem in which most vanish at the optimal solution. Mathematical programmers, however, are interested in being able to handle an arbitrary number of linear variables in addition to this moderate number of nonlinear variables. Let us therefore consider some methods appropriate to such problems.

A nonlinear problem can sometimes be reformulated in purely linear programming terms. Occasionally this can be done by changing the variables; an example is the problem discussed by Box [10]. This involves products of several variables y_j with a single other variable x_1. Mylander [11] points out that this becomes a linear programming problem if a new variable $z_j = y_j x_1$ is used instead of y_j. Such a transformation often has a simple physical interpretation. The variable x_1 may represent the total output of some process, and y_j may represent the production or use of some material per unit of total output. Then z_j represents the total production or use of the material. However, product terms cannot always be removed in this way.

Some nonlinear problems can be reformulated in linear terms by approximating a nonlinear feasible region by a polytope, which is a highbrow name for a polyhedron in n dimensions. For example, suppose that we wish to optimize the operation of a complex of chemical plants, and that there is some flexibility about the temperature and pressure under which one plant can be operated, but that there are complicated restrictions on the feasible combinations of temperature and pressure. We may then be able to use a linear programming model containing variables representing the lengths of time for which the plant is operated under specified extreme but feasible combinations of temperature and pressure. This approach implies that the costs, and also the amounts of materials produced and used, are approximately linear functions of temperature and pressure over the permissible combinations, so that a solution in which the plant is operated under a mixture of extreme conditions can be interpreted as operating continuously under appropriate intermediate conditions. Alternatively, if intermediate conditions are appreciably more favourable than this approach indicates, then further variables can be added to the linear programming model representing the lengths of time for which the plant is operated under suitable intermediate conditions.

Note that this formulation takes advantage of our ability to deal effectively with large numbers of variables, particularly if many of them have nonzero coefficients in only a few constraints. Normally the variables that represent operations under different conditions are all generated before the linear programming calculations start. But it is possible to defer the explicit genera-tion of these variables until they are required to enter the basis. This approach uses the *decomposition principle*, due to Dantzig and Wolfe [12] and discussed further in Dantzig [3] and Beale [6].

Nonlinear functions of single arguments can be represented in essentially the same way as nonlinear boundaries. If a function $f(z)$ of some argument z occurs in either the objective function or the constraints of an otherwise linear programming problem, we can take a set of points P_k ($k = 1, ..., K$) on the graph of this function, where say $z = a_k$ and $f(z) = b_k$. Then we introduce non-negative variables λ_k, representing weights attached to the points P_k, and the equations

$$\sum_{k=1}^{K} \lambda_k = 1 \qquad (8)$$

$$\sum_{k=1}^{K} a_k \lambda_k = z. \qquad (9)$$

The function $f(z)$ is then approximated by $\Sigma b_k \lambda_k$. In this representation the argument z may be either an original variable of the problem, or a linear function of other variables.

The intention behind this representation is to allow linear interpolation between adjacent points P_k and P_{k+1} by giving λ_k and λ_{k+1} positive values that sum to one. This produces a piecewise linear approximation to $f(z)$ between the selected K points. But the program may give positive values to two or more nonadjacent λ_k, which will in general correspond to values of z and $f(z)$ off the graph. In two important special circumstances this will not matter. If $f(z)$ is either a convex function that is part of the objective function being minimized, or a convex function on the left-hand side of a less-than-or-equal-to inequality, then the use of nonadjacent λ_k will always over-estimate $f(z)$, and this does not allow the linear programming model to gain any advantage from the artificial flexibility of the model.

In other circumstances the use of nonadjacent λ_k must be prevented. In Separable Programming, due to Miller [13], this is done by restricting the variables that are allowed to enter the basis at any iteration of the simplex method. The variables $\lambda_1 ... \lambda_K$ are known as a *set* (or group) of *special variables* of separable programming, and if two (or more) members of the set are in the basis then no others may be considered, while if one member, say λ_k, is in the basis then only its neighbors, λ_{k-1} and λ_{k+1}, may be considered. With these restrictions, the simplex method finds a local optimum solution to the problem, which may not be a global optimum. This possible failure to find a global optimum solution to a nonconvex problem is a defect that separable programming shares with hill-climbing methods. It can be overcome by the use of integer programming, but this takes us outside the scope of this conference. A simple procedure that will usually produce at least a good local optimum is to start by solving a separable programming problem as an ordinary linear programming problem, that is without the special restrictions of separable programming, to substitute the resulting values of the special variables into all equations of the form of Eqn (9) to define trial values of the arguments z of the nonlinear functions, and then to start a separable programming calculation from an initial basis containing the special variables appropriate to this z for each set. An automatic procedure for doing this within the LP system for the computer, and also for interpolating extra points on the curves defining nonlinear functions in relevant places, is described by Beale [5].

An alternative approach to separable programming is to define variables $\delta_1 \ldots \delta_{K-1}$ and to introduce the equations

$$0 \leqslant \delta_k \leqslant 1 \quad (k = 1, \ldots, K-1) \tag{10}$$

$$a_1 + \sum_{k=1}^{K-1} (a_{k+1} - a_k)\delta_k = z. \tag{11}$$

The function $f(z)$ is then approximated by $b_1 + \Sigma(b_{k+1} - b_k)\delta_k$.

This approach is of some interest when using a linear programming code that has special facilities for simple upper bounds, that is constraints of the form of Eqn (10), but not for generalized upper bounds, that is constraints of the form of Eqn (8). The special restrictions of separable programming are then that no δ_k may enter the basis unless δ_{k-1} is out of the basis at its upper bound, and no δ_{k-1} may be released from its upper bound unless δ_k is out of the basis at its lower bound. Two disadvantages of this approach are worth noting. One is that the approximation to a nonconvex problem obtained by solving it as an ordinary linear programming problem is more inaccurate. The other is that there is no analogue to the formulation obtained by replacing Eqn (8) by

$$\sum_{k=1}^{K} \lambda_k = x \tag{12}$$

and retaining Eqn (9). The expression $\Sigma b_k \lambda_k$ then represents $xf(z/x)$. This may not seem a particularly useful thing to be able to represent; but it is useful if x represents the amount of some product to be made, z represents the amount of some input material used, and f represents say the selling price per unit as a function of the concentration of this input material.

Separable programming deals primarily with nonlinear functions of single arguments. It can also be used to represent arbitrary quadratic functions, since these can be expressed as sums and differences of squares of linear functions of the variables. In particular, it can be used to represent simple product terms, since $uv = \frac{1}{4}(u+v)^2 - \frac{1}{4}(u-v)^2$. More general polynomials can be represented using logarithms and exponentials.

Another nonlinear programming problem that has been widely discussed in the literature is quadratic programming. This is the name given to the problem of maximizing a quadratic objective function of variables subject to linear constraints. Many methods have been devised for solving such problems on lines similar to the simplex method for linear programming. One method, described by Beale [14] in 1955 and elaborated by him later [5, 15], uses the product form of inverse and allows an arbitrary number of linear variables as well as a moderate number entering the quadratic part of the objective function. Nevertheless quadratic programming is not widely used; in many problems a quadratic objective function is only an approximation, and it is more convenient to use a more general piecewise linear approximation and separable programming. De Angelis [16] has studied the mechanics of the product form algorithm in detail to produce an efficient special procedure for such separable objective functions, whether or not they are convex. Another possible application for quadratic programming is to multiple regression problems when some of the regression coefficients are required to take non-negative values; but such problems can be solved by an extension of the optimum regression method of Beale *et al.* [17] and this approach allows us to explore alternative formulae with less than the maximum number of non-zero regression coefficients.

Another important nonlinear programming technique that is used on a wide variety of problems is the Method of Approximation Programming (or MAP) due to Griffith and Stewart [18]. This is based on the solution of a sequence of linear programming problems obtained by making linear approximations to all nonlinear functions which are valid in the neighbourhood of the current values of the nonlinear variables. It may be necessary to add upper and lower bounds on the values of the nonlinear variables to prevent the linear programming solution from reaching a point where the linear approximations are too inaccurate. The optimum values of the nonlinear variables are then used as the origin for new linear approximations. This method, and an extension using local quadratic programming approximations, was discussed by Beale [5].

5. The Relative Merits of Different Methods

Many optimization methods have been proposed recently, and comparisons must be made between them. This is not easy. In a comparative study of nonlinear programming codes, Colville [19] refrained from drawing sweeping conclusions from the performances of various codes on test problems that may not be typical of those faced by any particular organization, either in their content or their manner of presentation.

Note that there is an important difference in principle between the method of approximation programming and hill-climbing methods on the one hand, and the other methods discussed in this paper. Using the first group of methods we do not need to know anything about the problem outside the immediate neighbourhood of the current trial solution. This is obviously very useful if the problem involves functions that are difficult to calculate. However, it does mean writing special computer programs both to assemble data initially and to make new local approximations at intermediate points of the calculation. This is generally fairly easy when using a code that is designed to solve fairly small problems within the immediate access store of the computer. The standard optimization code simply calls a subroutine to make the approximations. On the other hand, if the standard optimization code is a linear programming code designed to solve large problems, the insertion of the subroutine is more complicated. The linear programming code probably uses all the available immediate access store, and the data needed by the subroutine will probably be packed in a way that is efficient for the linear programming code but may not be so conveniently accessible to the writer of the subroutine. This subroutine must not only revise the linear programming problem but also arrange for the linear programming code to invert to a specified basis that should be near-optimum in the light of the solution to the previous problem. For these reasons the method of approximation programming is a somewhat heavy-handed, although effective, approach to large problems.

The practical attractions of separable programming are associated with these points. Separable programming did not prove a very quick procedure in Colville's tests, and similar results were reported by Akeroyd [20] on a particular small problem—that described by Box [14] solved as a nonlinear problem. This slowness is largely because the implementations of separable programming used on both tests were in general-purpose linear programming codes designed primarily to solve large problems, with no special facilities for storing all relevant program and data within the immediate access store of the computer on small problems. Akeroyd's paper describes the steps involved in transforming product terms into a suitable form for separable programming in some detail. The original nonlinear problem had five independent variables, seven dependent variables, and upper and lower bounds on the values of seven variables. Akeroyd's separable programming formulation had forty rows and sixty-eight variables before any variables were added by interpolation. He points out that many of the variables and rows could be eliminated by substitution, but that this is not worthwhile on a problem as small as this when using a computer system designed for large problems.

There is a fair amount of work involved in preparing such a problem for the separable programming code, and one would normally want to write a special computer program to do this. But the writing of such a program, normally called a *matrix generator*, is in any case very desirable when tackling any problem involving a substantial amount of input data. The matrix generator is used to print out the input data in a format that enables the user to check it easily, and at the same time it can be programmed to set up the necessary special variables for separable programming. The whole program can be written in a high-level language, such as Fortran, and does not require any contact with the inside of the linear programming code itself. Note that the interpolation scheme described by Beale [5] is a purely numerical procedure based on the data already in the linear programming matrix, and does not require any support from special programs written for the particular application.

The conclusion is that separable programming, like the others considered in this paper, is a reliable technique, but that it is not very fast on small problems. Its real strength is that it remains reliable and convenient even if the problem is extended to include a large number of additional linear variables. This emphasis on large numbers of linear variables, and consequently on computational methods based on the product form of the inverse matrix version of the simplex method, is the feature that distinguishes mathematical programmers from other workers in the field of optimization.

References

1. G. B. Dantzig (1951). Maximization of a linear function of variables subject to linear inequalities. *In* "Activity Analysis of Production and Allocation" Chap XXI. (T. C. Koopmans, ed.) Wiley, New York.
2. G. E. P. Box and K. B. Wilson (1951). On the experimental attainment of optimum conditions. *Jl R. statist. Soc.* **B13**, 1-45.
3. G. B. Dantzig (1963). "Linear Programming and Extensions". Princeton University Press, Princeton, New Jersey.
4. P. Wolfe (1967). Methods of nonlinear programming. *In* "Nonlinear Programming" (J. Abadie, ed.) pp. 97-131. North-Holland Publishing Company, Amsterdam.
5. E. M. L. Beale (1967). Numerical Methods. *In* "Nonlinear Programming" (J. Abadie, ed.) pp. 135-205. North-Holland Publishing Company, Amsterdam.
6. E. M. L. Beale (1968). "Mathematical Programming in Practice". Pitmans, London.
7. J. Abadie (1969). In "Optimization" (R. Fletcher, ed.) chap. 4, this Volume. Academic Press, London.
8. G. Zoutendijk (1960). "Methods of Feasible Directions". Elsevier, Amsterdam.

9. G. B. Dantzig and W. Orchard Hayes (1954). The product form of the inverse in the simplex method. *Maths Comput.* **8**, 64-67.
10. M. J. Box (1965). A new method of constrained optimization and a comparison with other methods. *Comput. J.* **8**, 42-52.
11. W. C. Mylander (1966). Nonlinear programming test problems. Letter to the editor, *Comput. J.* **8**, 391.
12. G. B. Dantzig and P. Wolfe (1960). Decomposition principle for linear programming. *Ops Res.* **8**, 101-111.
13. C. E. Miller (1963). The simplex method for local separable programming. *In* "Recent Advances in Mathematical Programming" (R. L. Graves and P. Wolfe, ed.) pp. 89-100. McGraw Hill, New York.
14. E. M. L. Beale (1955). On minimizing a convex function subject to linear inequalities. *Jl R. statist. Soc.* **B 17**, 173-184.
15. E. M. L. Beale (1959). On quadratic programming. *Nav. Res. Logist. Q.* **6**, 227-243.
16. V. de Angelis (1969). *In* "Optimization" (R. Fletcher, ed.) chap. 6, this Volume. Academic Press, London.
17. E. M. L. Beale, M. G. Kendall and D. W. Mann (1967). The discarding of variables in multivariate analysis. *Biometrika* **54**, 357-366.
18. R. E. Griffith and R. A. Stewart (1961). A nonlinear programming technique for the optimization of continuous processing systems. *Mgmt. Sci.* **7**, 379-392.
19. A. R. Colville (1967). "A Comparative Study of Nonlinear Frogramming Codes." Paper presented at the International Symposium on Mathematical Programming, Princeton, New Jersey 14th-18th August, 1967.
20. A. J. Akeroyd (1966). An application of separable programming. *Comput. J.* **8**, 344-346.

Discussion

POWELL. I would like to ask Mr Beale a question concerning his statement that if one is using a LP formulation then it is very nice to do all the special purpose programming before one starts. In particular he suggests that where there are nonlinear constraint functions then by evaluating the constraint function at various points one can get an approximation to the function which is valuable. I was wondering about the extension of this to nonlinear constraint functions of several variables. I would think that under these circumstances it would not be possible to calculate the function at enough points in the space because the space becomes so large, in which case one would have to be prepared to enter some special purpose program during the running of the LP code. Do you agree?

BEALE. Yes. This approach is discussed in Miller's paper, including a mechanism for avoiding illegitimate combinations of points in nonconvex problems. I do not know if this method has been used in practice. The usual separable programming approach, which is effective in many but not all problems, is to avoid this difficulty by adding extra variables so as to transform all nonlinear functions into sums and differences of functions of single arguments, for example by turning quadratic functions into sums of squares or by using logarithms. But, as I have indicated, this approach is not always feasible, and when it is it can be cumbersome.

POWELL. I would like to see this tried for constraint functions of several variables with one doing the special purpose programming as one went along, although I am not saying that it would work.

BEALE. I think it might be better to use some other approach such as the method of approximation programming or POP (process optimization procedure) which is basically the same idea in which you make an honest linear programming approximation at the trial solution and put bounds on the amount by which the nonlinear variables can move away from that solution, and use the solution of this problem as the new trial solution.

ANON. Would you comment on the practical problems associated with this method.

BEALE. The method will deal with very general problems which cannot be put into separable form and you can still deal with a large number of linear variables. The method may look very heavy handed and must be supplemented by some means of adjusting the amounts by which the nonlinear variables move from their current values. If a variable is oscillating between its upper and lower bounds on successive LP approximations then the bounds should be reduced. Conversely if the variable keeps to say its upper bound then it is likely that more rapid progress will be made by increasing the bound. Philip Wolfe was contrasting this method with the cutting plane method in his presentation; the cutting plane method is a more elegant method in a sense but has two important disadvantages. One is that it may not work for nonconvex problems, although this can be overcome. The other is that to my knowledge no one has succeeded in using it effectively on real problems because it is inherently ill-conditioned. As one gets nearer and nearer the solution so the cutting planes become more nearly parallel and the numerical difficulties increase. Although this is probably manageable on a two variable problem it is not so as the number of variables increases.

VAJDA. I would like to point out that I did not find any reference to a point which I think is rather significant—that is the application of intermediate smaller problems before you continue with the simplex method. This happens in the way in which Gilmore and Gomory treat the trim-loss problem where they introduce an intermediate knapsack problem. Do you think this of significance or just a phase in the development of the subject?

BEALE. No I do not think this is a passing phase but something which is likely to have wider application. Philip Wolfe mentioned decomposition as a special case of these general ideas, of adding columns as you go along. One advantage of the revised simplex method is that it makes this sort of thing rather easy, because you can add a column as you go along and work with the original form of the constraints, so that there is negligible extra work involved in incorporating the extra variables, other than in working out its coefficient in all the rows.

VAJDA. I am not sure now whether this is true in the Gilmore and Gomory method for the knapsack problem, an integer linear programming problem. I am not sure whether this is helped by the product form.

BEALE. It is significant that they do use the inverse matrix methods rather than a straight tableau method to do this. It is easy enough to generate the coefficients of your new variable in the original equations, but it is less easy to generate the coefficients of any new variable in a tableau.

WOLFE. I think that automatic column generation is one of the four or five truly practically important contributions to optimization in this field. Many of the problems where people have had success have used precisely this idea. The trouble is that it is not an easy idea but has very general application and is probably best described in a paper of Gomory's which very few people have read, that is: R. E. Gomory (1963). Large and nonconvex problems in linear programming. *Proc. Sympos. appl. Math.* **XV,** 125-139. American Mathematical Society, Providence, Rhode Island.

ANON. Are there any methods which use a quadratic programming approximation rather than LP approximation?

BEALE. Yes there are. There is a subroutine called SOLVER by Robert Wilson at Stanford. In these methods one has to do something about getting the quadratic part of the active constraints into the objective function. It has been done however for problems of up to seventy variables. The limit is caused by the linear variables due to the method of quadratic programming which he was using. I have a method for quadratic programming which allows an arbitrary number of linear variables but have not worked this up into a general nonlinear programming method. One of the difficulties that prevents this from being widely used is that one must estimate the quadratic terms both in the objective function and in the expressions for all the active constraints. This may lead to a fair amount of work.

4. Generalization of the Wolfe Reduced Gradient Method to the Case of Nonlinear Constraints

J. Abadie and J. Carpentier

Electricité de France and Institut Statistique de de l'Université de Paris, and Electricté de France and Public Power Corporation of Greece

Introduction

The GRG algorithm in this paper extends the reduced gradient method, due to Wolfe [1, 2], to the case where not only the objective function but also the constraints are nonlinear.

The principle of the current iteration is as follows:

(1) Partition the variables into dependent (i.e. basic) variables, y, and independent (i.e. non-basic) variables, x; the dependent variables are implicitly determined by the independent ones, which makes the objective function a function $F(x)$ of only the x variables.

(2) Compute the reduced gradient in the Wolfe sense, that is dF/dx.

(3) Determine the direction of progression of the independent variables by making zero the components of the reduced gradient corresponding to an unfeasible modification; modify these variables accordingly.

(4) Modify the dependent variables in order to verify the constraints.

Major differences with the Wolfe method in the case of linear constraints are:

(a) The computation of the reduced gradient is changed in order to avoid frequent reinversions of the basis, otherwise necessary: some duality properties due to Beale are related to this approximation.

(b) The modification of the dependent variables is obtained in solving a nonlinear system by Newton's method

A first experimental code has been written and tried on two different types of problems: test-problems where the solutions are known exactly, and industrial problems. Numerical experiments are reported.

37

1. The GRG algorithm numerically solves mathematical programming problems in the following standard form:

$$\text{Maximize} \quad f(X)$$

$$\text{subject to:} \; g(X) = 0, \quad A \leqslant X \leqslant B,$$

where $X \in R^N$ (the real N-dimensional space), g is a continuously differentiable mapping of R^N into R^m whose domain is the parallelotope $A \leqslant X \leqslant B$, and f is a continuously differentiable function defined on this parallelotope. We call P this (generalized) parallelotope, and V the manifold defined in R^N by $g(X) = 0$. Notice that any mathematical program can be put into the standard form by introduction of non-negative slack variables, if there are inequalities among the constraints, and by allowing some of the bounds A_i, B_i to be $-\infty$ or $+\infty$ if necessary.

2. Assume that we know some feasible solution X^0, that is some $X^0 \in P \cap V$. We make a *nondegeneracy assumption*, which can be stated as follows: the vector X^0 can be split into two components:

$$X^0 = (x^0, y^0), \quad y^0 \in R^m, \quad x^0 \in R^n, \quad n = N - m,$$

and correspondingly for A, B:

$$A = (\alpha, \alpha'), \quad B = (\beta, \beta'),$$

such that:

(1) $\alpha' < y^0 < \beta'$;

(2) the square $m \times m$ matrix $(\partial g / \partial y)_{x = x^0}$ is nonsingular.

Restating the mathematical programming problem as:

$$\text{maximize } f(x, y)$$

subject to:

$$g(x, y) = 0 \tag{1}$$

$$\alpha \leqslant x \leqslant \beta, \tag{2}$$

$$\alpha' \leqslant y \leqslant \beta', \tag{3}$$

the Kuhn–Tucker conditions for optimality at (x^0, y^0) are the existence of $u \in R_m$ (the notation R_m designates the dual space to R^m) and the existence of $v \in R_n$ such that:

$$v_i \begin{cases} \leqslant 0 \text{ if } x_i{}^0 = \alpha_i \\ \geqslant 0 \text{ if } x_i{}^0 = \beta_i \\ = 0 \text{ if } \alpha_i < x_i{}^0 < \beta_i \end{cases} \tag{4}$$

$$\frac{\partial f}{\partial x^0} - u \frac{\partial g}{\partial x^0} = v \tag{5}$$

$$\frac{\partial f}{\partial y^0} - u \frac{\partial g}{\partial y^0} = 0. \tag{6}$$

By substitution, we find from Eqns (5) and (6):

$$u = \frac{\partial f}{\partial y^0} \left(\frac{\partial g}{\partial y^0} \right)^{-1} \tag{7}$$

$$v = \frac{\partial f}{\partial x^0} - \frac{\partial f}{\partial y^0} \left(\frac{\partial g}{\partial y^0} \right)^{-1} \frac{\partial g}{\partial x^0}.$$

If v, so computed, satisfies Eqn (4), then X^0 is a stationary point (an optimal solution of the program if the initially given program has the proper convexity).

If v does not satisfy Eqn (4), then the value of x will be changed according to the following formula:

$$x = x^0 + \theta h, \tag{8}$$

where θ is a positive number, and the nonzero $h \in R^n$ is defined by:

$$h_i = 0 \begin{cases} \text{if } x_i^0 = \alpha_i \text{ and } v_i < 0 \\ \\ \text{or if } x_i^0 = \beta_i \text{ and } v_i > 0 \end{cases} \tag{9}$$

$$h_i = v_i \quad \text{or } 0 \text{ otherwise.} \tag{10}$$

We identify here R^n and R_n; moreover, the choice between the values v_i or 0, when possible, depends on the variant chosen for the method, and will be explained in section 7.2.

From Eqn (8), we define, at least in a neighbourhood of (x^0, y^0), a continuous curve Γ by the formulae:

$$x = x^0 + \theta h, \quad \theta \geqslant 0 \tag{11}$$

$$y = y(\theta), \quad y(0) = y^0 \tag{12}$$

$$g(x^0 + \theta h, \quad y(\theta)) = 0. \tag{13}$$

The tangent L to Γ at (x^0, y^0) is the (half) line:

$$x = x^0 + \theta h, \quad \theta \geqslant 0 \tag{11}$$

$$y = y^0 + \theta k \tag{14}$$

$$\frac{\partial g}{\partial x^0} h + \frac{\partial g}{\partial y^0} k = 0. \tag{15}$$

The following formula is easily checked:

$$(d/d\theta) f(x^0 + \theta h, y(\theta)) = (d/d\theta) f(x^0 + \theta h, y^0 + \theta k) = \theta |h|^2 \tag{16}$$

where $|h|$ designates the euclidean norm

$$|h| = \left(\sum_i (h_i)^2 \right)^{\frac{1}{2}}.$$

3. A value of X, called $\tilde{X}^1 = (x^1, \tilde{y}^1)$ will be chosen on L. First, θ must be restricted to those values for which:

$$\alpha \leqslant x^0 + \theta h \leqslant \beta, \quad \alpha' \leqslant y^0 + \theta k \leqslant \beta';$$

this gives for θ some non-zero interval, which is possibly infinite:

$$0 \leqslant \theta \leqslant \bar{\theta}.$$

We determine now θ_1 as maximizing $f(x^0 + \theta h, y^0 + \theta k)$, where $0 \leqslant \theta \leqslant \bar{\theta}$, from which $\tilde{X}^1 = (x^1, \tilde{y}^1)$ is computed as:

$$x^1 = x^0 + \theta_1 h, \quad \tilde{y}^1 = y^0 + \theta_1 k.$$

The point $\tilde{X}^1$, does not, in general, belong to V. We deduce from $\tilde{X}^1$ a point $X^1 = (x^1, y^1) \in V$, which is in fact $X^1 \in \Gamma$, by solving the following equation, where y is the unknown and x^1 is kept fixed,

$$g(x^1, y) = 0.$$

This is done by Newton's method, starting with $\tilde{y}^1$, the current iteration for Newton's method being as follows:

$$\Delta y^t = - \left(\frac{\partial g}{\partial y} \right)^{-1} g(x^1, \tilde{y}^t)$$

$$\tilde{y}^{t+1} = \tilde{y}^t + \Delta y^t,$$

where $\partial g / \partial y$ should be computed at $(x^1, \tilde{y}^t)$.

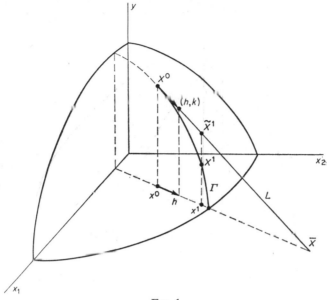

FIG. 1.

Figure 1 corresponds to a case where $m = 1$ (one constraint), $n = 2$, $A = 0$, $B = +\infty$; the point $\bar{X} \in L$ corresponds to $\theta = \bar{\theta}$, while $\tilde{X}^1$ corresponds to $\theta = \theta_1$.

Assuming that Newton's method converges to $X^1 \in V \cap P$, there are two possible ends for the current iteration.

3.1. If $\alpha' < y^1 < \beta'$, then we are in a position to repeat on X^1 what has been done on X^0 (this is the case in Fig. 1).

3.2. If, on the contrary, $y_r^1 = \alpha_r'$ or β_r' for some index r, then the y_r variable must become an x variable, while some x_s variable becomes a y variable in order to replace y_r. This operation, called the "change of basis", is performed as in the Simplex Method for linear programming, using the derivatives $f'(X^1)$, $g'(X^1)$ instead of the usual constant cost row and constraints matrix. This case is illustrated in Figs 2 and 3.

4. Newton's method can be unsuccessful in many different ways.

(1) It can fail to converge. For instance after, say, twenty iterations $(x^1, \tilde{y}^t)$ is not on Γ or some neighborhood of Γ; or after, say, five iterations, some norm $\|g(x^1, \tilde{y}^t)\|$ is greater than $\|g(x^1, \tilde{y}^1)\|$.

(2) For some iterate $\tilde{y}^t$, we have $f(x^1, \tilde{y}^t) < f(x^0, y^0)$.

(3) For some iterate $\tilde{y}^{t+1}$, the point $(x^1, \tilde{y}^{t+1})$ is outside the parallelotope P.

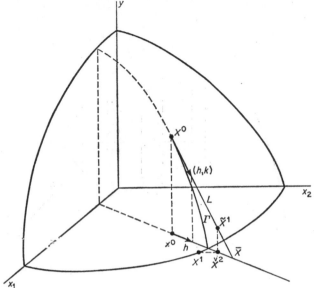

FIG. 2.

In cases (1) and (2), we reduce θ_1 (taking for instance, $\theta_1/2$ or $\theta_1/10$) and repeat Newton's method for the new point thus obtained on L.

In case (3) let $(x^1, \tilde{y}^t) \in P$; then the segment joining $(x^1, \tilde{y}^t)$ to $(x^1, \tilde{y}^{t+1})$ intersects the boundary of P at a point $(x^1, \check{y}^{t+1})$ such that $\check{y}_r^{t+1} = \alpha_r'$ or β_r' for some index r: we make a change of basis, as in section 3.2, and continue Newton's method with the new splitting thus obtained for X. Figure 2 explains such a situation, where we assume that neither case (1) nor case (2) is met: starting from $\tilde{X}^1 \in L$, case (3) is immediately encountered, so that $\check{X}^2$ is $(x^1, \check{y}^{t+1})$. Let us suppose now that the change of basis replaces y by x_2: Newton's method then leads to $X^1 \in V$, which is the next starting point for initiating the algorithm, as described in section 3. Still another possible realization of the method is shown in Fig. 3, where the iterates of Newton's method, starting from

$$\tilde{X}^1 = (x_1^1, x_2^1, \tilde{y}^1)$$

are

$$\check{X}^2 = (x_1^1, x_2^1, 0)$$
$$\check{X}^3 = (x_1^1, 0, 0),$$

from which Newton's method gives $X^1 \in V$. (Here also we assume that cases (1) and (2) are not met on the iterates $\check{X}^2, \check{X}^3, X^1$.)

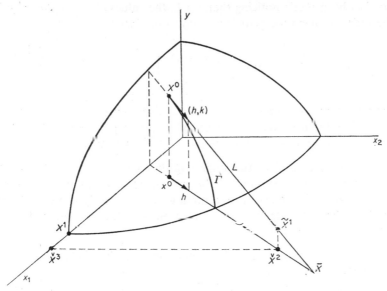

FIG. 3.

5. It is rather long, in practice, to compute $(\partial g/\partial y)^{-1}$ at each successive point X met during the computation. Instead, let us suppose that the inverse $(\partial g/\partial y^0)^{-1}$ has been computed at X^0: one can use it for all other points up to X^1, which is the next point on V, and even at X^1 and some of the subsequent points. We use also the following approximate value of $(\partial g/\partial y)^{-1}$:

$$\left(\frac{\partial g}{\partial y}\right)^{-1} \cong \left(\frac{\partial g}{\partial y^0}\right)^{-1} - \left(\frac{\partial g}{\partial y^0}\right)^{-1}\left(\frac{\partial g}{\partial y} - \frac{\partial g}{\partial y^0}\right)\left(\frac{\partial g}{\partial y^0}\right)^{-1}$$

that is

$$\left(\frac{\partial g}{\partial y}\right)^{-1} \cong 2\left(\frac{\partial g}{\partial y^0}\right)^{-1} - \left(\frac{\partial g}{\partial y^0}\right)^{-1}\frac{\partial g}{\partial y}\left(\frac{\partial g}{\partial y^0}\right)^{-1}.$$

By some easily derived check, such a series of approximations of the inverse can be stopped when necessary, and a true inverse is then computed.

It is worthwhile to mention here that an interesting duality property, due to Beale, is related to this approximation, although this point will not be discussed here.

Another point of practical importance is to avoid too large values of θ: we could put an upper bound on θ or, better, on $\theta|h|$. However, another device has been used: noticing that, on L, the values of $g_i(x^0 + \theta h, y^0 + \theta k)$ is of the order of magnitude $O(\theta^2)$ when g_i is of class C^2, we set

$$K = \max_{1 \leqslant i \leqslant m} \sqrt{\left[\frac{|g_i(\theta_1)|}{\varepsilon_i} \right]},$$

where ε_i is some (not necessasily small) positive number. If $K > 1$, we replace θ_1 by θ_1/K in the formulae giving $\tilde{X}^1$, while we keep θ_1 as it is if $K \leqslant 1$; this device approximately keeps the values of $g_i(\tilde{X}^1)$ within some bounds.

6. We have assumed since the beginning that we know a feasible solution. If this were not the case, any of the procedures used as a phase 1 of the simplex method for linear programming can be easily converted into a phase 1 for the GRG algorithm. If such a phase 1 does not succeed, then the constraint set is empty or nonconvex.

7. There are many variants of the GRG method. They differ mostly in two main points.

7.1. Differences with respect to the memorization of $(\partial g/\partial y^0)^{-1}$ are as follows.

7.1.1. In the "full-inverse" variant, the inverse $(\partial g/\partial y^0)^{-1}$ is kept in the memories of the machine and is modified, when a change of basis occurs, in the same way as in the corresponding simplex method for linear programming.

7.1.2. In the "product form of the inverse" variant, we retain in the memories of the machine at each change of basis, the so-called eta-vector (or η-vector) (see again the corresponding simplex method for linear programming).

7.1.3. In the "tableau" variant, the memories contain $(\partial g/\partial y^0)^{-1} (\partial g/\partial x^0)$, as in the simplex method computed by successive tableaux.

Other variants can be introduced for large structured problems, for instance block triangular or block angular structure of $g(X)$. It is worthwhile to notice that whatever variant is chosen, the computation of v by Eqn (7) in section 2 is analogous to the computation of the reduced cost or δ_j, or $c_j - z_j$ (depending on the various authors), in the simplex method for linear programming.

7.2. A second way of varying the method lays in the application of Eqn (10).

7.2.1. *GRG Method.* We choose in Eqn (10) $h_i = v_i$ for all possible indices i, that is for all indices i for which Eqn (9) does not compel us to set $h_i = 0$. In the case where $g(x) = Ax - b$ (which is the case of linear constraints), the

GRG method coincides with the reduced gradient method due to Wolfe [2]: this is why we call this method the Generalized Reduced Gradient method. In the terminology of Wolfe, which we adopt here, v is the reduced gradient. The reduced gradient method (for linear constraints) has been coded and extensively studied by Faure and Huard [3].

7.2.2. *GRGS Method.* We define an index s by:

$$|v_s| = \max_{i \in I} |v_i|.$$

where I is the set of all indices for which Eqn (10) holds, and we set:

$$h_i = 0, \quad \text{if } i \neq s$$
$$h_s = v_s.$$

This variant coincides with the simplex method of Dantzig in the case of a linear programming problem ($f(x) = cx$ and $y(x) = Ax - b$). One theoretical advantage of this simplex-like method is that the direction L does not depend entirely on the units on the axes of R^N—which is not the case for the GRG method—and another is that the amount of computations for the direction X is much less than in the GRG method.

7.2.3. *GRGC Method.* This cyclic method consists in setting

at the 1st iteration: $h_1 = v_1$, $h_1 = 0$ if $i \neq 1$;

at the 2nd iteration: $h_2 = v_2$, $h_i = 0$ if $i \neq 2$;

and so on, returning to $h_1 = v_1$ after N steps, and omitting the kth iteration each time k is the index of some y_k variable or an index such that Eqn (9) holds. This method does not depend at all on the units on the axes of R^N; moreover, it posseses some interesting continuity property of the operator $X^0 \rightarrow X^1$. The GRGC method coincides in the case where there are no constraints $g(X) = 0$ (that is where we maximize $f(X)$ simply subject to $X \in P$) with a method due to d'Esopo [4].

8. Experimental codes have been written and experiments have been conducted on two machines: IBM 7094 model 2 and CAE 510. Four main series of experiments have been run.

1st Series. These were problems with known solutions, found in the literature; among them were problems due to Roubault, van de Panne, Rosen and Suzuki [5–9]. These problems are quite small and were solved during the debugging process.

2nd Series. Some problems from chemical industry, sized ten constraints and thirty-four variables for a typical one, have been solved. The solutions were not known in advance: however, the results were the same as those given by the method of Centres of Huard [10], which was used for checking.

3rd Series. Colville (IBM, New York Scientific Center) [11] has created a kind of club whose purpose is to conduct numerical experiments in solving nonlinear programming problems. In mid 1968 the club had thirty-nine members, using codes based on various techniques, which are experimented on various machines. The real machine-time is divided by a factor, depending on the particular machine, in order to eliminate the influence of the speed of that machine. Ten problems are at the moment on the list of the test problems offered by the club to be solved by each member. The overall results show that the GRG method seems to behave quite satisfactorily, as it is leading in this friendly competition. For the description of these test problems and discussion of the results, readers are referred to Colville's report [11].

4th Series. A code has been written whose input is m, n, and whose output is a randomly generated convex program with a quadratic objective function in n variables and m quadratic inequality constraints in those variables (the bounds are simply $x \geqslant 0$). The optimal solution is unique and exactly known in advance. Typical of this series are four problems where $(m, n) = (5, 10)$, $(10, 20), (30, 50), (50, 50)$. Readers are referred to [12].

9. A second code is written for the CDC 6600. Here, we use the Fletcher and Reeves [13] approach to the conjugate gradient method in order to accelerate the convergence of the maximization of $F(x)$ (for the definition of $F(x)$, see the Introduction).

Acknowledgements.

It is our pleasure to thank Dr Philip Wolfe for his stimulating comments during his visit to Electricité de France, and Messrs, Hensgen and Niederlander whose help was invaluable in writing the first version of the code. We thank also Mr J. Guigou who is writing a second version which is different in principle from the first and very much more elaborate; the latter code appears to attain much better achievement both in precision and computing time. A report is due to appear as soon as numerical experiments are completed.

References

1. P. Wolfe (1967). Methods for linear constraints. *In* "Nonlinear Programming" (J. Abadie, ed.) pp. 120-124. North-Holland Publishing Company, Amsterdam.
2. P. Wolfe (1963). Methods of nonlinear programming. *In* "Recent Advances in Mathematical Programming", (Graves and Wolfe, eds.) pp. 67-86, especially section 6. McGraw-Hill, New York.
3. P. Faure and P. Huard (1965). "Résolution des Programmes Mathématiques à Fonction Nonlinéaire par la Méthode du Gradient Réduit". Revue Française de Recherche Opérationnelle 9, 167-205.
4. D. A. d'Esopo (1959). A convex programming procedure. *Nav. Res. Logist. Q.* 6, No. 1, 33-42.
5. M. C. Roubault (1963). "Etude d'un Algorithme de Résolution pour les Programmes Nonlinéaires Convexes". Thése, Paris, 1963.
6. C. van de Panne (1964). "Programming with a Quadratic Constraint", Informal discussion paper, University of Birmingham. 1964.
7. J. B. Rosen (1960). The gradient projection method for nonlinear programming. Part I: linear constraints. *J. Soc. ind. appl. Math.* 8, 181-217.
8. J. B. Rosen (1961). The gradient projection method for nonlinear programming. Part II: nonlinear constraints. *J. Soc. ind. appl. Math.* 9, 514-532.
9. J. B. Rosen and S. Suzuki (1965). Construction of nonlinear programming test problems. *Communs. Ass. comput. Mach.* 8, 113.
10. P. Huard (1967). Resolution of mathematical programming with nonlinear constraints by the method of centres. *In* "Nonlinear Programming" (J. Abadie, ed.) pp. 209-219, North-Holland Publishing Company, Amsterdam.
11. A. R. Colville (1968). "A Comparative Study on Nonlinear Programming Codes." IBM New York Scientific Center. Report 320-2949.
12. J. Abadie and J. Carpentier (1967). "Some Numerical Experiments with the GRG Method for Nonlinear Programming". Paper HR 7422, Electricité de France.
13. R. Fletcher and C. M. Reeves (1964). Function minimization by conjugate gradients. *Comput. J.* 6, 163 *et seq.*

Discussion

ANON. What can be done in the method described by Professor Abadie when the object function is not convex?

ABADIE. The method works for nonconvex functions provided that the derivatives are continuous, but the solution obtained may only be a local optimum. A global study, or some other information, would be required to guarantee a global minimum of a function.

VAJDA. (University of Birmingham) Will Professor Abadie tell us which of the methods mentioned by Dr Fletcher are incorporated in his code?

ABADIE. Many of the methods such as Davidon's and Powell's could be incorporated, but we have chosen Fletcher and Reeves' method of conjugate gradients for many reasons, but primarily because of low storage requirements. For example a hundred equation problem would require 10,000 memories with a matrix method, which we could not afford. In applying the method, the conjugate gradient is used in place of the gradient, except when the method reaches a boundary, in which case the cycle is restarted.

5. Programmation Mathématique Convexe

P. HUARD

Conseiller Scientifique E.D.F. Service I.M.A

1. Introduction

Nous établissons dans ce qui suit, avec deux variantes, un algorithme de résolution pour les programmes mathématiques convexes, du type:

maximiser $f(x)$ sous les conditions

$$g_i(x) \geqslant 0, \quad i = 1, 2, \dots m$$

où les fonctions numériques f et g_i, définies dans R^n, sont concaves et continuement différentiables. Nous supposons de plus qu'il existe un point $x^0 \in R^n$ tel que $g_i(x^0) > 0$, pour l'ensemble des fonctions g_i non affines.

L'intérêt de cet algorithme est de conduire à la résolution d'une suite, théoriquement infinic mais finie pratiquement, de programmes linéaires analogues, c'est-à-dire dont le nombre de contraintes est constant. Plus précisément, ces contraintes linéaires sont obtenues en linéarisant les diverses fonctions f et g_i, en des points tendant vers la solution optimale du problème donné.

Si certaines fonctions g_i sont de plus affines, les contraintes correspondantes des programmes linéaires ont leurs expressions inchangées, ce qui fait que cet algorithme ne détruit pas la partie linéaire du problème: à la limite, si le problème donné est un programme linéaire, les calculs se ramènent à la méthode simpliciale, avec paramétrisation d'un second membre.

En fait, cet algorithme, sous sa première variante, n'est qu'une application de la méthode des centres telle qu'elle est définie dans [1].

On désigne par cette expression une famille d'algorithmes, permettant de résoudre les programmes mathématiques non linéaires sous des hypothèses assez larges (par exemple: la variable x peut être un élément d'un espace topologique quelconque et la convexité n'est pas nécessaire), chaque algorithme étant défini par le choix d'une fonction devant satisfaire à quelques conditions simples, le choix de cette fonction demeurant par ailleurs assez arbitraire. C'est en utilisant les propriétés de convexité et de différentiabilité du programme envisagé ici que nous avons pu choisir une telle fonction s'adaptant bien aux calculs par linéarisation.

Nous donnons en Section 2 des rappels concernant la méthode des centres, en nous limitant au minimum nécessaire, et nous établissons en Section 3 l'adaptation au problème posé ici. On remarquera qu'une partie des calculs de l'algorithme (partie que nous désignons par "algorithme partiel") consiste à maximiser une fonction concave non continuement différentiable sur un polyèdre, et peut être considérée comme une généralisation de la méthode de Franck et Wolfe [2]. Enfin, la partie Section 4 est consacrée à quelques considérations pratiques.

Cet article était rédigé lorsque nous avons eu connaissance de celui de Topkis et Veinott [3], dont le passage consacré aux problèmes de maximisation sous contraintes contient un des algorithmes présentés ici. En fait, la variante dite du premier ordre décrite dans [3] utilise—implicitement—la méthode des centres. C'est pourquoi nous avons ajouté 3.5.5. afin de préciser cette liaison.

Notations:

Si $A \subset R^n$, on désigne par A^0 l'intérieur de A et par Fr (A) la frontière de A.

Si $f : R^n \to R$ est une fonction numérique continuement différentiable, on désigne par $\nabla f(x)$ la valeur du gradient de f calculé au point x.

Si $a, b \in R^n$, leur produit scalaire est noté simplement $a.b$.

2. Rappels Sur La Méthode des Centres

2.1. *Problème envisagé*

On considère le programme mathématique P suivant:

$$P: \begin{cases} \text{Maximiser } f(x) \text{ sous les conditions} \\ x \in A \subset R^n \\ x \in B \subset R^n \end{cases}$$

où $f : R^n \to R$ est une fonction numérique continue, bornée supérieurement sur $A \cap B$, telle que

$$\{x|f(x) = \lambda\} = Fr\{x|f(x) \geqslant \lambda\}, \forall \lambda \in R \qquad (1)$$

et où A est un ensemble de R^n vérifiant l'hypothèse suivante:

$$A^0 \neq \emptyset \text{ et } Fr(A) = Fr(A^0) \qquad (2)$$

Aucune hypothèse n'est faite à présent sur B.

On désignera par $E(\lambda)$ un "tronçon" de A, c'est-à-dire:

$$E(\lambda) = \{x | x \in A, \; f(x) \geqslant \lambda\} \;\; (\lambda \in R)$$

2.2. *Fonction F-distance*

On apelle *F*-distance compatible avec f toute fonction numérique $d: R^n \times R \to R$ satisfaisant aux conditions:

(i) $d(x, \lambda) > 0, \;\;\;\; \forall \lambda \in R$ et $\forall x \in E^0(\lambda)$

(ii) $d(x, \lambda) = 0, \;\;\; \forall \lambda \in R$ et $\forall x \in Fr\,[E(\lambda)]$

(iii) $\forall \lambda, \lambda' \in R$, on a:

$$x \in E(\lambda) \text{ et } \lambda > \lambda' \Rightarrow d(x, \lambda) \leqslant d(x, \lambda')$$

(iiii) (compatibilité avec f)

$\forall$ la suite $\lambda_k \in R, k = 1, 2, 3, \ldots$ monotone non décroissante, et $\forall$ la suite $x^k \in R^n, k = 1, 2, 3, \ldots$ telle que

$$x^k \in Fr\,[E(\lambda_k)], \forall k,$$

on a $\;\;\;\;\;\;\;\;\;\;\; f(x^{k+1}) - f(x^k) \to 0 \;\;$ quand $k \to \infty$

$$\Rightarrow d(x^{k+1}, \lambda_k) \to 0 \;\;$$ quand $k \to \infty$

2.3. *ε-centre d'un tronçon*

Etant donné un tronçon $E(\lambda)$ et $\varepsilon \in R$ tel que $0 \leqslant \varepsilon < \sup\,\{d(x, \lambda) | x \in E(\lambda)\}$, on appelle ε-centre de $E(\lambda)$ tout point $\bar{x}$ de $E(\lambda)$ tel que:

$$d(\bar{x}, \lambda) \geqslant \sup\,\{d(x, \lambda) | x \in E(\lambda)\} - \varepsilon.$$

Un 0-centre, s'il existe, est appelé un centre.

2.4. *Méthode des centres*

Cette méthode consiste, après avoir choisi une *F*-distance d compatible avec f, et une valeur initiale λ_0 convenable, c'est-à-dire telle que:

$$\lambda_0 < \sup\,\{f(x) | x \in A \cap B\},$$

à résoudre la suite de programmes mathématiques suivante, pour $k = 0, 1, 2, 3, \ldots$:

$$Q_k: \begin{cases} \text{Maximiser } d(x, \lambda_k), \text{ à } \varepsilon_k \text{ près, sous les conditions} \\ x \in E(\lambda_k) \\ x \in B \end{cases}$$

En d'autres termes, la résolution de Q_k consiste à déterminer un "ε_k — centre relatif" de $E(\lambda_k)$ dans B. Soit x^{k+1} ce point. La suite des valeurs λ_k est obtenue par:

$$\lambda_k = f(x^k), \quad k = 1, 2, 3, \ldots$$

La suite des ε_k doit être monotone décroissante, convergente vers 0 et satisfaire à

$$0 \leqslant \varepsilon_k < \sup \{d(x, \lambda_k) | x \in E(\lambda_k) \cap B\}.$$

Les ε_k-centres relatifs étant des points intérieurs aux tronçons $E(\lambda_k)$, la contrainte $x \in E(\lambda_k)$ est généralement inutile *sur le plan pratique des calculs*.

On obtient une suite de points x^k, $k = 1, 2, 3, \ldots$, généralement infinie, qui converge vers une solution optimale $\hat{x}$ du problème P. Les tronçons $E(\lambda_k)$ diminuent par inclusion et tendent vers une limite $E(\hat{\lambda})$ telle que $E^0(\hat{\lambda}) \cap B = \emptyset$. Les valeurs de

$$\sup \{d(x, \lambda_k) | x \in E(\lambda_k) \cap B\}$$

tendent vers 0 quand $k \to \infty$, et cette remarque permet de définir les ε_k de la façon suivante:

$$\varepsilon_k = (1 - \rho) \sup \{d(x, \lambda_k) \mid x \in E(\lambda_k) \cap B\}$$

avec $\rho \in \,]0, 1]$, constante indépendante de k. Dans ces conditions, les points x^{k+1} vérifient la relation ci-dessous, qui nous sera utile par la suite:

$$x^{k+1}: d(x^{k+1}, \lambda_k) \geqslant \rho \sup \{d(x, \lambda_k) \mid x \in E(\lambda_k) \cap B\}, \quad \rho \in \,]0, 1]. \qquad (3)$$

3. Résolution du Problème P

3.1. *Adaption de la méthode des centres*

Nous nous proposons, dans ce qui suit, d'appliquer la méthode des centres, telle qu'elle est décrite en Section 2, dans les conditions particulières suivantes:

(a) I est un ensemble fini d'indices. Par exemple $I = \{1, 2, \ldots, m\}$:

$$f, g_i : R^n \to R, \quad i \in I, \qquad (4)$$

sont des fonctions concaves continuement différentiables.

(b) f satisfait à (1) et les g_i à une condition semblable.

(c) $A = \{x \mid g_i(x) \geqslant 0, \ \forall i \in I\}$. $\qquad (5)$

A est donc un ensemble convexe fermé de R^n, ainsi que les tronçons $E(\lambda)$.

(d) $A^0 \neq \emptyset$ $\qquad (6)$

Par suite, puisque A est convexe, on a bien $Fr(A) = Fr(A^0)$ et l'hypothèse (2) est satisfaite (Section 1).

(e) B est un polyèdre linéaire fermé de R^n,   (7)

c'est-à-dire qu'il est défini par un système d'inégalités et d'égalités linéaires. Nous supposons de plus qu'il est borné, donc compact, ce qui est toujours possible en ajoutant éventuellement la condition $x \in C$, où C est un pavé de R^n, assez grand pour ne pas modifier la solution du problème P, supposée à distance finie.

(f) $d(x, \lambda) = \min \{f(x) - \lambda, g_i(x) \mid i \in I\}$.   (8)

Il s'agit bien d'une F-distance compatible avec f [1], p. 69, fonction concave en x pour tout λ fixé, prenant des valuers $\leqslant 0$ pour tout $x \notin E(\lambda)$. Elle atteint son maximum par rapport à x, sur chaque ensemble $E(\lambda) \cap B$, tel que $E^0(\lambda) \cap B \neq \emptyset$, en un point intérieur à $E(\lambda)$.

(g) Les différents points x^{k+1} fournis par la méthode des centres seront déterminés d'après le critère (3), c'est-à-dire en résolvant le problème suivant:

$$Q_k: \begin{cases} \text{Trouver } x^{k+1}: d(x^{k+1}, \lambda_k) \geqslant \rho \max \{d(x, \lambda_k) \mid x \in E(\lambda_k) \cap B\} \\ \text{avec } \rho \in]0, 1] \text{ constante indépendante de } k \\ \lambda_k = f(x^k) \end{cases}$$

La suite des problèmes Q_k, $k = 1, 2, 3, \ldots$ représente l'essentiel des calculs à effectuer.

3.2. Linéarisation

Soit $g_i': R^n \times R^n \to R$, $\forall i \in I$, des fonctions numériques définies par:

$$g_i'(x; y) = g_i(y) + \nabla g_i(y) \cdot (x - y)$$ (9)

et soit $f': R^n \times R^n \to R$ la fonction numérique définie par:

$$f'(x; y) = f(y) + \nabla f(y) \cdot (x - y).$$ (10)

Les g_i' et f' sont des fonctions affines de x. Posons par ailleurs

$$d'(x, \lambda; y) = \min \{f'(x; y) - \lambda, \ g_i'(x; y) \mid i \in I\}$$ (11)

d' est une fonction concave et affine par morceaux de x.

Nous avons la relation classique, liée à la concavité des fonctions envisagées:

$$d'(x, \lambda; y) \geqslant d(x, \lambda) \quad \forall x, y \in R^n, \quad \forall \lambda \in R.$$ (12)

Considérons le programme mathématique $Q'(\lambda; y)$ suivant:

$$Q'(\lambda; y): \begin{cases} \text{Maximiser } d'(x, \lambda; y) \text{ sous la condition} \\ \quad x \\ x \in B \end{cases}$$

λ et y étant fixés. Ce problème peut se formuler sous la forme d'un programme linéaire en introduisant une variable supplémentaire μ:

$$Q''(\lambda; y) \begin{cases} \text{Maximiser } \mu \text{ sous les conditions} \\ \quad g_i'(x; y) - \mu \geqslant 0, \quad \forall i \in I \\ \quad f'(x; y) - \mu \geqslant \lambda \\ \quad x \in B \end{cases}$$

λ et y étant fixés.

3.3. *Algorithme partiel*

Notons $z(\lambda; y)$ une solution optimale du programme linéaire $Q''(\lambda; y)$. Considérons l'algorithme suivant, au cours duquel λ demeure constant.

(1) Choisir une valeur de départ $y^0 \in B$. Faire $h = 0$.

(2) Résoudre $Q''(\lambda; y^h)$ — Soit $z^h = z(\lambda; y^h)$ une solution.

(3) Déterminer $y^{h+1}: d(y^{h+1}, \lambda) = \max \{d(x, \lambda) \mid x \in [y^h, z^h]\}$.

(4) Aller en (2) avec $h+1$ au lieu de h.

On obtient ainsi une suite infinie de points y^h, avec éventuellement à partir d'un certain rang, des points identiques: c'est le cas si l'on trouve $y^{h+1} = y^h$ pour un certain h, ce qui entraine alors que tous les problèmes $Q''(\lambda; y^h)$ ultérieurs sont identiques.

La suite des valeurs $d(y^h, \lambda)$, $h = 0, 1, 2, \dots$ est monotone non décroissante d'après la partie (3) de l'algorithme.

3.4. *Convergence de l'algorithme partiel*

Soit S la suite infinie des indices h. On a, $\forall h \in S$, $z^h \in B$ par définition, et $y^{h+1} \in B$ également, car B est convexe, $y^{h+1} \in [y^h, z^h] \subset B$ et $y^0 \in B$. Puisque B est compact, $B \times B$ l'est aussi, et il existe une sous-suite $S' \subset S$ d'indices h tels que

$$(y^h, z^h) \to (\bar{y}, \bar{z}) \text{ quand } h \to \infty, \ h \in S'$$

avec $\qquad\qquad\qquad \bar{y} \in B, \ \bar{z} \in B.$

D'une part, on peut écrire, $\forall h \in S$:

$$d'(z^h, \lambda; y^h) \geqslant d'(y^h, \lambda; y^h) \text{ car } z^h \text{ maximise } d' \text{ sur } B$$

$$= d(y^h, \lambda) \quad \text{d'après les définitions (9) et (10).}$$

Si $h \to \infty$, $h \in S'$, on a à la limite:

$$d'(\bar{z}, \lambda; \bar{y}) \geqslant d(\bar{y}, \lambda) \tag{13}$$

D'autre part, soit $\theta \in [0, 1]$. On a, $\forall h, h' \in S'$, $h' > h$:

$$d[y^h + \theta(z^h - y^h), \lambda] \leqslant \max \{d(x, \lambda) \mid x \in [y^h, z^h]\}$$

$$= d(y^{h+1}, \lambda) \text{ par définition}$$

$$\leqslant d(y^h, \lambda) \text{ car } h' \geqslant h+1, \text{ et la suite des valeurs}$$

$$d(y^h, \lambda) \text{ est monotone non décroissante}$$

Par suite, quand $h \to \infty$, $h \in S'$, en passant à la limite, avec θ fixé, on obtient:

$$d[\bar{y} + \theta(\bar{z} - \bar{y}), \lambda] \leqslant d(\bar{y}, \lambda) \quad \forall \theta \in [0, 1] \tag{14}$$

La relation (14) entraîne avec (13):

$$d'(\bar{z}, \lambda; \bar{y}) = d(\bar{y}, \lambda) \tag{15}$$

En effet, du fait que f et les g_i sont contiuement différentiables, (14) entraîne (16) ou (17): ʹ

$$\exists i_0 \in I: \quad \left| \begin{array}{l} g_{i0}(\bar{y}) = d'(\bar{y}, \lambda; \bar{y}) = d(\bar{y}, \lambda) \\ \nabla g_{i0}(\bar{y}) \cdot (\bar{z} - \bar{y}) \leqslant 0 \end{array} \right. \tag{16}$$

$$\left| \begin{array}{l} f(\bar{y}) - \lambda = d'(\bar{y}, \lambda; \bar{y}) = d(\bar{y}, \lambda) \\ \nabla f(\bar{y}) \cdot (\bar{z} - \bar{y}) \leqslant 0 \end{array} \right. \tag{17}$$

et puisque $d'(x, \lambda; y) \leqslant \min \{f'(x, y) - \lambda, g''_{i0}(x; y)\}$, (16) ou (17), avec (13), entraîne (15).

Montrons que $\bar{y}$ maximise $d(x, \lambda)$ sur $E(\lambda) \cap B$.

Soit un point quelconque $x \in B$. On a, $\forall h \in S$:

$$d(x, \lambda) \leqslant d'(x, \lambda; y^h) \text{ d'après (12)} \tag{18}$$

$$\leqslant d'(z^h, \lambda; y^h) \text{ car } z^h \text{ maximise } d'(x, \lambda; y^h) \text{ sur } B \tag{19}$$

A la limite, quand $h \to \infty$, $h \in S'$:

$$d(x, \lambda) \leqslant d'(\bar{z}, \lambda; \bar{y}) = d(\bar{y}, \lambda) \text{ d'après (15)}$$

En définitive

$$d(x, \lambda) \leqslant d(\bar{y}, \lambda), \quad \forall x \in B \tag{20}$$

Puisque la suite des valeurs $d(y^h, \lambda)$, $h \in S$, est monotone non décroissante, elle converge vers $d(\bar{y}, \lambda)$.

3.5. Remarques sur la résolution de Q_k

3.5.1—Nous venons d'établir en Sections 3.3 et 3.4 un algorithme, dit partiel, qui permet de résoudre le problème Q_k, dans le cas $\rho = 1$, par une séquence infinie de résolutions de programmes linéaires Q'' $(\lambda_k; y^h)$, sous l'hypothèse de concavité des fonctions f et g_i envisagées.

En fait, il est possible de reprendre la démonstration de la convergence, établie en Section 3.4, en n'utilisant plus cette hypothèse: le résultat, plus faible, ainsi obtenu, est que l'algorithme converge vers un point stationnaire (*cf.* Section 3.5.2)

On peut également, lors de la maximisation de $d(x, \lambda)$ sur le segment $[y^h, z^h]$, ne pas aller jusqu'au bout de cette maximisation, comme il est indiqué en Sections 3.5.3, sans détruire la convergence de l'algorithme partiel.

Enfin, la méthode des centres n'exige pas pour chaque tronçon de calculer un centre, (c'est-à-dire de prendre $\rho = 1$ dans la définition de Q_k), mais seulement un ε-centre ($\rho \in \,]0, 1]$). Nous donnons en Sections 5.3.4. un moyen de contrôler si le point y^h trouvé est bien une solution de Q_k, c'est-à-dire vérifiant la condition (3).

Nous montrons en Sections 5.3.5. que le premier point y^h trouvé dans l'algorithme partiel peut toujours être considéré comme un ε-centre, c'est-à-dire que l'on peut théoriquement se contenter d'un seul programme linéaire par itération de la méthode des centres.

3.5.2. Abandon de la concavite

Dans l'établissement de la convergence, en Section 3.4, la concavité des fonctions f et g n'est utilisée qu'à la fin, quand on fait appel à la relation (12): on peut établir ainsi que $\bar{y}$ est un maximum global de $d(x, \lambda)$ sur B. Si on abandonne l'hypothèse de concavité, la relation (15) demeure par contre valable.

Soit $x \in B$ et $\varepsilon = \|x - \bar{y}\|$ — Si ε est assez petit, et $\forall h \in S'$ assez grand (c'est-à-dire y^h suffisamment voisin de $\bar{y}$), on a au moins l'un des résultats suivants (21) et (22):

$$\exists i_0 \in I: \quad \left| \begin{array}{l} d(y^h, \lambda) = g_{i0}\,(y^h) \\[1ex] d'(x, \lambda; y^h) = g_{i0}\,(y^h) + \nabla g_{i0}\,(y^h) \cdot (x - y^h) \end{array} \right. \qquad (21)$$

$$\left| \begin{array}{l} d(y^h, \lambda) = f(y^h) - \lambda \\[1ex] d'(x, \lambda; y^h) = f(y^h) - \lambda + \nabla f(y^h) \cdot (x - y^h). \end{array} \right. \qquad (22)$$

Par ailleurs,

$$d'(x, \lambda; y^h) \leqslant d'(z^h, \lambda, y^h) \text{ car } z^h \text{ maximise } d' \text{ sur } B. \qquad (23)$$

En passant à la limite quand $h \to \infty$, $h \in S'$, les relations (21) à (23) entraînent, en tenant compte de (15), au moins l'un des deux résultats suivants, correspondant respectivement aux situations (21) et (22):

$$\nabla g_{i0}\,(\bar{y}) \cdot (x - \bar{y}) \leqslant 0 \qquad (24)$$

$$\nabla f(\bar{y}) \cdot (x - \bar{y}) \leqslant 0 \qquad (25)$$

ce qui montre que la limite $\bar{y}$ est un point stationnaire pour la maximisation de $d(x, \lambda)$ dans B.

3.5.3. *Maximisation approchée sur un segment*

Si on remplace l'ordre (3) de l'algorithme partiel par l'ordre suivant:

(3)   Déterminer y^{h+1}:

$$d(y^{h+1}, \lambda) \geqslant d(y^h, \lambda) + \rho'[\max \{d(x, \lambda) \mid x \in [y^h, z^h]\} - d(y^h, \lambda)]$$

avec $\rho' \in \,]0, 1]$, ρ' constante indépendante de h

l'algorithme partiel converge encore. En effet, le raisonnement qui aboutit à la relation (13) demeure valable, et celui qui aboutit à la relation (14) donne dans ces conditions:

$$d[y^h + \theta(z^h - y^h), \lambda] \leqslant d(y^h, \lambda) + (1/\rho')[d(y^{h'}, \lambda) - d(y^{h'}\lambda)] \quad \forall \theta \in [0, 1]$$

et à la limite, quand $h \to \infty$, $h \in S'$, on retrouve (14).

Ce résultat est intéressant sur le plan pratique, car la maximisation de $d(x, \lambda)$ sur un segment est toujours approchée.

3.5.4. *Reconnaissance d'un ε-centre*

Etant donné $\rho \in \,]0, 1]$, le fait que pour un rang h^* fini, on ait $d(y^{h^*}, \lambda_k) \geqslant$ $\rho\, d(\bar{y}, \lambda_k)$ est évident. Il reste à pouvoir reconnaître que cette inégalité est satisfaite; autrement dit, il faut définir un critère d'arrêt.

Pour cela, calculons une borne supérieure de $d(\bar{y}, \lambda)$ en prenant $x = \bar{y}$ et $\lambda = \lambda_k$ dans la relation (19), ce qui donne:

$$d(\bar{y}, \lambda_k) \leqslant d'(z^h, \lambda_k; y^h), \quad \forall h \in S \tag{26}$$

Par suite, on peut arrêter l'algorithme partiel quand on a obtenu un point y^{h^*} tel que:

$$d(y^{h^*}, \lambda_k) \geqslant \rho\, d'(z^{h^*}, \lambda_k; y^{h^*}) \; \rho \in \,]0, 1] \text{ donné.} \tag{27}$$

Ce point y^{h^*} est solution de Q_k et l'on pose

$$x^{k+1} = y^{h^*}.$$

3.5.5. *La première étape est un ε-centre*

Si, dans la résolution de Q_k, on s'arrête systématiquement à la première solution y^1 trouvée, que l'on pose $x^{k+1} = y^1$ pour passer au problème Q_{k+1} suivant, et si $x^0 \in B \cap E(\lambda_0)$, on obtient encore une méthode des centres. Il suffit, pour le montrer, d'établir que la première solution de Q_k est un ε_k-centre relatif pour le tronçon $E(\lambda_k)$, c'est-à-dire que $x^k \in B \cap E(\lambda_k)$, $\forall k$, et que la séquence infinie des ε_k tend vers zéro quand $k \to \infty$.

L'algorithme ainsi défini conduit à une séquence S d'itérations qui se résument comme suit:

Etant donnés $x^k, \lambda_k = f(x^k)$, déterminer z^k, puis x^{k+1}, tels que:

$$d'(z^k, \lambda_k; x^k) = \max \{d'(x, \lambda_k; x^k) \mid x \in B\} \tag{28}$$

$$d(x^{k+1}, \lambda_k) = \max \{d(x, \lambda_k) \mid x \in [x^k, z^k]\}. \tag{29}$$

Dans ces conditions on vérifie aisément de proche en proche, à partir de x^0, que $x^k \in B \cap E(\lambda_k)$ et que $d(x^k, \lambda_k) = f(x^k) - \lambda_k = 0$. D'ou l'on remarque que:

$$d(x^{k'}, \lambda_k) \geqslant d(x^{k+1}, \lambda_k), \qquad \forall k,\ k' \in S,\ k' > k. \tag{30}$$

Par ailleurs, λ_k étant borné supérieurement, il existe une sous-suite $S' \subset S$ telle que:

$$(x^k, z^k, \lambda_k) \to (\bar{x}, \bar{z}, \bar{\lambda}) \text{ quand } k \to \infty, k \in S'.$$

On en déduit, puisque $0 = d(x^k, \lambda_k)$, $\forall k$, et en passant à la limite:

$$d(\bar{x}, \bar{\lambda}) = 0. \tag{31}$$

En reprenant le raisonnement de Section 3.4, mais en remplaçant l'indice d'itération h par k, en tenant compte de ce que λ prend une valeur différente λ_k à chaque itération, et en utilisant les remarques (30) et (31):

$$d'(\bar{z}, \bar{\lambda}; \bar{x}) = 0. \tag{32}$$

Exprimons une borne supérieure de l'écart ε_k entre la valeur trouvée $d(x^{k+1}, \lambda_k)$ et la valeur de d en un centre relatif c^k du tronçon $E(\lambda_k)$ dans B, sous l'hypothèse de concavité des fonctions f et g_i:

$$\varepsilon_k = d(c^k, \lambda_k) - d(x^{k+1}, \lambda_k) \geqslant 0 \tag{33}$$

$d(c^k, \lambda_k) \leqslant d'(c^k, \lambda_k; x^k)$ d'après (12), en prenant $x = c^k$, $\lambda = \lambda_k$ et $y = x^k$

$\leqslant d'(z^k, \lambda_k; x^k)$ car z^k maximise d' sur B. $\tag{34}$

Par ailleurs, d'après (30):

$$d(x^{k+1}, \lambda_k) \geqslant d(x^k, \lambda_k) = 0, \quad \forall k \in S \tag{35}$$

L'expression (33) de l'écart devient, avec (34) et (35):

$$\varepsilon_k \leqslant d'(z^k, \lambda_k; x^k) \quad \forall k \in S'. \tag{36}$$

En passant à la limite, et d'après (32), la borne supérieure tend vers zéro: il en est de même pour l'écart $\varepsilon_k \geqslant 0$.

3.6. *Troncatures par excès et par défaut*

Dans la méthode des centres, telle qu'elle est décrite en Section 2.4, les tronçons sont déterminés par les solutions x^k. Plus précisément, une fois trouvé, dans B, un ε-centre relatif x^{k+1} du tronçon $E(\lambda_k)$, le nouveau tronçon est pris égal à $E[f(x^{k+1})]$.

Une idée assez naturelle consiste à envisager un procédé de troncature différent, conduisant à l'algorithme suivant:

(1) Prendre $\alpha = -\alpha_0$ et $\beta = +\beta_0$, α_0 et $\beta_0 > 0$ assez grands.

Choisir $\lambda_0 \in [\alpha, \beta]$ arbitraire. Aller en (2) avec $k = 0$.

(2) Déterminer $T_k = E(\lambda_k) \cap B$

Si $T_k \neq \emptyset$, prendre $\alpha = \lambda_k$

Si $T_k = \emptyset$, prendre $\beta = \lambda_k$

Aller en (3).

(3) Prendre $\lambda_{k+1} = (\alpha+\beta)/2$.

Aller en (2) avec $k+1$ au lieu de k.

Cet algorithme opère avec des valeurs de troncature λ_k ajustées par excès et par défaut et convergeant vers la valeur $\bar{\lambda} = f(\hat{x})$ cherchée.

Pour vérifier ai un tronçon $E(\lambda)$ a, avec B, une intersection $T(\lambda)$ vide ou non, il suffit de maximiser $d(x,\lambda)$, définie par (8), pour $x \in B$. S'il est possible de trouver un point $\bar{x}$ tel que $d(\bar{x},\lambda) \geqslant 0$, alors $T(\lambda) \neq \varnothing$. Sinon, c'est-à-dire si max $\{d(x,\lambda) \mid x \in B\} < 0$, on en déduit que $T(\lambda) = \varnothing$.

Soit $\bar{x}(\lambda)$ un point qui maximise $d(x,\lambda)$ sur B. Posons $\bar{\mu}(\lambda) = d[\bar{x}(\lambda),\lambda]$. Cette fonction numérique de λ est un critère de vacuité pour $T(\lambda)$:

$$T(\lambda) \neq \varnothing \Leftrightarrow \bar{\mu}(\lambda) \geqslant 0$$

$$T(\lambda) = \varnothing \Leftrightarrow \bar{\mu}(\lambda) < 0.$$

Pour chaque valeur λ proposée, on peut utiliser l'algorithme partiel décrit en Section 3, afin de déterminer $\bar{\mu}(\lambda)$. On calcule par dichotomies successives la valeur de λ qui annule $\bar{\mu}(\lambda)$, en utilisant toute technique d'interpolation utile. En particulier, en remarquant que $\bar{\mu}(\lambda)$ est une fonction concave (résultat classique sur la paramétrisation du second membre des programmes mathématiques convexes), la valeur de λ donnée par l'interpolation linéaire

$$\lambda' = \frac{\beta\bar{\mu}(\alpha) - \alpha\bar{\mu}(\beta)}{\bar{\mu}(\alpha) - \bar{\mu}(\beta)}$$

donne une estimation par défaut de λ.

Par ailleurs, la dualité permet d'obtenir une estimation par excès de $\hat{\lambda}$ à chaque résolution de programme linéaire $Q''(\lambda_k; y^h)$, si cette dernière nous fournit, avec la solution optimale, les multiplicateurs de Kuhn et Tucker correspondants. Les conditions d'optimalité de $Q''(\lambda_k; y^h)$ donnent, entre autres relations, en posant $B = \{x \mid B_l x > b_l, l \in L\}$:

$$\exists u^i, i \in I; \ u^0; v^l, \ l \in L \text{ tels que:}$$

$$u^i \geqslant 0, \quad \forall i \in I$$

$$u^0 \geqslant 0$$

$$v^l \geqslant 0, \quad \forall l \in L$$

$$\sum_{i \in I} u^i \nabla g_i(y^h) + u^0 \nabla f(y^h) + \sum_{i \in I} v^l B^l = 0$$

$$\sum_{i \in I} u^i \qquad + u^0 = 1.$$

Si $u^0 \neq 0$ (il est toujours possible de se placer dans ce cas, pour y^h fixé, en prenant λ_k assez grand), les valeurs u^i/u^0, $i \in I$, et v^l/u^0, $l \in L$, fournissent avec y^h, une solution réalisable pour le dual D de P. Rappelons que ces deux programmes s'écrivent :

$$P: \begin{cases} \text{Maximiser } f(x) \text{ sous les conditions} \\ g_i(x) \geqslant 0, \quad \forall i \in I \\ B_l x \geqslant b_l, \quad \forall l \in L. \end{cases}$$

$$D: \begin{cases} \text{Minimiser } \sum_{i \in I} u^i g^i(x) + f(x) + \sum_{l \in L} v^l(B_l x - b_l) \text{ sous les conditions :} \\ \\ u^i \geqslant 0, \quad \forall i \in I \\ v^l \geqslant 0, \quad \forall l \in L \\ \sum_{i \in I} u^i \nabla g_i(x) + \nabla f(x) + \sum_{l \in L} v^l B_l = 0. \end{cases}$$

On a la relation classique :

$$\hat{\lambda} \leqslant f(y^h) + \sum_{i \in I} u^i g_i(y^h) + \sum_{l \in L} v^l(B_l y^h - b_l).$$

4. Aspects Pratiques—Paramétrisation

4.1. *Aspect fini des calculs*

Nous avons vu, au Section 3, que la détermination d'un ε-centre revient à résoudre une suite finie de programmes linéaires, séparés par la recherche du maximum d'une fonction concave sur un segment de droite. Les programmes linéaires peuvent être résolus, par exemple, par la méthode simpliciale. Si les maximisations sur un segment sont réalisées de façon approchée, comme il est indiqué dans la remarque 3.5.3, une méthode classique par dichotomies conduira en définitive à un nombre fini de calculs de valeurs de d.

Enfin, bien que le nombre d'itérations de la méthode des centres, c'est-à-dire le nombre d'ε-centres x^k à calculer, est théoriquement infini en général, il devient fini si on détermine non pas une solution optimale exacte $\hat{x}$ du problème P, mais une valeur approchée $\hat{x}'$ telle que

$$f(\hat{x}') \geqslant f(\hat{x}) - \varepsilon'$$

avec les conditions

$$\hat{x}' \in A \cap B \text{ et } \varepsilon' > 0 \text{ donné}$$

(il en est toujours ainsi en pratique).

Par suite, le calcul d'une solution approchée $\hat{x}'$ peut se faire à l'aide d'une suite finie d'opérations (en particulier de résolutions de programmes linéaires).

4.2. *Cas linéaire.—Paramétrisation*

Si l'on envisage le cas d'un programme P entièrement linéaire, c'est-à-dire où les fonctions f et g_i, $i \in I$, sont affines, le calcul d'un centre se réduit à la résolution d'un simple programme linéaire, car on a

$$g_i'(x;y) = g_i(x), \qquad \forall i \in I \quad \text{et} \quad \forall y$$
$$f'(x;y) = f(x), \qquad \forall y$$

et ce programme s'écrit :

$$(Q_k) : \begin{cases} \text{Maximiser } \mu \text{ sous les conditions} \\ g_i(x) - \mu \geqslant 0, \quad \forall i \in I \\ f(x) - \mu \geqslant \lambda_k \\ x \in B. \end{cases}$$

Lorsque l'on a déterminé la solution optimale (x^{k+1}, μ^{k+1}) de ce programme linéaire, on peut alors faire varier continuement la valeur de la "troncature" λ, à partir de λ_k, par valeurs croissantes. La valeur de la solution optimale varie avec λ, et en utilisant l'algorithme classique de la méthode simpliciale paramétrée, on peut déterminer une solution optimale extrême $[\bar{x}(\lambda), \bar{\mu}(\lambda)]$, qui varie linéairement par morceaux avec λ. La méthode simpliciale paramétrée fournit directement la suite des valeurs de λ qui déterminent les intervalles de variation linéaire. Il est aisé de voir comme il a été indiqué en Section 3.6 que $\bar{\mu}(\lambda)$ est une fonction concave décroissante de λ, linéaire par morceaux.

En ajoutant la condition $\mu \geqslant 0$, la paramétrisation s'arrête quand on ne peut plus augmenter λ. On a alors $\hat{x}(\lambda) = \hat{x}$, et $\lambda = f(\hat{x})$.

4.3. *Paramétrisation dans le cas non linéaire*

Le procédé de paramétrisation décrit en Section 4.2 n'a qu'un intérêt théorique, puisqu'il n'est défini que pour le cas linéaire et qu'il utilise ... la méthode simpliciale. Néanmoins, on peut envisager de l'adapter au cas non linéaire, lorsqu'on détermine des ε-centres définis par la condition (3), ou pratiquement, par la condition (27).

En effet, supposons que l'on soit à l'étape k, avec la solution correspondante x^k, et la troncature $\lambda_k = f(x^k)$. Après résolution d'un premier programme linéaire $Q''(\lambda_k; x^k)$, on obtient les points correspondants z^1 et

y^1. Si la condition (27), où l'on remplace h^* par 1, est satisfaite, on peut conserver la linéarisation au point x^k, c'est-à-dire ne pas modifier $Q''(\lambda_k; x^k)$, mais paramétrer ce programme linéaire par rapport à λ, à partir de λ_k. On obtient ainsi une suite de points z^h, et pour chacun d'entre eux, on détermine le point, y^h qui maximise $d(x, \lambda_h')$ sur $[x^k, z^h]$, en désignant par λ_h' la valeur du paramètre λ correspondant à la solution z^h. Quand la condition (27), où l'on remplace h^* par h et λ_k par λ_h', n'est plus satisfaite pour un certain rang h, on choisit pour nouveau point x^{k+1} de linéarisation le point y^{h-1}, et l'on prend pour nouvelle valeur de troncature $\lambda = f(x^{k+1})$. On résoud alors $Q''(\lambda; x^{k+1})$, et on essaie de nouveau la paramétrisation.

L'intérêt pratique de ce procédé de paramétrisation n'est pas évident, et doit dépendre des caractéristiques du problème P traité: fonctions faiblement non linéaires, beaucoup de contraintes actives à l'optimum, etc. ... Dans le cas d'un programme linéaire, on retrouve la méthode décrite en Section 4.2, la linéarisation devenant sans objet, et les points y^h sont de vrais centres. D'un autre côté, dans le cas d'un programme fortement non linéaire (variations importantes des ∇q_i), on retrouvera pratiquement la méthode des centres décrite en Section 3.1 si, pour tout x^k, on doit linéariser en tous les points y^h.

4.4. *Démarrage des programmes linéaires*

La résolution d'un problème $Q''(\lambda_k; x^k)$ par la méthode simpliciale nécessite de satisfaire aux conditions initiales posées par la technique de cette méthode. En particulier, il faut avoir une solution réalisable extrême, dite "de base", comme solution de départ.

En introduisant des variables d'écart y_0, y_i, $i \in I$, la matrice des contraintes du PL à traiter peut se présenter sous la forme suivante:

x	μ		y			y_0		
	1	+1						
	1		+1					
A	1			+1		0	=	a
	1				+1			
A_0	1		0			+1	=	a_0
B	0		0			0	=	b

où
$$A_i = -\nabla g_i(x^k), \quad a_i = -\nabla g_i(x^k).x^k + g_i(x^k), \quad i \in I$$
$$A_0 = -\nabla f(x^k), \quad a_0 = -\nabla f(x^k).x^k + f(x^k) - \lambda_k.$$

On suppose ici que le polyèdre B est représenté par $Bx = b$, $x \geqslant 0$.

Si x^0 est une solution extrême (solution de base K) du polyèdre B, et si l'on possède le tableau simplicial correspondant, il suffit de rendre nuls les éléments des lignes A_0, A_i situés dans les colonnes d'indices $j \in K$, par les procédés classiques (combinaisons linéaires des lignes entre elles ou pré-multiplication matricielle). La base du PL complet est obtenue en réunissant à K les indices de l'ensemble $I + \{0\}$.

La solution de départ a pour valeur:

$$\begin{cases} x = x^0 \\ \mu = \min \{f'(x^0;x^k) - \lambda_k, \ g_i'(x^0;x^k) \mid i \in I\} \\ y_i = g_i'(x^0;x^k) - \mu, \ i \in I \\ y_0 = f'(x^0;x^k) - \lambda_k - \mu. \end{cases}$$

On peut vérifier que, à l'exclusion de μ, toutes ces composantes sont bien $\geqslant 0$. Il faut alors faire entrer μ dans la base.

References

1. B. Trong Lieu et P. Huard (1966). La méthode des Centres dans un espace topologique. *Num. Math.* **8**, 56-57,

2. M. Franck et P. Wolfe (1956). An algorithm for quadratic programming. *Nav. Res. Logist. Q.* mars-juin. 1956. Voir également C. Berge et A. Ghouila-Houri. *Programmes, jeux et réseaux de transport.* pp. 90-91. Dunod, Paris.

3. D. M. Topkis et A. F. Jr. Veinott (1967). On the convergence of some feasible direction algorithms for nonlinear programming. *J. Soc. ind. appl. Math. Control* **5**, (2), 268-279.

6. On the Solution of a Structured Linear Programming Problem in Upper Bounded Variables

V. DE ANGELIS*

University of Rome, Rome, Italy

1. Introduction

The problem considered is:

$$
\left.
\begin{aligned}
&\min \sum_{j=1}^{n} f_j(x_j) + \sum_{l=1}^{p} h_l y_l \\[2mm]
&\quad \sum_{j=1}^{n} a^j x_j + \sum_{l=1}^{p} g^l y_l = e\dagger \\[2mm]
&\quad l_j \leqslant x_j \leqslant u_j \qquad (j = 1, ..., n) \\[2mm]
&\quad y_l \geqslant 0 \qquad\qquad (l = 1, ..., p)
\end{aligned}
\right\}
\tag{1}
$$

A convenient approach to problem (1), which is nonlinear because the $f_j(x_j)$ are nonlinear, is to replace every function $f_j(x_j)$ by a piece-wise linear approximation, with vertices in

$$(l_{j1}, f_{j1}) = (l_j, f_j(l_j)),$$

$$(l_{jp_j}, f_{jp_j}) = (u_j, f_j(u_j))$$

and in other points

$$(l_{j2}, f_{j2}) = (l_{j2}, f_j(l_{j2})), ..., (l_{j, \, p_j-1}, f_{j, \, p_j-1}) = (l_{j, \, p_j-1}, f_j(l_{j, \, p_j-1}))$$

---

* Also at University of Birmingham, Birmingham England.
† $^T a^j = [a_{1j}, ..., a_{mj}]$, where $^T a^j$ stands for the transpose of a^j; $^T g^l = [g_1{}^l, ..., g_m{}^l]$; $^T e = [e_1, ..., e_m]$.

V. DE ANGELIS

chosen arbitrarily; as shown in Fig. 1.

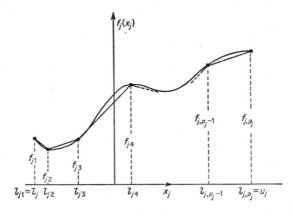

FIG. 1

Approximate solutions to problem (1) are then given by the solutions of the linear programming problem in upper bounded variables:

$$
\min \sum_{j=1}^{n} \sum_{r=1}^{p_j-1} c_{jr} x_{jr} + \sum_{l=1}^{p} h_l y_l + \sum_{j=1}^{n} f_{j1}
$$

$$
\sum_{j=1}^{n} a^j \left(l_{j1} + \sum_{r=1}^{p_j-1} x_{jr} \right) + \sum_{l=1}^{p} g^l y_l = e
$$

$$
0 \leqslant x_{jr} \leqslant l_{j,r+1} - l_{jr} \qquad (j=1,...,n; \ r=1,...,p_j-1)
$$

$$
y_l \geqslant 0 \qquad (l=1,...,p)
$$

(2)

where

$$
c_{jr} = \left(\frac{f_{j,r+1} - f_{jr}}{l_{j,r+1} - l_{jr}} \right) \qquad (j=1,...,n; \ r=1,...,p_{j-1}),
$$

provided that no $x_{jr} > 0$ unless $x_{j1}, ..., x_{j,r-1}$ are equal to their upper bound.*

The constant term $\sum_{j=1}^{n} f_{j1}$ need not be considered in the solution of

---

* For a different formulation of the approximating problem, see the paper by Beale [1], and also [2].

problem (2) and will therefore be omitted hereafter. By putting:

$$b = e - \sum_{j=1}^{n} a^j l_{j1}$$

problem (2) can be reformulated on:

$$
\begin{aligned}
& \min \sum_{j=1}^{n} \sum_{r=1}^{p_j-1} c_{jr} x_{jr} + \sum_{l=1}^{p} h_l y_l \\
& \sum_{j=1}^{n} \sum_{r=1}^{p_j-1} a^j x_{jr} + \sum_{l=1}^{p} g^l y_l = b \\
& 0 \leqslant x_{jr} \leqslant l_{j,\,r+1} - l_{jr} \qquad (j = 1, ..., n; r = 1, ..., p_j - 1) \\
& y_l \geqslant 0 \qquad\qquad\qquad (l = 1, ..., p)
\end{aligned}
\right\} \quad (3)
$$

From the optimal solutions of Eqn (3), the optimal values of the variables x_j are derived as:

$$x_j = l_{j1} + \sum_{r=1}^{p_j-1} x_{jr} \qquad (j = 1, ..., n)$$

and the approximate value of the objective function as:

$$\sum_{j=1}^{n} f_{j1} + \sum_{j=1}^{n} \sum_{r=1}^{p_j-1} c_{jr} x_{jr} + \sum_{l=1}^{p} h_l y_l.$$

If the functions $f_j(x_j)$ $(j = 1, ..., n)$ are all convex, no particular device need be used in order to obtain an optimal solution of (3) not having any $x_{jr} > 0$ unless all variables $x_{j1}, ..., x_{j,\,r-1}$ are equal to their upper bound. The optimal solutions of Eqn (3) will satisfy such property automatically and will provide approximate global optima of problem (1).

If the functions $f_j(x_j)$ $(j = 1, ..., n)$ are not all convex, however, particular rules should be used to solve problem (3) and the optimal solutions are more likely to provide only local optima.

In the latter case, a feasible starting solution of problem (3), satisfying the condition that no $x_{jr} > 0$ unless $x_{j1}, ..., x_{j,\,r-1}$ are equal to their upper bound, can easily be found by calculating a feasible solution $\bar{x}_1, ..., \bar{x}_n, \bar{y}_1, ..., \bar{y}_l$ of the system:

$$\sum_{j=1}^{n} a^j x_j + \sum_{l=1}^{p} g^l y_l = b$$

$$l_j \leqslant x_j \leqslant u_j \qquad (j = 1, ..., n)$$

$$y_l \geqslant 0 \qquad (l = 1, ..., p)$$

and then if:

$$l_{jr_j} \leqslant \bar{x}_j < l_{j, r_j+1}$$

by putting:

$$\left. \begin{array}{ll} x_{j1} & = l_{j2} - l_{j1} \\ \quad\vdots & \\ x_{j, r_j-1} = l_{jr_j} - l_{j, r_j-1} \\ \qquad\quad = \bar{x}_j - l_{jr_j} \\ x_{jr_j} \\ x_{j, r_j+1} = 0 \\ \quad\vdots & \\ x_{j, p_j-1} = 0 & (j = 1, ..., n) \\ y_l \qquad = \bar{y}_l & (l = 1, ..., p) \end{array} \right\} \quad (4)$$

Once (4) has been calculated, problem (3) can then be solved by restricting the choice of the nonbasic variable* to be introduced into the basis at each iteration to those variables only whose increase or decrease does not alter the property of the current basic solution of having $x_{jr} > 0$, only if $x_{j1}, ..., x_{jr, -1}$ are equal to their upper bound, for $j = 1, ..., n$. Therefore the choice will be restricted to those nonbasic variables $x_{ks} = 0$ such that

$$x_{k, s-1} = l_{ks} - l_{k, s-1},$$

and to those nonbasic variables

$$x_{ks} = l_{k, s+1} - l_{ks}$$

---

* Basis is here defined as a set of m variables such that the matrix of their coefficients in the m constraints is not singular.

such that $x_{k,\,s+1} = 0$.* No restriction applies to the introduction into the basis of any variable y_l ($l = 1, ..., p$).

A particular routine, which makes use of the property of the variables x_{jr} ($r = 1, ..., p_j$) having the same vector of coefficients a_j ($j = 1, ..., n$), will be described in the following sections to accelerate the solution of problem (3). Before describing such a routine, it will be recalled briefly what is involved in performing a standard iteration for a linear problem in upper bounded variables by using the product form of the inverse matrix method.

2. Standard Solution of Problem (3).

Solving problem (3) by means of the product form of the inverse matrix method involves rewriting the objective function as:

$$x_0 + \sum_{j=1}^{n} \sum_{r=1}^{p_j} c_{jr}\, x_{jr} + \sum_{l=1}^{p} h_l\, y_l = 0$$

(x_0 to be maximised) and having available, for the current basic feasible solution, the inverse B^{-1} of the $(m+1) \times (m+1)$ matrix B of the elements of x_0 and of the m basic variables, in the objective function (which is here assumed to be the first row of B, or row 0) and in the m constraints, as a product of elementary matrices. Then an iteration is performed in the following steps.

Step 1. Form the first row of the inverse, called "pricing vector", by post-multiplying the row unit vector u_1, whose first element is unity and all other elements are zero, by all elementary matrices whose product forms the inverse of the basis, in the opposite order to that in which they were generated. This operation is called the backward transformation.

Step 2. Calculate the reduced cost of the nonbasic variables eligible to be introduced into the basis, by postmultiplying the row vector calculated in step 1 by the column of their coefficients.

If all nonbasic variables equal to zero have a positive reduced cost and all nonbasic variables equal to their upper limit have a negative reduced cost, the current solution is optimal. Otherwise, select the variable with greatest reduced cost, in modulus, among those equal to zero which have a negative reduced cost and those equal to their upper bound which have a positive reduced cost.

Increase the selected nonbasic variable if it is equal to zero, decrease it if it is equal to its upper bound.

---

*In linear programming with upper bounded variables, the nonbasic variables are set equal either to their upper or to their lower bound [3, chapter 18].

Step 3. Calculate the updated coefficients α_i $(i = 0, 1, ..., m)$ of the chosen variable by premultiplying the column of its original coefficients by all elementary matrices whose product forms the inverse of the basis, in the order in which they were generated. Such operation is called the forward transformation.

Step 4. Perform a ratio test in order to determine a candidate pivot. If the nonbasic variable is to be increased, the candidate pivot is in the row where the minimum occurs of the following ratios:

$$\frac{\beta_i}{\alpha_i} \quad \text{for } \alpha_i < 0 \text{ and } i \geqslant 1$$

$$\frac{\beta_i - u_i}{\alpha_i} \text{ for } \alpha_i > 0 \text{ and } i \geqslant 1$$

where β_i $(i = 0, 1, ..., m)$ stands for the current value of the ith basic variable and u_i for the upper limit of the ith basic variable.

If the nonbasic variable is to be decreased, the candidate pivot is in the row where the minimum occurs of the following ratios:

$$\left| \frac{\beta_i}{\alpha_i} \right| \quad \text{for } \alpha_i < 0 \text{ and } i \geqslant 1$$

$$\left| \frac{\beta_i - u_i}{\alpha_i} \right| \text{ for } \alpha_i > 0 \text{ and } i \geqslant 1$$

In both cases, let θ be the value of the minimum ratio and r the index of the row where it occurs.

Step 5. Compare θ with the upper bound u of the variable to be made basic and put $\theta' = \min \{\theta, u\}$.

Step 5a. If $\theta' < u$, the chosen nonbasic variable is to be introduced into the basis at the place of the rth basic variable. In this case, update the set of basic variables and the right-hand side, calculate another elementary matrix which, postmultiplied by all other elementary matrices previously calculated, in the opposite order to that in which they were generated, gives the inverse of the new basis.

Step 5b. If $\theta' = u$, the nonbasic variable is to be made equal to either its upper or lower bound, according to whether it is being increased or decreased, but no change of basis occurs. Register its new value and update the value of the basic variables.

3. Special Features of Problem (3)

Let $\bar{c}_{ks}$ be the reduced cost of a variable x_{ks}, given a current basis.

Once $\bar{c}_{ks}$ is calculated, the reduced cost of any variable x_{kt} $(t = 1, ..., p_{k-1})$ can easily be calculated as:

$$\bar{c}_{kt} = \bar{c}_{ks} + (c_{kt} - c_{ks}). \tag{5}$$

Formula (5) holds since the pricing vector has unity as first element and the vector of coefficients of the variable x_{kt} is equal to the sum of the vector of the coefficients of the variable x_{ks} and the vector having $(c_{kt} - c_{ks})$ as first element and all the other elements equal to zero:

$$\begin{bmatrix} c_{kt} \\ a_{1k} \\ a_{2k} \\ \vdots \\ a_{mk} \end{bmatrix} = \begin{bmatrix} c_{ks} \\ a_{1k} \\ a_{2k} \\ \vdots \\ a_{mk} \end{bmatrix} + \begin{bmatrix} c_{kt} - c_{ks} \\ 0 \\ 0 \\ \vdots \\ 0 \end{bmatrix}$$

Let x_{qz} be the rth basic variable in the current basis and B be the matrix of the coefficients of x_0 and the m current basic variables.

If we consider a basis different from the current one only in the fact that the variable x_{qz} is replaced by a variable x_{qv} $(v = 1, ..., p_q - 1)$, the reduced cost $\bar{c}_{ks}$ of x_{ks} corresponding to such a basis is:

$$\bar{c}_{ks} = \bar{c}_{ks} - (c_{qv} - c_{qz})\alpha_r^{ks}. \tag{6}$$

where α_r^{ks} is the rth element of the vector

$$\alpha^{ks} = B^{-1} \begin{bmatrix} c_{ks} \\ a^k \end{bmatrix}.$$

The relation (6) holds because the matrix $\bar{B}$ of the coefficients of x_0 and of the m basic variables in the new basis can be rewritten as:

$$\bar{B} = B \begin{bmatrix} 1 & 0 & \cdots & (c_{qv} - c_{qz}) & \cdots & 0 \\ 0 & 1 & \cdots & 0 & \cdots & 0 \\ & \vdots & & & & \\ 0 & 0 & \cdots & 1 & \cdots & 0 \\ & \vdots & & & & \\ 0 & 0 & \cdots & 0 & \cdots & 1 \end{bmatrix} \begin{matrix} \\ \\ \\ \leftarrow r\text{th row} \\ \\ \end{matrix}$$
$$\uparrow$$
$$r\text{th column}$$

Therefore:

$$\bar{c}_{ks} = [1 \quad 0 \quad \dots \quad 0]\, \bar{B}^{-1} \begin{bmatrix} c_{ks} \\ a^k \end{bmatrix}$$

$$= [1 \quad 0 \quad \dots \quad 0] \begin{bmatrix} 1 & 0 & \dots & -(c_{qv}-c_{qz}) & \dots & 0 \\ 0 & 1 & \dots & 0 & \dots & 0 \\ & \vdots & & & & \\ 0 & 0 & \dots & 1 & \dots & 0 \\ & & & & & \\ 0 & 0 & \dots & 0 & \dots & 1 \end{bmatrix} B^{-1} \begin{bmatrix} c_{ks} \\ a_{1k} \\ \vdots \\ a_{rk} \\ \vdots \\ a_{mk} \end{bmatrix}$$

$$= [1 \quad 0 \quad \dots \quad 0] \begin{bmatrix} 1 & 0 & \dots & -(c_{qv}-c_{qz}) & \dots & 0 \\ 0 & 1 & \dots & 0 & \dots & 0 \\ & \vdots & & & & \\ 0 & 0 & \dots & 1 & \dots & 0 \\ & & & & & \\ 0 & 0 & \dots & 0 & \dots & 1 \end{bmatrix} \begin{bmatrix} \bar{c}_{ks} \\ \alpha_1^{ks} \\ \vdots \\ \alpha_r^{ks} \\ \vdots \\ \alpha_m^{ks} \end{bmatrix}$$

$$= [1 \quad 0 \quad \dots \quad 0]^T \,[(c_{ks}-(c_{qv}-c_{qz})\alpha_r^{ks}),\, \alpha_1^{ks},\, \dots,\, \alpha_m^{ks}]$$
$$= \bar{c}_{ks} - (c_{qv}-c_{qz})\,\alpha_r^{ks}.$$

Relations (5) and (6) will enable us to perform some special iterations which will reduce the amount of calculations and the time required to solve problem (3).

4. Modified Procedure to Solve Problem (3)

Let the variable to be introduced into the current basis, chosen by means of steps 1 and 2 described in Section 2, be one of the variables

$$x_{jr}\ (j = 1, \dots, n;\ r = 1, \dots, p_j),$$

say x_{ks}.†

If the value of the variable x_k corresponding to the new value of x_{ks}, obtained by means of step 5 through steps 3 and 4, is less than u_k, in case of

---

† A similar argument applies where the variable to be introduced into the basis is a y_l $(l = 1, \dots, p)$ variable. In this Case, however, only Case a can occur.

an increase of x_{ks}, or greater than l_k, in case of a decrease of x_{ks}, it may well be worthwhile to try and increase (or decrease) x_k further. This is very likely to produce a further decrease in the objective function, particularly if the functions $f_j(x_j)$ $(j = 1, ..., n)$ are smooth so that there is not much difference between the values of a pair $c_{jr}, c_{j, r+1}$ $(j = 1, ..., n; r = 1, ..., p_j - 2)$.

We shall now examine the increase (decrease) θ' given to x_{ks} and study in detail the two cases where $\theta' < (l_{k, s+1} - l_{ks})$ and $\theta' = (l_{k, s+1} - l_{ks})$.

Case a: $\theta' < (l_{k, s+1} - l_{ks})$.

We focus our attention on the rth basic variable, that is on the variable selected by step 4 as the variable to be made nonbasic.

If such variable is a y_l variable $(l = 1, ..., p)$, say y_z, x_{ks} cannot be increased (decreased) any further because y_z would become negative. Variable x_{ks} should be introduced into the basis and step 5a performed.

If the rth basic variable is an upper bounded variable, x_{qz}, attention must be paid to the value it takes by giving x_{ks} an increase (decrease) equal to θ'.

If x_{qz}, upon giving x_{ks} an increase (decrease) equal to θ', becomes equal to its upper bound and $l_{q, z+1} = u_q$, then x_k cannot be increased (decreased) any further because x_q would exceed its upper bound. Variable x_{ks} should be introduced into the basis and step 5a performed.

Similarly, if x_{qz} becomes equal to its lower bound and $l_{qz} = l_q$, then x_k cannot be increased (decreased) any further because x_q would fall below its lower bound. Variable x_{ks} should be introduced into the basis and step 5a performed.

If, on the contrary, x_{qz} becomes equal to its upper bound and $l_{q, z+1} < u_q$ or x_{qz} becomes equal to its lower bound and $l_{qz} > l_q$, a further increase (decrease) of x_k is feasible because it does not make any of the original variables x_j $(j = 1, ..., n)$, y_l $(l = 1, ..., p)$ exceed their bounds.

In this last case, we do not perform either step 5a or 5b, but, through formula (6), we calculate the reduced cost $\bar{c}_{ks}$ of x_{ks} corresponding to the basis obtained from the current one by substituting x_{qz} by x_{qv}; where $v = z+1$ if x_{qz} has become equal to its upper bound, $v = z-1$ if x_{qz} has become equal to its lower bound upon the increase (decrease) of x_{ks} equal to θ'.

If $\bar{c}_{ks} \geqslant 0$ in the case where x_{ks} is being increased and $\bar{c}_{ks} \leqslant 0$ in the case where x_{ks} is being decreased, it is not profitable to increase (decrease) x_{ks} any more. Variable x_{ks} should be introduced into the basis and step 5a performed.

If, on the contrary, $\bar{c}_{ks} < 0$ in the case where x_{ks} is being increased or $\bar{c}_{ks} > 0$ in the case where x_{ks} is being decreased, it is profitable to increase (decrease) x_{ks} further. Variable x_{ks} should not be introduced into the basis but variable x_{qv} should be introduced instead, in order to make it possible to increase x_{ks} further.

Step 5c should be performed.

Step 5c. Put the nonbasic variable x_{ks} equal to θ' or to $(l_{k\ s+1} - l_{ks}) - \theta'$, according to whether it is being increased or decreased, and leave it as a nonbasic variable.* Update the value of the current basic variable by putting it equal to

$$\beta \mp \alpha^{ks} \theta' \tag{7}$$

where $-$ or $+$ holds according to whether x_{ks} is being increased or decreased.

Exchange the basic variable x_{qz} with the nonbasic variable x_{qv}, and reset the rth component of Eqn (7) equal to the value of x_{qv}.

Calculate the elementary matrix

$$\begin{bmatrix} 1 & \cdots & -(c_{qv}-c_{qz}) & \cdots & 0 \\ 0 & \cdots & 0 & \cdots & 0 \\ \vdots & & & & \\ 0 & \cdots & 1 & \cdots & 0 \\ \vdots & & & & \\ 0 & \cdots & 0 & \cdots & 1 \end{bmatrix}$$

which, postmultiplied by all elementary matrices previously calculated, will give the inverse of the matrix of the coefficients of the new current basis.

Reset the value of $\alpha_0{}^{ks}$ equal to $\bar{c}_{ks}$.

Once step 5c has been performed, the next iteration, which involves increasing (decreasing) x_{ks} further, will start from step 4.†

By increasing (decreasing) x_{ks} further, the possibility is therefore created of performing iterations which do not involve either a backward or a forward transformation. Since such transformations usually take a considerable amount of time, the modified procedure suggested here enables us to cut substantially the solution time of problem (3). Also, the non-unit vector in Eqn (7) has only two non-zero entries and can be stored very easily.

Case b: $\theta' = (l_{k,\ s+1} - l_{ks})$.

If the value of x_k corresponding to the new value of x_{ks} is less than u_k in the case of an increase of x_{ks}, or greater than l_k in the case of a decrease of x_{ks}, x_k can easily be increased (decreased) further by simply increasing $x_{k,\ s-1}$ (decreasing $x_{k,\ s-1}$).

---

* The solution obtained through step 5c is not basic. Such a solution, however, is only an intermediate solution which eventually leads to a basic solution. The fundamental feature of the simplex method of dealing with basic solutions is lost only temporarily.

† The value θ calculated through step 4 must be considered in this case as a $\Delta\theta$. The new value of ϑ will be $\theta = \theta' + \Delta\theta$.

Whether such increase (decrease) is profitable, is easily seen through formula (5), by calculating

$$\bar{c}_{k, s+1} = \bar{c}_{ks} + (c_{k, s+1} - c_{ks})$$

(or $\bar{c}_{k, s-1} = \bar{c}_{ks} + (c_{k, s-1} - c_{ks})$).

If $\bar{c}_{k, s+1}$ is non-negative (or $\bar{c}_{k, s-1}$ is non-positive), x_k cannot be profitably increased (decreased) any more. Step 5b should be performed.

If $\bar{c}_{k, s+1}$ is negative (or $\bar{c}_{k, s-1}$ is positive), step 5d should be performed.

Step 5d. Increase (decrease) x_{ks} by an amount equal to its upper bound. Update the value of the basic variables. Calculate $\alpha^{k, s-1}$ ($\alpha^{k, s-1}$) as:

$$\alpha^{k, s+1} = \alpha^{ks} + \begin{bmatrix} c_{k, s+1} - c_{ks} \\ 0 \\ \vdots \\ 0 \end{bmatrix}$$

$$\alpha^{k, s-1} = \alpha^{ks} + \begin{bmatrix} c_{k, s-1} - c_{ks} \\ 0 \\ \vdots \\ 0 \end{bmatrix}$$

Once step 5d has been performed, the next iteration will start from step 4.

Again, the possibility is offered of performing an iteration which does not involve either a backward or a forward transformation.

The following numerical examples will help to clarify the procedure described above.

5. Numerical Examples and Conclusions*

Example 1.

$$\min z = \sum_{j=1}^{2} f_j(x_j)$$

$$5 x_1 - 8 x_2 = 4$$

$$x_1 \geqslant 0, x_2 \geqslant 0$$

---

* The numerical examples are taken from [4].

where the functions $f_1(x_1), f_2(x_2)$ are approximated by the polygonal functions described below:

$$f_1(x_1) \simeq \begin{cases} -x_1 & 0 \leqslant x_1 \leqslant 3 \\ -\frac{3}{2} - \frac{1}{2}x_1 & 3 \leqslant x_1 \leqslant 8 \\ -\frac{23}{2} + \frac{3}{4}x_1 & 8 \leqslant x_1 \leqslant 16 \end{cases}$$

$$f_2(x_2) \simeq \begin{cases} \frac{1}{3}x_2 & 0 \leqslant x_2 \leqslant 2 \\ -\frac{4}{3} + x_2 & 2 \leqslant x_2 \leqslant 6 \end{cases}$$

In terms of upper bounded variables, the problem is formulated as:

$$\min z = -x_{11} - \frac{1}{2}x_{12} + \frac{3}{4}x_{13} + \frac{1}{3}x_{21} + x_{22}$$

$$5x_{11} + 5x_{12} + 5x_{13} - 8x_{21} - 8x_{22} = 4$$

$$0 \leqslant x_{11} \leqslant 3, \quad 0 \leqslant x_{12} \leqslant 5, \quad 0 \leqslant x_{13} \leqslant 8$$

$$0 \leqslant x_{21} \leqslant 2, \quad 0 \leqslant x_{22} \leqslant 4$$

The matrix of the coefficients is:

	x_0	x_{11}	x_{12}	x_{13}	x_{21}	x_{22}	b
row 0	1	-1	$-\frac{1}{2}$	$\frac{3}{4}$	$\frac{1}{3}$	1	0
row 1	0	5	5	5	-8	-8	4

Let the starting basic feasible solution be:

$$\begin{cases} x_0 = \frac{4}{5} & x_{11} = \frac{4}{5} & \text{basic variables} \\ x_{12} = x_{13} = x_{21} = x_{22} = & 0 \text{ nonbasic variables} \end{cases}$$

The elementary matrix which gives the inverse of the basis is:

$$E_1 = \begin{bmatrix} 1 & \frac{1}{5} \\ 0 & \frac{1}{5} \end{bmatrix}$$

The current basic solution and steps 1, 2, 3, are described in the following tableaux:

First solution (basic solution)*

$$\text{pricing vector} = \begin{bmatrix} 1 & \frac{1}{5} \end{bmatrix}$$

---

* Only the nonbasic variables marked with * can be introduced into the basic.

b. v.	β	n. b. v.	their value	their red. cost	α^{21}
		x_{12}	0		
x_0	$\frac{4}{5}$	x_{13}	0		$-\frac{19}{15}$
x_{11}	$\frac{4}{5}$	$*x_{21}$	0	$-\frac{19}{15}$	$-\frac{8}{5}$
		x_{22}	0		

The ratio test (step 4) gives:

$$\theta = \min \begin{bmatrix} \frac{\frac{4}{5}-3}{-\frac{8}{5}} \end{bmatrix} = \frac{11}{8}$$

and $r = 1$ (number of the pivotal row).

The value θ' is

$$\theta' = \min \{\tfrac{11}{8}, 2\} = \tfrac{11}{8} < 2.$$

The basic variable x_{11} becomes equal to its upper bound. Since the corresponding value of x_1 ($x_1 = 3$) is less than the upper bound of x_1, x_{21} can be increased further.

In order to find out whether such increase if profitable, we calculate:

$$\bar{c}_{21} = -\tfrac{19}{15} - (-\tfrac{1}{2}+1)(-\tfrac{8}{5}) = -\tfrac{7}{15}.$$

Since $\bar{c}_{21} < 0$, x_{21} can be profitably increased further. Its further increase will cause x_1 to increase further, and therefore can only be achieved by allowing x_{12} to become basic instead of x_{11}.

Step 5c must be performed, that is we calculate the elementary matrix E_2 whose postmultiplication by E_1 gives the inverse of the basis x_0, x_{12}, and represent the solution obtained by increasing x_{21} by an amount equal to θ' as a nonbasic solution where x_{21} is a nonbasic variable.

We obtain:

$$E_2 = \begin{bmatrix} 1 & -\tfrac{1}{2} \\ 0 & 1 \end{bmatrix}$$

and the following.

Second solution (nonbasic solution).

b. v.	β	n. b. v.	their value	α^{21}
x_0	$\frac{61}{24}$	x_{11}	3	$-\frac{7}{15}$
x_{12}	0	x_{13}	0	$-\frac{8}{5}$
		x_{21}	$\frac{11}{8}$	
		x_{22}	0	

The ratio test relative to a further increase of x_{21} gives:

$$\Delta\theta = \min \left[\frac{0-5}{-\frac{8}{5}} \right] = \frac{25}{8}$$

and $r = 1$.

The total increase for x_{21} is therefore:

$$\theta = \frac{11}{8} + \frac{25}{8} = \frac{9}{2}.$$

The value θ' is:

$$\theta' = \min \left\{ \frac{9}{2}, 2 \right\} = 2.$$

By an increase equal to 2, x_{21} becomes equal to its upper bound. Since the corresponding value of x_2 ($x_2 = 2$) is less than the upper bound of x_2, we can consider increasing x_{22}.
We therefore calculate:

$$\bar{c}_{22} = -\frac{7}{15} + (1 - \frac{1}{3}) = \frac{1}{5}.$$

Since $\frac{1}{5} > 0$, we cannot increase x_2 profitably any further. We therefore set x_{21} equal to its upper bound (no change of basis occurs with respect to the second solution) and perform step 5b.

Third solution (basic solution).

$$\text{pricing vector} = \left[1 \quad \frac{1}{10} \right]$$

b. v.	β	n. b. v.	their value	their red. cost
x_0	$\frac{17}{6}$	x_{11}	3	
x_{12}	1	x_{13}	0	
		*x_{21}	2	$-\frac{7}{15}$
		*x_{22}	0	$\frac{1}{5}$

Since the reduced cost of x_{21} is negative and the reduced cost of x_{22} is positive, the current basic solution is optimal and the solution of the problem is $x_1 = 4$, $x_2 = 2$, $z = -\frac{17}{6}$.

The solution of this numerical example has taken two backward and one forward transformations, compared with three backward and two forward transformations of the standard procedure of solution [4].

Example 2.

$$\min z = \sum_{j=1}^{3} f_j(x_j)$$

$$3x_1 + 2x_2 - 2x_3 = 18$$

$$x_1 - x_2 + 2x_3 = 8$$

$$x_1 \geqslant 0, \; x_2 \geqslant 0, \; x_3 \geqslant 0$$

where the functions $f_j(x_j)$ are approximated by the polygonal functions described below:

$$f_1(x_1) \simeq \begin{cases} \frac{1}{2}x_1 & 0 \leqslant x_1 \leqslant 2 \\ -\frac{1}{2} + \frac{3}{4}x_1 & 2 \leqslant x_1 \leqslant 4 \\ -\frac{3}{2} + x_1 & 4 \leqslant x_1 \leqslant 10 \end{cases}$$

$$f_2(x_2) \simeq \begin{cases} -x_2 & 0 \leqslant x_2 \leqslant 3 \\ -3 & 3 \leqslant x_2 \leqslant 5 \\ -13 + 2x_2 & 5 \leqslant x_2 \leqslant 17 \end{cases}$$

$$f_3(x_3) \simeq \begin{cases} -\frac{1}{2}x_3 & 0 \leqslant x_3 \leqslant 4 \\ -6 + x_3 & 4 \leqslant x_3 \leqslant 6 \\ -9 + \frac{3}{2}x_3 & 6 \leqslant x_3 \leqslant 8 \end{cases}$$

In terms of upper bounded variables, the problem is formulated as:

$$\min z = \tfrac{1}{2}x_{11} + \tfrac{3}{4}x_{12} + x_{13} - x_{21} + 0x_{22} + 2x_{23} - \tfrac{1}{2}x_{31} + x_{32} + \tfrac{3}{2}x_{33}$$

$$3x_{11} + 3x_{12} + 3x_{13} + 2x_{21} + 2x_{22} + 2x_{23} - 2x_{31} - 2x_{32} - 2x_{33} = 18$$

$$x_{11} + x_{12} + x_{13} - x_{21} - x_{22} - x_{23} + 2x_{31} + 2x_{32} + 2x_{33} = 8$$

$$0 \leqslant x_{11} \leqslant 2, \; 0 \leqslant x_{12} \leqslant 2, \; 0 \leqslant x_{13} \leqslant 6$$

$$0 \leqslant x_{21} \leqslant 3, \; 0 \leqslant x_{22} \leqslant 2, \; 0 \leqslant x_{23} \leqslant 12$$

$$0 \leqslant x_{31} \leqslant 4, \; 0 \leqslant x_{32} \leqslant 2, \; 0 \leqslant x_{33} \leqslant 2$$

The matrix of the coefficients is:

	x_0	x_{11}	x_{12}	x_{13}	x_{21}	x_{22}	x_{23}	x_{31}	x_{32}	x_{33}	b
row 0	1	$\frac{1}{2}$	$\frac{3}{4}$	1	-1	0	2	$-\frac{1}{2}$	1	$\frac{3}{2}$	0
row 1	0	3	3	3	2	2	2	-2	-2	-2	18
row 2	0	1	1	1	-1	-1	-1	2	2	2	8

We start from the basic feasible solution:

$$x_0 = -11 \quad x_{23} = 5 \quad x_{33} = 1 \qquad\qquad \text{basic variables}$$
$$\left. \begin{array}{lll} x_{11} = 2 & x_{12} = 2 & x_{13} = 0 \quad x_{21} = 3 \\ x_{22} = 2 & x_{31} = 4 & x_{32} = 2 \end{array} \right\} \text{nonbasic variables}$$

The elementary matrices whose product gives the inverse of the current basis are:

$$E_1 = \begin{bmatrix} 1 & -1 & 0 \\ 0 & \frac{1}{2} & 0 \\ 0 & \frac{1}{2} & 1 \end{bmatrix} \qquad E_2 = \begin{bmatrix} 1 & 0 & -\frac{7}{2} \\ 0 & 1 & 1 \\ 0 & 0 & 1 \end{bmatrix}$$

First solution (basic solution).

$$\text{pricing vector} = [1 \quad -\tfrac{11}{4} \quad -\tfrac{7}{2}]$$

b. v.	β	n. b. v.	their value	their red. cost	α^{13}
x_0	-11	x_{11}	2		$-\frac{43}{4}$
x_{23}	5	$*x_{12}$	2	-11	4
x_{33}	1	$*x_{13}$	0	$-\frac{43}{4}$	$\frac{5}{2}$
		x_{21}	3		
		x_{22}	2		
		x_{31}	4		
		x_{32}	2		

The ratio test gives:

$$\theta = \min \{\tfrac{5}{4}, 1/\tfrac{5}{2}\} = \min \{\tfrac{5}{4}, \tfrac{2}{5}\} = \tfrac{2}{5}$$

and $r = 2$.

The value θ' is:

$$\theta' = \min \{\tfrac{2}{5}, 6\} = \tfrac{2}{5} < 6.$$

The basic variable x_{33} becomes equal to its lower bound. Since the corresponding value of x_3 ($x_3 = 6$) is greater than the lower bound of x_3, x_{13} can be increased further. We calculate:

$$\bar{c}_{13} = -\tfrac{43}{4} - (1 - \tfrac{3}{2})\tfrac{5}{2} = -\tfrac{19}{2}.$$

Since $\bar{c}_{13} < 0$, x_{13} can be profitably increased further, therefore we replace x_{33} in the basis by x_{32}. We obtain:

$$E_3 = \begin{bmatrix} 1 & 0 & \tfrac{1}{2} \\ 0 & 1 & 0 \\ 0 & 0 & 1 \end{bmatrix}$$

Second solution (nonbasic solution).

b. v.	β	n. b. v.	their value	α^{13}
x_0	$-\tfrac{67}{10}$	x_{11}	2	$-\tfrac{19}{2}$
x_{23}	$\tfrac{17}{5}$	x_{12}	2	4
x_{32}	2	x_{13}	$\tfrac{2}{5}$	$\tfrac{5}{2}$
		x_{21}	3	
		x_{22}	2	
		x_{31}	4	
		x_{33}	0	

The ratio test relative to a further increase of x_{13} gives:

$$\Delta\theta = \min \{\tfrac{17}{5}/4,\ 2/\tfrac{5}{2}\} = \tfrac{4}{5}$$

(i.e. $\theta = \tfrac{6}{5}$) and $r = 2$.

The value θ' is:

$$\theta' = \min \{\tfrac{6}{5},\ 6\} = \tfrac{6}{5}.$$

The variable x_{32} becomes equal to its lower bound. Since the corresponding value of x_3 ($x_3 = 4$) is greater than the lower bound of x_3, x_{13} can be increased further.

We put $\bar{c}_{13} = -\tfrac{19}{2}$ and calculate a new value of $\bar{c}_{13}$:

$$\bar{c}_{13} = -\tfrac{19}{2} - (-\tfrac{1}{2} - 1)\tfrac{5}{2} = -\tfrac{23}{4}.$$

Since $\bar{c}_{13} < 0$, x_{13} can be profitably increased further, and therefore we replace x_{32} in the basis by x_{31}. We obtain:

$$E_4 = \begin{bmatrix} 1 & 0 & \frac{3}{2} \\ 0 & 1 & 0 \\ 0 & 0 & 1 \end{bmatrix}$$

Third solution (nonbasic solution).

b. v.	β	n. b. v.	their value	α^{13}
x_0	$\frac{9}{10}$	x_{11}	2	$-\frac{23}{4}$
x_{23}	$\frac{1}{5}$	x_{12}	2	4
x_{31}	4	x_{13}	$\frac{6}{5}$	$\frac{5}{2}$
		x_{21}	3	
		x_{22}	2	
		x_{32}	0	
		x_{33}	0	

The ratio test gives:

$$\Delta\theta = \min\{\tfrac{1}{5}/4, \tfrac{4}{5}/2\} = \tfrac{1}{20}$$

(i.e. $\theta = 5$) and $r = 1$.

The value θ' is:

$$\theta' = \min\{\tfrac{5}{4}, 6\} = \tfrac{5}{4}.$$

The variable x_{23} becomes equal to its lower bound. Since the corresponding value of x_2 ($x_2 = 5$) is greater than the lower bound of x_2, x_{13} can be increased further.

We put $\bar{c}_{13} = -\frac{23}{4}$ and calculate:

$$\bar{c}_{13} = -\tfrac{23}{4} - (0-2)\,4 = \tfrac{9}{4}.$$

Since $\bar{c}_{13} > 0$, it is not profitable to increase x_{13} any further. We therefore perform step 5a, that is we perform a standard iteration by replacing x_{23} in the basis by x_{13}. We obtain:

$$E_5 = \begin{bmatrix} 1 & \frac{23}{16} & 0 \\ 0 & \frac{1}{4} & 0 \\ 0 & -\frac{5}{8} & 1 \end{bmatrix}$$

Fourth solution (basic solution).

pricing vector $= [1 \quad -\frac{5}{16} \quad -\frac{1}{16}]$

b. v.	β	n. b. v.	their value	their red. cost
r_0	$\frac{19}{8}$	x_{11}	2	
x_{13}	$\frac{5}{4}$	x_{12}	2	
x_{31}	$\frac{31}{8}$	x_{21}	3	
		*x_{22}	2	$-\frac{9}{16}$
		*x_{23}	0	$\frac{23}{16}$
		x_{32}	0	
		x_{33}	0	

Since the reduced cost of x_{22} is <0 and the reduced cost of x_{23} is >0, the fourth solution is optimal and the solution of the problem is

$$x_1 = \tfrac{21}{4}, \; x_2 = 5, \; x_3 = \tfrac{31}{8}, \; z = -\tfrac{19}{16}.$$

The solution of this numerical example has taken two backward and one forward transformations, against four backward and three forward transformations in the standard procedure of solution.

Acknowledgement

The author wishes to thank Professor S. Vajda, of Birmingham University, for helpful comments.

References

1. J. Abadie (ed). (1967). "Nonlinear Programming", North Holland Publishing Company, Amsterdam.
2. V. de Angelis. Minimization of a separable function subject to linear constraints using the product form of the inverse matrix method. To be published.
3. G. B. Dantzig (1963). "Linear Programming and Extensions", Princeton University Press, Priceton, New Jersey.
4. V. de Angelis (1965). Ricerca del minimo valore di una funzione convessa separabile sottoposta a vincoli lineari, "Giornale dell' Istituto Italiano degli Attuari", Anno XXVIII, 1, No. 1, 92-133.

7. Decomposition of a Nonlinear Convex Separable Economic System in Primal and Dual Directions*

T. O. M. KRONSJÖ

Faculty of Commerce and Social Science, University of Birmingham, Birmingham, England

1. Introduction

The following decompositional method by the author may be used to achieve optimal coordination of a nonlinear convex economic system which is assumed to be separable in terms of three types of variables and of the following general structure.

$$
\begin{array}{ll}
\min_{x_1, x_2, x_3} & \left\{ f_1(x_1) + f_2(x_2) + f_3(x_3) \quad \middle| \quad \text{Dual variables} \right. \\
& g_{11}(x_1) + g_{12}(x_2) + g_{13}(x_3) \leqslant 0 \qquad u_1 \\
& \qquad\qquad g_{22}(x_2) + g_{23}(x_3) \leqslant 0 \qquad u_2 \\
& \qquad\qquad\qquad\qquad g_{33}(x_3) \leqslant 0 \qquad u_3 \\
& x_1 \geqslant 0 \quad x_2 \geqslant 0 \quad x_3 \geqslant 0 \right\}
\end{array}
\tag{1}
$$

where x_j denotes a *vector* of variables, $f_j(x_j)$ denotes a convex *scalar* function, $g_{ij}(x_j)$ denotes a convex *column vector* function. The above functions are, however, assumed to be of such an internal structure that:

(1) feasible solutions may always be found to the first, second and third types of constraint by using only the x_1, x_2 and x_3 variables, respectively, independently of the values of the other variables; and

(2) no infinite solution exists to the overall problem and to the following subproblems

$$
\min_{x_2, x_3} \{ f_2(x_2) + u_1 g_{12}(x_2) + f_3(x_3) + u_1 g_{13}(x_3) \mid g_{22}(x_2) + g_{23}(x_3) \leqslant 0,
$$
$$
g_{33}(x_3) \leqslant 0, x_2 \geqslant 0, x_3 \geqslant 0 \}
$$

$$
\min_{x_3} \{ f_3(x_3) + u_1 g_{13}(x_3) + u_2 g_{23}(x_3) \mid g_{33}(x_3) \leqslant 0, x_3 \geqslant 0 \}
$$

irrespective of the values given to the u variables.

The functioning of the method may be summarized by the following illustration.

* This research was supported by the Swedish Council for Social Science Research, Stockholm, Sweden.

85

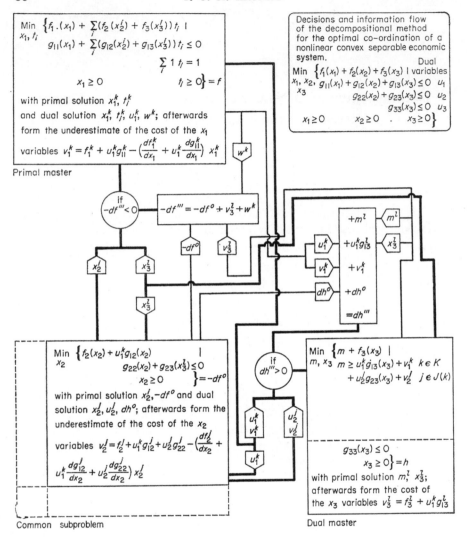

The method offers an interesting possibility for optimal coordination of a very large economic system in which the first type of constraints and the third type of variables are comparatively few and in which the second type of constraints and variables may be very many but with a computationally simple structure of the common subproblem

$$\min_{x_2} \{ f_2(x_2) + u_1^k\, g_{12}(x_2) \mid g_{22}(x_2) + g_{23}(x_3^l) \leqslant 0,\ x_2 \geqslant 0 \}.$$

Among such computationally favourable structures is the one when the common subproblem consists of n independent problems, each one of which may be solved independently by one computer. The decompositional method may then enable $n+2$ computers to operate almost all in parallel to achieve the optimal coordination of the economic system.

In the nonlinear case the proposed decomposition method will normally lead to an infinite process. As at every iteration the current primal objective function and an underestimate of its optimal value are available, it is possible to discontinue the iterations when the difference between these two values is less than or equal to ε to obtain an ε-optional solution.

Some previous acquaintance with nonlinear decompositional and duality theory [1, 2] may provide useful background knowledge for the detailed reading of parts of this paper.

2. The Primal Decomposition

The problem Eqn (1) may be solved by the decomposition method of Dantzig for a nonlinear convex program* [1] which results in the following *primal master*:

$$
\left.
\begin{array}{ll}
\min\limits_{\substack{x, t \\ 1; i}} \left\{ f_1(x_1) + \sum_i (f_2(x_2{}^i) + f_3(x_3{}^i))t_i \right. & \text{Dual variables or} \\
& \text{minus simplex} \\
& \text{multipliers} \\[2mm]
g_{11}(x_1) + \sum_i (g_{12}(x_2{}^i) + g_{13}(x_3{}^i))t_i \leqslant 0 & u_1 \\[2mm]
\sum_i 1 \, t_i = 1 & w \\[2mm]
x_1 \geqslant 0 & \left. t_i \geqslant 0 \right\} = f
\end{array}
\right\}
\tag{2}
$$

and the *primal subproblem*:

$$
\left.
\begin{array}{ll}
\min\limits_{x_2, x_3} \left\{ f_2(x_2) + u_1{}^k g_{12}(x_2) + f_3(x_3) + u_1{}^k g_{13}(x_3) + w^k \right. & \text{Dual} \\
& \text{variables} \\[2mm]
g_{22}(x_2) \qquad\qquad + g_{23}(x_3) \leqslant 0 & u_2 \\[2mm]
g_{33}(x_3) \leqslant 0 & u_3 \\[2mm]
x_2 \geqslant 0 \qquad\qquad\qquad \left. x_3 \geqslant 0 \right\} = -df
\end{array}
\right\}
\tag{3}
$$

The values of the variables of the primal master at iteration k may be denoted by the superscript k. The objective function of the master may then be expressed as

$$
f^k = f_1(x_1{}^k) + \sum_i (f_2(x_2{}^i) + f_3(x_3{}^i))t_i{}^k = f_1(x_1{}^k) + u_1{}^k g_{11}(x_1{}^k) - w^k
\tag{4}
$$

---

* Alternatively, a nonlinear master may be obtained by using the nonlinear decomposition method of Huard [3].

where the latter expression may be obtained by considering $x_1{}^k$ as constants and formulating Eqn (1) in the form

$$\min_{t_i} \left\{ \sum_i (f_2(x_2{}^i)+f_3(x_3{}^i))t_i \right. \qquad\qquad \text{Simplex multipliers}$$

$$-\sum_i (g_{12}(x_2{}^i)+g_{13}(x_3{}^i))t_i \geqslant g_{11}(x_1{}^k) \qquad\qquad u_1{}^k$$

$$-\sum_i 1\, t_i = -1 \qquad\qquad w^k$$

$$t_i \geqslant 0 \qquad \left.\right\} +f_1(x_1{}^k)$$

and then using the relationship that the objective function equals the product of the simplex multipliers and the right-hand constants [4], (footnote on p. 68).

A cost term $v_1{}^k$ dependent on the x_1 variables:

$$v_1{}^k = f_1{}^k + u_1{}^k g_{11} - \left(\frac{df_1{}^k}{dx_1} + u_1{}^k \frac{dg_{11}^k}{dx_1} \right) x_1{}^k \tag{5}$$

will later be required. An economic interpretation is given in [8].

3. The Dual Decomposition

Upon the basis of the author's nonlinear decompositional theory [2] the problem may also be decomposed into a dual master and a dual subproblem. The minimization process is then seen as the two-stage minimization:

$$\min_{x_3} \left\{ f_3(x_3) + \min_{x_1, x_2} \left\{ f_1(x_1)+f_2(x_2) \right.\right. \qquad\qquad \left.\right\}$$

$$g_{11}(x_1)+g_{12}(x_2)+g_{13}(x_3)\leqslant 0$$

$$g_{22}(x_2)+g_{23}(x_3)\leqslant 0$$

$$x_1\geqslant 0 \ \ x_2\geqslant 0 \qquad \left.\right\} \tag{6}$$

$$g_{33}(x_3)\leqslant 0$$

$$x_3 \geqslant 0 \Big\}$$

where the inner minimization is undertaken for a particular value of the vector x_3. As a consequence of the assumptions made in Section 1 the outer minimization problem will always involve feasible and finite x_3 values leading to finite values of the corresponding objective function, and the inner minimization problem will always possess a feasible x_1, x_2 solution with a finite optimal solution, no matter which x_3 values have been chosen.

It is therefore possible to substitute the inner minimization problem involving x_1, x_2 variables by its dual [5, 6] to obtain the equivalent problem:

$$
\begin{aligned}
\min_{x_3} \Big\{ f_3\,(x_3) + \max_{\substack{x_1,\,x_2 \\ u_1,\,u_2}} \{ f_1\,(x_1) + f_2\,(x_2) + u_1\,(g_{11}\,(x_1) + g_{12}\,(x_2) + \\
+ g_{13}\,(x_3)) + u_2\,(g_{22}\,(x_2) + g_{23}\,(x_3)) - \\
\left(\frac{df_1}{dx_1} + u_1\,\frac{dg_{11}}{dx_1} \right) x_1 - \\
\left(\frac{df_2}{dx_2} + u_1\,\frac{dg_{12}}{dx_2} + u_2\,\frac{dg_{22}}{dx_2} \right) x_2 \\
\frac{df_1}{dx_1} + u_1\,\frac{dg_{11}}{dx_1} \geqslant 0 \\
\frac{df_2}{dx_2} + u_1\,\frac{dg_{12}}{dx_2} + u_2\,\frac{dg_{22}}{dx_2} \geqslant 0 \\
x_1 \geqslant 0 \quad x_2 \geqslant 0 \quad u_1 \geqslant 0 \quad u_2 \geqslant 0 \\
g_{33}\,(x_3) \leqslant 0 \\
x_3 \geqslant 0 \Big\}
\end{aligned}
\tag{7}
$$

All the feasible solutions (usually an infinite number) to the dual constraints above may be assumed to be known and denoted by $x_1{}^k, x_2{}^j, u_1{}^k, u_2{}^j$ where $k = j = 1, ..., \infty$. Minimizing a function containing the maximum of the dual objective function over all feasible dual solutions is equivalent to

$$
\begin{aligned}
\min_{m,x_3} \Big\{ m + f_3\,(x_3) \Big| \\
m \geqslant f_1{}^k + f_2{}^j + u_1{}^k (g_{11}^k + g_{12}^j + g_{13}\,(x_3)) + u_2{}^j(g_{22}^j + g_{23}\,(x_3)) - \\
\left(\frac{df_1{}^k}{dx_1} + u_1{}^k\,\frac{dg_{11}^k}{dx_1} \right) x_1{}^k - \left(\frac{df_2{}^j}{dx_2} + u_1{}^k\,\frac{dg_{12}^j}{dx_2} + \right. \\
\left. u_2{}^j\,\frac{dg_{22}^j}{dx_2} \right) x_2{}^j \quad (k = j = 1, ..., \infty) \\
g_{33}\,(x_3) \leqslant 0 \\
x_3 \geqslant 0 \Big\} = h
\end{aligned}
\tag{8}
$$

where $f_1{}^k \equiv f_1\,(x_1{}^k)$, and so on.

The above problem may be named the full dual master. Instead of considering the full dual master it may be preferable to consider a *restricted* dual master which contains only some of the above constraints, that is $k = j = 1, ..., r$ and has the optimal solution $m = m^l, x_3 = x_3{}^l$.

The optimal solution of the full dual master (8) would give the optimal solution of the original problem, Eqn (1), that is the minimal value of the objective function. As some constraints of the full dual master have been omitted in the restricted dual master, it follows that an optimal solution of the restricted dual master will provide an (under) estimate of the cost which is less than or equal to the minimal.

The constraint of the full dual master which would be active when $x_3 = x_3{}^l$ may be obtained by solving the *dual subproblem*:

$$
\begin{aligned}
\max_{x_1, x_2, u_1, u_2} \quad & \Bigg\{ f_1(x_1) + f_2(x_2) + u_1(g_{11}(x_1) + g_{12}(x_2) + g_{13}(x_3{}^l)) + \\
& u_2(g_{22}(x_2) + g_{23}(x_3{}^l)) - \left(\frac{df_1}{dx_1} + u_1 \frac{dg_{11}}{dx_1} \right) x_1 - \\
& \left(\frac{df_2}{dx_2} + u_1 \frac{dg_{12}}{dx_2} + u_2 \frac{dg_{22}}{dx_2} \right) x_2 - m^l \\
& \frac{df_1}{dx_1} + u_1 \frac{dg_{11}}{dx_1} \geqslant 0 \\
& \frac{df_2}{dx_2} + u_1 \frac{dg_{12}}{dx_2} + u_2 \frac{dg_{22}}{dx_2} \geqslant 0 \\
& x_1 \geqslant 0 \quad x_2 \geqslant 0 \quad u_1 \geqslant 0 \quad u_2 \geqslant 0 \Bigg\} = dh
\end{aligned}
\tag{9}
$$

the optimal solution of which is denoted by $x_1{}^k, x_2{}^j, u_1{}^k, u_2{}^j$ and is used to form an additional constraint of the restricted dual master.

A cost term $v_3{}^l$ dependent upon the x_3 variables

$$
v_3{}^l = f_3{}^l + u_1{}^k g_{13}^l
\tag{10}
$$

will later be required.

4. The Common Subproblem

By inserting temporarily constant values $x_3 = x_3{}^l$ into the primal sub-problem disregarding some constant terms, a related problem may be obtained which may be named the *primal common subproblem*:

$$
\begin{aligned}
\min_{x_2} \quad & \Big\{ f_2(x_2) + u_1{}^k g_{12}(x_2) & & \text{Dual variables} \\
& g_{22}(x_2) + g_{23}(x_3{}^l) \leqslant 0 & & u_2 \\
& x_2 \geqslant 0 \qquad \Big\} = -df^0
\end{aligned}
\tag{11}
$$

Similarly, by inserting temporarily constant values $x_1 = x_1{}^k$, $u_1 = u_1{}^k$ into the dual subproblem and disregarding some constant terms, a related problem may be obtained which may be named the *dual common subproblem*:

$$
\left.
\begin{array}{l}
\max\limits_{x_2,\, u_2} \left\{ f_2\,(x_2) + u_1{}^k g_{12}\,(x_2) + u_2\,(g_{22}\,(x_2) + g_{23}\,(x_3{}')) - \right. \\[2mm]
\left. \left(\dfrac{df_2}{dx_2} + u_1{}^k \dfrac{dg_{12}}{dx_2} + u_2 \dfrac{dg_{22}}{dx_2} \right) x_2 \right| \\[4mm]
\dfrac{df_2}{dx_2} + u_1{}^k \dfrac{dg_{12}}{dx_2} + u_2 \dfrac{dg_{22}}{dx_2} \geqslant 0 \\[4mm]
x_2 \geqslant 0 \quad u_2 \geqslant 0 \qquad\qquad \left. \right\} = dh^0
\end{array}
\right\}
\tag{12}
$$

It is of importance to observe that Eqn (12) is the dual of Eqn (11). As a consequence of the assumptions made in Section 1 and the results of non-linear duality theory the optimal value of (12) must equal that of (11):

$$
-df^0 = dh^0.
\tag{13}
$$

Having obtained an optimal or improved feasible solution of the dual common subproblem, a cost term $v_2{}^j$ dependent upon the x_3 variables is formed as follows for later use:

$$
v_2{}^j = f_2{}^j + u_1{}^k g_{12}^j + u_2{}^j g_{22}^j - \left(\frac{df_2{}^j}{dx_2} + u_1{}^k \frac{dg_{12}^j}{dx_2} + u_2{}^j \frac{dg_{22}^j}{dx_2} \right) x_2{}^j.
\tag{14}
$$

5. The Information Transfer Decision

The primal solution of the common subproblem and of part of the dual master forms a feasible solution of the primal subproblem. The corresponding value of the objective function may be denoted by $-df''''$ which using Eqns (3), (10), (11) may be expressed as

$$
-df'''' = -df^0 + v_3{}^l + w^k.
\tag{15}
$$

If $-df'''' < 0$ then the corresponding primal subproblem solution may decrease the primal master and it will therefore be useful to transfer information concerning $x_2{}^l, x_3{}^l$ to the primal master in the form of the normally more compact column vector

$$
\begin{pmatrix} f_2\,(x_2{}^l) + f_3\,(x_3{}^l) \\[2mm] g_{12}\,(x_2{}^l) + g_{13}\,(x_3{}^l) \end{pmatrix}
\tag{16}
$$

Similarly, the dual solution of the common subproblem and of part of the primal master forms a feasible solution of the dual subproblem. The corresponding value of the objective function may be denoted by dh''' which using Eqns (5), (9), (12) may be expressed as

$$dh''' = dh^0 + v_1{}^k + u_1{}^k g_{13}^l - m^l. \tag{17}$$

Similarly, if $dh''' > 0$ then the corresponding dual subproblem may increase the dual master and it will therefore be useful to transfer information concerning $u_1{}^k, v_1{}^k, u_2{}^j, v_2{}^j$ in the form of the normally more compact function

$$u_1{}^k g_{13}(x_3) + u_2{}^j g_{23}(x_3) + v_1{}^k + v_2{}^j. \tag{18}$$

If only a feasible rather than an optimal primal common subproblem solution $x_2{}^i$ for the partially primal feasible dual master solution $x_3{}^l$ has been obtained, then the value of the objective function of the primal subproblem may be denoted by $-df'$ instead of $-df'''$, where

$$-df' \geqslant -df'''. \tag{19}$$

Similarly, if only a feasible instead of an optimal dual common subproblem solution $u_2{}^j, x_2{}^j$ corresponding to the partially dual feasible primal master solution $u_1{}^k, x_1{}^k$ has been obtained, then the value of the objective function of the dual subproblem may be denoted by dh' instead of dh''', where

$$dh' \leqslant dh'''. \tag{20}$$

Finally, it may be convenient to denote the possible improvement of the primal and the dual common subproblems by $-df''$ and dh'' respectively, giving the relationships

$$-df' - df'' = -df'''$$

and $$dh' + dh'' = dh'''.$$

It may also be useful to introduce absolute overestimates of $-df''$ and dh'' and denote these by $-df{**}$ and $dh{**}$, respectively. These values may then be determined and used to facilitate the computational work in a way fully analogous to that outlined in an earlier paper by the author ([4], §§3, 4).

6. The Functioning of the Decomposition Method

6.1. *Initiation*

The solution process is initiated by reading the data including $\varepsilon \geqslant 0$, formulating the initial parts of the common subproblem, the primal and the dual master, setting $f = w = \infty$, $h = m = -\infty$, choosing some arbitrary

$$x_1 = x_1{}^k, \quad u_1 = u_1{}^k \tag{21}$$

which satisfy part of the dual constraints of the primal master problem, that is

$$\frac{df_1}{dx_1} + u_1 \frac{dg_{11}}{dx_1} \geq 0 \qquad x_1 \geq 0, \quad u_1 \geq 0 \tag{22}$$

and some arbitrary

$$x_3 = x_3{}^l \tag{23}$$

which satisfy part of the primal constraints of the dual master problem, that is

$$g_{33}(x_3) \leq 0, \qquad x_3 \geq 0. \tag{24}$$

6.2. General Control of the Process

If $f - h \geq \varepsilon$ then the following three steps 6·3–6·5 are solved in parallel or in some sequence, otherwise the last step (6.6). To ensure convergence the results of the following theorem should be taken into account.

6.3. Common Subproblem and Information Transfer Decision

On the basis of the latest information received concerning $u_1{}^k$, $x_3{}^l$, the modified objective function and the modified constants of the common subproblem, Eqns (11), (12), are obtained.

The common subproblem is then solved for (i) a primal, (ii) a dual, or (iii) a primal and dual feasible solution. The decision as to which type of solution is desired may be made in the course of the solution process in order to minimize the computational work necessary to produce a primal or dual solution, or both, which may improve the primal or the dual master, or both. The optimization of the primal or dual or both solution(s) of the common subproblem must at least be continued until (i) $-df' < 0$, (ii) $dh' > 0$, or (iii) $-df' < 0$ and $dh' > 0$.

If $-df' < 0$, information concerning the vector (16) of t_i coefficients corresponding to the achieved feasible solution $x_2{}^i$, $x_3{}^l$ of the primal common subproblem is sent to the primal master.

If $dh' > 0$, information concerning the function (18) of x_3 corresponding to the achieved feasible solution $u_1{}^k$, $x_1{}^k$, $u_2{}^j$, $x_2{}^j$ is sent to the dual master.

6.4. Primal Master

An optimal or improved primal feasible and partially dual feasible solution of the primal master is determined.

The solution is required to be dual feasible with respect to the x_1 and u_1 variables so that the dual constraints (22) are satisfied. The primal master solution so determined is denoted by a superscript k, viz:

$$x_1 = x_1{}^k, \quad t = t^k, \quad u_1 = u_1{}^k, \quad w = w^k \quad \text{and} \quad f = f^k. \tag{25}$$

An underestimate $v_1{}^k$ of the cost of the x_1 variables is formed using Eqn (5). The information $u_1{}^k$, w^k and $v_1{}^k$ is then sent to the common subproblem.

6.5. *Dual Master*

An optimal or improved dual feasible and partially primal feasible solution of the dual master is determined. The solution is required to be primal feasible with respect to the primal constraints (24). The dual master solution so determined is denoted by a superscript l, viz:

$$m = m^l, \quad x_3 = x_3{}^l \quad \text{and} \quad h = h^l. \tag{26}$$

A cost estimate of the x_3 variables is formed using Eqn (10). The information m^l, $x_3{}^l$ and $v_3{}^l$ is then sent to the common subproblem.

6.6. *The Final Solution*

An ε-optimal primal solution is obtained from the primal master by setting

$$x_1 = x_1{}^k, \quad x_2 = \sum_i x_2{}^i t_i{}^k, \quad x_3{}^i = \sum_i x_3{}^i t_i{}^k \tag{27}$$

and
$$f = f_1(x_1) + f_2(x_2) + f_3(x_3).$$

7. The Convergence of the Process

THEOREM. *If there is available an optimal solution to the primal master with $f = f^k$, an optimal solution to the dual master with $h = h^l$ for which $f^k > h^l$, and an optimal primal and dual solution of the corresponding common subproblem, then the corresponding primal and dual subproblem solutions inserted into the primal and dual master problems may improve either or both of the master problems.*

Proof. The optimal value of the primal subproblem Eqn (3) for the constant value of $x_3 = x_3{}^l$ is given by $-df'''$. The optimal value of the dual subproblem Eqn (9) for the constant value of $x_1 = x_1{}^k$, $u_1 = u_1{}^k$ is given by dh'''.

A positive value of $df''' + dh'''$ means that either df''' or dh''' or both are positive and hence that the corresponding primal and dual subproblem solutions may improve either the primal or the dual or both master problems.

Thus it is of importance to consider the value of

$$df''' + dh''' =$$

$$
\begin{aligned}
&= -(-df^0 + f_3^l + u_1{}^k g_{13}^l + w^k) &&\text{from Eqns (3) and (11)} \\
&\quad + (dh^0 + f_1{}^k + u_1{}^k (g_{11}^k + g_{13}^l)) - &&\text{from Eqns (9) and (12)} \\
&\quad - \left(\frac{df_1{}^k}{dx_1} + u_1{}^k \frac{dg_{11}^k}{dx_1} \right) x_1{}^k - m^l = \\[4pt]
&= f_1{}^k + u_1{}^k y_{11}^l - w^k &&= f^k \quad \text{cf Eqn (4)} \\
&\quad - (m^l + f_3^l) + &&= -h^l \text{ cf Eqn (8)} \\
&\quad + df^0 + dh^0 - &&= 0 \quad \text{cf Eqn (13)} \\[4pt]
&\quad - \left(\frac{df_1{}^k}{dx_1} + u_1{}^k \frac{dg_{11}^k}{dx_1} \right) x_1{}^k = &&= 0 \quad \text{cf the} \\
& &&\text{assumption of an} \\
&= f^k - h^l = &&\text{optimal solution of} \\
&> 0 &&\text{the primal master}
\end{aligned}
\tag{28}
$$

This leads to the conclusion that the corresponding primal and dual subproblem solutions may improve either the primal or the dual or both master problems.

7.1. *The cases in which the optimal primal and the optimal dual common subproblem solution may not improve one of the masters.*

It follows from Eqn (28) that

$$\text{if} \qquad\qquad df''' \geqslant f - h > 0 \ \text{ then } dh''' \leqslant 0 \tag{29}$$

in which case it would be of no avail to obtain an improved feasible or optimal dual common subproblem solution as neither would be able to improve the dual master.

Similarly, it follows from Eqn (28) that

$$\text{if} \qquad\qquad dh''' \geqslant f - h > 0 \ \text{ then } df''' \leqslant 0 \tag{30}$$

in which case it would be of no avail to obtain an improved feasible or optimal primal common subproblem solution as neither would be able to improve the primal master.

8. Illustrations of Programs with the Assumed Properties
8.1. *A Nonlinear Case*

The assumptions made in Section 1 are satisfied by, for instance, the nonlinear convex program

$$\min_{a,\,b,\,c,\,x,\,y,\,z} \left\{ \bar{s}a + \bar{u}b + \bar{v}c + A(x) + B(y) + C(z) \right.$$

$$
\left.
\begin{array}{llll}
Ia & H(x) + K(y) + L(x) & \geqslant P \\
& Ib & + D(y) + M(z) & \geqslant Q \\
& & Ic & + N(z) & \geqslant R \\
& & -Ix & \geqslant -\bar{x} \\
& & -Iy & \geqslant -\bar{y} \\
& & -Iz & \geqslant -\bar{z} \\
\multicolumn{4}{c}{a \geqslant 0 \quad b \geqslant 0 \quad c \geqslant 0 \quad x \geqslant 0 \quad y \geqslant 0 \quad z \geqslant 0}
\end{array}
\right\}
\tag{31}
$$

where

$$
\left.
\begin{array}{l}
a, b, c, x, y, z \quad \text{denote vectors of variables, and} \\[1em]
\bar{s}, \bar{u}, \bar{v}, \bar{x}, \bar{y}, \bar{z} \quad \text{vectors of non-negative constants}
\end{array}
\right\}
\tag{32}
$$

or, for instance, by the same program but assuming that

$$
\left.
\begin{array}{ll}
a, b, c & \text{denote scalar variables} \\
x, y, z & \text{as before, vectors of variables} \\
\bar{s}, \bar{u}, \bar{v}, \bar{x}, \bar{y}, \bar{z} & \text{scalar non-negative constants} \\
I = (1, 1, ..., 1) & \text{column vector of unit coefficients} \\
-I = (-1, -1, ..., -1) & \text{row vector of minus unit coefficients}
\end{array}
\right\}
\tag{33}
$$

The detailed structure of the nonlinear program (see Eqn 1) would then be

$$
\left\{
\begin{array}{l}
f_1\ (x_1) \equiv [\quad \bar{s}a + A(x)]\ f_2\ (x_2) \equiv [\quad \bar{u}b + B(y)]\ f_3\ (x_3) \equiv [\quad \bar{v}c + C(z) + 0] \\[1em]
g_{11}\ (x_1) \equiv \begin{bmatrix} -Ia - H(x) \\ Ix \end{bmatrix} g_{12}(x_2) \equiv \begin{bmatrix} -K(y) \end{bmatrix} g_{13}\ (x_3) \equiv \begin{bmatrix} -L(z) + P \\ -\bar{x} \end{bmatrix} \\[2em]
\qquad\qquad g_{22}\ (x_2) \equiv \begin{bmatrix} -Ib - D(y) \\ Iy \end{bmatrix} g_{23}\ (x_3) \equiv \begin{bmatrix} -M(z) + Q \\ -\bar{y} \end{bmatrix} \\[2em]
\text{(34)} \qquad\qquad\qquad\qquad\qquad\qquad g_{33}\ (x_3) \equiv \begin{bmatrix} -Ic - N(z) + R \\ Iz - \bar{z} \end{bmatrix}
\end{array}
\right.
$$

8.2 A Nonlinear–Linear Case

Among possible nonlinear–linear cases, a computationally rather pleasant case may be obtained by assuming the structure of Eqns (33) and (34), linear functions with respect to the variables x and z, and a computationally favourable structure of the common subproblem. The corresponding primal and dual master programs will then become linear programs, with a nonlinear common subproblem. The overall nonlinear programming problem could then be comparatively simple to solve.

8.3. *A Linear Case*

An interesting special case occurs when the substructure of the program is as in Eqns (32)–(34) but all functions $A(x), H(x)$, and so on, are assumed to be linear and hence may be written as Ax, Hx, and so on, where A, H, and so on, denote vectors and matrices. The method may then be given a completely symmetric formulation as derived in another paper by the author [4] which generalizes and proves the convergence of a decomposition method proposed by Pigot [7].

9. Conclusion

The method developed seems to be of considerable theoretical interest as it allows one to take advantage of computationally favourable structures of a nonlinear convex separable economic system.

References

1. G. B. Dantzig (1963). "Linear Programming and Extensions", pp. 471-478, section 24-1. Princeton University Press, Princeton, New Jersey.
2. T. O. M. Kronsjö (1967). Decomposition of an Economic System Consisting of a Convex and a Nonconvex Section. University of Birmingham, CREES, Discussion Papers, Series RC/A, No. 9, Birmingham, England, 25th September, 1967.
3. P. Broise, P. Huard and J. Sentenac (1968). "Decomposition des Programmes Mathématiques", section 3-3. Dunod, Paris.
4. T. O. M. Kronsjö (1968). Centralization and Decentralization of Decision Making, The Decomposition of Any Linear Programme in Primal and Dual Directions to Obtain a Primal and a Dual Master Solved in Parallel with One or More Common Subproblems. Revue Française d'Informatique et de Recherche Operationelle, 2e année, No. 10, pp. 73-114.
5. A. Whinston (1967). Some applications of the conjugate function theory to duality. *In* (J. Abadie, ed.) "Nonlinear Programming", chap. V. North-Holland Publishing Company, Amsterdam.
6. P. Huard (1964). Programme dual. *In* (P. Huard, ed.) "Mathématique des programmes économiques", pp. 13-17. Dunod, Paris.
7. D. Pigot (1964). Double décomposition d'un programme linéaire. *In* Proceedings of the 3rd International Conference on Operational Research pp. 72-78 Oslo, July 1963. Dunod, Paris.
8. T. O. M. Kronsjö(1969). International and interregional economic cooperation and planning by linked computers. *In* (A. J. Scott, ed.) "Studies in Regional Science." Pion Ltd., London.

8. Large Step Gradient Methods for Decomposable Nonlinear Programming Problems

LAWRENCE E. SCHWARTZ*

Center for Naval Analyses of The University of Rochester, Arlington, Virginia, U.S.A.

1. Introduction

Several new algorithms for solving large-scale nonlinear programming problems which have a special structure are presented here. These problems occur in economics and in certain related contexts [1]. The constraint set in problems of the type to be considered contains both nonlinear and linear functions, the linear ones having already been reduced to the usual canonical block diagonal or staircase form—the latter being reducible to the former [1]. Many otherwise independent linear blocks are linked together by a relatively small number c of nonlinear connecting relations as follows:

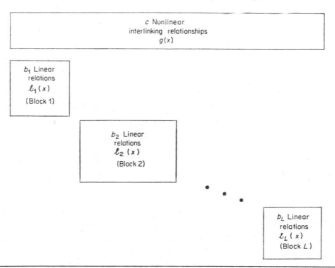

---

* This paper does not necessarily represent the views of the Center for Naval Analyses or of the U.S. Navy.

The methods to be presented below are extensions to these structured programs of some ideas of Zoutendijk [2, 3], as modified by ideas needed to eliminate separate anti-zigzagging procedures due to Topkis and Veinott [4] and McCormick [5]. Simplex-based Generalized Upper Bounding Techniques of Dantzig and Van Slyke [6, 7], as generalized by Sakarovitch and Saigal [8], are the basis of this extension. The methods to be presented here solve a sequence of relatively simple direction-finding or step-size subproblems, each of significantly lower dimension.

These methods differ from such competing algorithms as Rosen's partition programming [9–11] in that only a single problem need be solved and no interlinking variables need be introduced into a problem. Thus, dimensionality and computation time are much reduced.

The first procedure, in several variant algorithms, involves linearizing nonlinear constraints with a convenient method, for example with adjoined cutting planes or with numerical differentiation of the nonlinear constraints in the neighborhood of the current trial solution as in MAP [12]. The second procedure, and its variants, incorporate all nonlinear constraints in the objective. These interior point methods are members of a class of procedures due to Huard [13, 14].

Powerful new tools for structured nonlinear programming problems which have a large linear constraint matrix are the result of the integration of these already proved algorithms. Sketches of proofs of convergence for algorithms of both types will be given, assuming concave objectives† for the linearized constraint procedures and convex constraints for both methods. Computational results, although preliminary, will also be discussed.

2. Problem

The problem addressed by these procedures is of the primal form: find a possibly local optimum x^*, if it exists, of $f(x)$ a real-valued continuous function over a closed subset S of E^n, euclidean n-space

$$\text{minimize } \{f(x): x \in S\}. \tag{1}$$

The form of the feasible set of m constraints S is

$$S = \bigcap_j \{x: g_j(x) \leqslant 0, \wedge j \in J_1, l_j(x) \leqslant 0, \wedge j \in J_2, h_j(x) \leqslant 0, \wedge j \in J_3\}. \tag{2}$$

The $l_j(x)$ are linear constraints with the finite indexing set J_2, having a relatively large number of linear blocks, which are completely independent of each other because they contain only their own decision variables and not those of any other block. The $g_j(x)$ are a relatively small number of

---

† The assumption that x is an element in some arbitrary topological space is sufficient for the interior point method, see Lieu and Huard [15]. However, this paper is specialized to euclidean n-space.

nonlinear constraints, with a finite indexing set J_1, expressing nonlinear interrelationships between all the decision variables x in all blocks. In addition, nonlinear equalities $h_j(x)$ have finite indexing set J_3. The set S is assumed to be compact: and $f(x)$, $g(x)$ and $h(x)$ are assumed to be twice continuously differentiable.

There are two general types of problems: (1) problems of the form of Eqn (1) with a few "nearly linear" constraints $g(x)$, $h(x)$, and (2) problems of the form of Eqn (1) with many highly nonlinear constraints $g(x)$. Consider firstly a very natural approach to type 1 problems, which involves linearizing the $g(x)$.

3. Algorithms

3.1. *"Nearly Linear" Constraints* $g(x)$, $h(x)$

The procedures of this section are designed for problems of the form (1) for which there are only a relatively few nonlinear constraints $g(x)$ and $h(x)$ and these few are "nearly linear", that is the $g(x)$ and $h(x)$ have neither too much curvature nor too many changes in curvature in the neighborhoods of the iterative trial solutions x^k. If this is the case, the $g(x)$ and $h(x)$ would be well-approximated locally in the neighborhood of the current best trial solution x^k by a cutting plane or other convenient linearization such as a finite difference [16] approximation of the partial derivatives of the $g(x)$, for example as in MAP.

Qualitative characterization of the class of nonlinearities for which these algorithms are efficient is not now possible at a theoretical level. It is, however, clear that some characterization in terms of measure theory is possible, and such a characterization will appear later. From a practical point of view, however, extremely slow convergence of an algorithm using the cutting plane method, and either slow convergence or convergence to a suboptimal solution when numerical differentiation is used, usually signals nonlinearities in the constraint set which cannot be successfully attacked with either of these linearizations. It is efficient when this happens to use the best trial solution found, with the method of this section as the starting point for one of the algorithms of the next section designed with highly nonlinear constraints in mind.

For convenience, assume that the objective $f(x)$ is concave and that the constraints $g(x)$ and $h(x)$ are convex. Modification of the procedure for nonconvex problems on a connected subset S of E^n will also be described for an algorithm using the cutting plane method.

Optimization calculations start from a feasible initial point x^0. This preliminary step of the method is common to all variations of the algorithm discussed here. If no such point is available construct, as described in [3], a slightly modified problem for which an interior solution always exists.

In the iterated step of these algorithms, where k is the iteration number, linear subproblems

$$P_k = \min \{f(x^k): l(x^k) \leq 0\} \tag{3}$$

are solved by repeated application of one of several alternative large step gradient procedures to be described below.

In all first-order gradient procedures, it is necessary to determine x^{k+1} from the current trial solution x^k by the use of the usual recursion

$$x^{k+1} = x^k + t^k d^k \tag{4}$$

in which d^k is the direction of fastest decrease of the objective $f(x)$, $-\partial f(x^k)/\partial x_i$ at the best current solution x^k, while t^k is the length of admissible travel from x^k along the direction d^k. Notice that it can be assumed without loss of generality that x^{k+1} lies on the boundary of the feasible region S, since the boundary can be reached from any interior point by increasing t^k.

There are several first-order gradient procedures which lead to convenient algorithms for structured problems, at least in practice. Zoutendijk [3] suggests the use of his cauchy-based large step gradient procedure $P2$ [2] to optimize the linear subproblems (3), since it does not require additional anti-zigzagging precautions. In this procedure, determination of x^{k+1} requires first the solution of the direction-finding problem at each iteration

$$d^k = \min \{\nabla f(x^k)^T d: A^T d \leq 0\}. \tag{5}$$

Using matrix A from $l(x)$ and a linear norm a simplex method can be applied. Because these subproblems are large structured linear programs a variant simplex algorithm, due to Sakarovitch and Saigal [8], should be integrated into this part of the algorithm to take advantage of the special structure of the linear constraints. Two problems are not required as in the Dantzig-Wolfe decomposition algorithm.

Using the fact that the number of blocks L in the constraint matrix greatly exceeds the dimension c of the coupling block, a much smaller working basis is obtained. Consequently both computation and core storage requirements are reduced, since all computations are done on that working basis, which is of much lower dimension than the original basis.

The reduced system is defined by deleting subsystems $a_i x_i = b_i$ where a set S_i of columns is "inessential", that is it has no more than the number m_i of variables in the ith subsystem. The set of "essential" basic columns is the working basis B, which forms a basis for the reduced system (see [8] for proof). The value of the "inessential" basic variables is determined by solving almost trivial low order subsystems of the form $Bx_i = b_i$. Sakarovitch and Saigal provide revised simplex rules for finding each new working basis from its predecessor. They also point out that because of the special structure of the matrix B, Dantzig's compact basis triangularization algorithm should be more efficient computationally. Details of that algorithm are available in [8] and [17].

A stepsize problem is then solved at each iteration in one of several ways, for instance by finding the smallest positive root of the equations

$$\psi_p(x^k + td^k) = 0, \quad (p = 1, 2, \dots m) \tag{6}$$

with, for example, Newton's method. These systems are easy to solve if they are broken down into subsystems corresponding to the "essential" and "inessential" columns in the final iteration of the algorithm of Sakarovitch and Saigal used to solve the direction-finding problem at each iteration.

To apply this method successfully in n dimensions, two stabilizing devices are useful. The first requires a one dimensional search on t for a minimum on the line joining the current best solution x^k and the proposed new solution x^{k+1}, that is

$$\min_t \{f(x^k) + td^k\}. \tag{7}$$

The new point x^{k+1} is then $x^k + td^k$. The second stabilizing device [18] adds the Levenberg parameter λ to the diagonal elements of the Hessian matrix $H(x^k)$ in the usual Newton's method recursion

$$x^{k+1} = x^k + t^k H(x^k) \nabla f(x^k) \tag{8}$$

where
$$H(x^k) = \{\nabla^2 f(x)\}^{-1}.$$

Another first-order gradient procedure, a modified cauchy algorithm which does not require an anti-zigzagging routine is due to McCormick [5, CM2], and it should be computationally efficient to integrate it into this algorithm instead of Zoutendijk's $P2$. It should be more efficient than Zoutendijk's $P2$ because it requires fewer calculations of the objective function gradient. McCormick's algorithm proceeds as follows. As suggested by Wolfe [19], the reduced gradient method described in [20] is used to reduce the problem

$$\min \{f(x): \ A - xb \leqslant 0\} \tag{9}$$

to the problem

$$\min \{f(x): \ x \leqslant 0\}. \tag{10}$$

Minimization proceeds from the current point x^k along the direction d^k of the fastest decrease of $f(x)$ evaluated at the best current solution x^k. Motion in the direction d^k continues until a constraint of the form of that in the reduced problem (10) is encountered or until $f(x)$ is locally minimized along the ray connecting the current best solution x^k and the proposed new one x^{k+1}. If a boundary of the reduced problem (10) is encountered, the appropriate component is set equal to zero; that is

$$x_i^k = \max \ (0, x_i^k + t^k d^k), \qquad (i = 1, 2, ..., n) \tag{11}$$

with t^k the smallest local minimizing point of the subproblem

$$\min \{f(x^k): \ t \geqslant 0, \ l(x) \leqslant 0\}. \tag{12}$$

The subproblems to be solved when McCormick's algorithm is used thus involve step length rather than direction as in Zoutendijk's P2. The algorithm of Sakarovitch and Saigal should be used to solve these subproblems to take advantage of the special structure of the constraints. However, a first-order Taylor's series approximation at x^k must first be obtained.

After a local minimum along the ray connecting x^k and x^{k+1} has been found new gradient information is used to generate the next step size t^{k+1}.

A third possible first-order gradient method for this algorithm has been suggested by Topkis and Veinott [4]. This method is a simple variant of Zoutendijk's procedure P1 [2], but does not require additional anti-zigzagging steps. The direction subproblems are once again more efficiently solved if the Sakarovitch and Saigal algorithm is used.

While the available literature is limited to these three variants of the cauchy method of steepest descent under constraint, there is no problem in principle with using more sophisticated gradient approaches. For example, convergence should be accelerated if the Partan method of Shah and co-workers [21] is used in these algorithms for structured problems instead of the cauchy procedure.

It is possible that one of the first-order procedures described above can be used until an optimum is reached. If, however, the first-order procedure becomes very inefficient in the vicinity of the optimum, the first-order gradient procedure should be stopped after a few long steps. A natural approximation to the objective at this point in these algorithms is a quadratic one. Consequently, it is more efficient to switch to a second-order procedure for one or a few steps until the minimum is reached. Quadratic programming, for example Wolfe's algorithm [21a], can then be used to solve the step-size or direction-finding subproblems, respectively.

When a constrained version of Newton's method such as a modification of that in [22] or [4] is used, the second derivatives are not usually available. As a consequence, the method of Davidon [23], in which successive values of $\nabla f(x^k)$ are used to build an estimate of $\nabla^2 f(x^k)$, and thus of the Hessian matrix $H(x^k)$ required in Newton's method, would greatly reduce the computational effort. In order to perform convex programming to solve the respective step size or step direction subproblems with the Sakarovitch and Saigal algorithm, a linear programming technique, the objective of the step size or direction-finding problem, respectively, can be linearized. This can be done by introducing a new variable x_{n+1} and then minimizing it instead of $\nabla f(x^k)^T d - \frac{1}{2} d^T H(x^k) d$, using adjoined cutting planes to linearize the new constraint, for example,

$$x_{n+1} - \nabla f(x^k)^T d - \frac{1}{2} d^T H(x^k) d. \tag{13}$$

Alternatively, numerical differentiation could be used to linearize the new constraint. A numerical approximation of the first partials of each non-linear constraint $g(x)$, $h(x)$ can then be substituted for $g(x)$ or $h(x)$ itself, assuming that Δx is properly chosen. However, caution must be used when selecting the increment in each element of x, for too small a value produces large round-off errors while too large a value approximates the slope by a chord instead of the tangent. Whether cutting planes or numerical differentiation is used, the linearized constraints will be of the form

$$\nabla g(x^k)^T x^k \leqslant \nabla g(x^k)^T x^k. \tag{14}$$

If an interior minimum is found on the ray connecting x^k and x^{k+1}, it is taken as the successor point x^{k+1}. If an interior minimum is not found, linearize for example $g(x) \leqslant 0$ with respect to the boundary points x^k by adding cutting planes.

The major difference between these two procedures is that cutting planes need not be added to the constraint set if numerical differentiation is used. This may provide a substantial computational advantage if there are very many $g(x)$, $h(x)$ by reducing the dimensionality of the constraint set.

If some $g(x^k) < 0$ in the constraint set of problem (1) is non-convex at the current trial solution, say, constraints which were introduced when the current trial solution was far from the present one may cut off a portion of the feasible set S which should not be cut off. To insure that no portion of S is definitely cut off take out, as suggested by Zoutendjik [3], any active linearization of that $g(x)$ in the linear subproblem. The test need not be made at each step, but only occasionally, judged emphatically on results of the computation. As pointed out by Beale [24], the basis may become ill-conditioned unless high precision is carried in all computations near the minimum. Thus, speed of computation may be adversely affected. Finally, add these inequalities to the constraints of the linear subproblem P giving the linear subproblem P_{k+1}.

Various acceleration devices can be introduced. For example, a constraint utilizing the conjugate gradient idea

$$\{\nabla f(x^{k+1}) - \nabla f(x^k)\}^T (x^k - x^{k+1}) = 0 \tag{15}$$

can be added whenever x^{k+1} is an interior minimum of $f(x)$ on the ray connecting x^k and x^{k+1}. Such an additional constraint makes the method converge in a finite number of steps.

In this way, a sequence of points x^k with decreasing values $f(x^{k+1}) < f(x^k)$ is generated, converging to a global minimum of $f(x)$ on S if $f(x)$ is convex and the $g(x)$ are also convex. If the Kuhn-Tucker conditions

$$\nabla f(x^*) = \sum_{j=1}^{m} \pi_j \nabla g_j(x^*) + \delta \text{ and } \pi_j \geqslant \mu \tag{16}$$

(where δ and μ are arbitrary pre-established tolerances) are satisfied, stop. Otherwise, repeat all of the iterated steps of any of the several variant algorithms presented above until the global minimum of $f(x)$ is reached.

If objectives are nonconcave global optima cannot be found with these large step gradient methods. The best that these procedures can guarantee is a sequence of local optima, since constrained gradient methods use only local information. The only known method for finding global optima in this circumstance is to exhaustively search all points of S. A random sampling procedure can find a global optimum with given probability levels if large enough samples are taken [25]. The sampling procedures, in contrast to gradient procedures, accomplish only a separation of starting points into sets of points with good and poor potential for producing gradient directions which lead eventually to lower minima of $f(x)$.

Because random sampling procedures make no use of local information about an unknown objective $f(x)$, they are inefficient in locating local minima. Consequently, it is natural to combine these two methods into a single procedure for finding global minima. Such a procedure, which is useful when Bayesian prior probability functions can be fitted as information about $f(x)$ obtained through exploration by gradient procedures, was developed by the author in [26] for a somewhat different problem. Rules for switching from one kind of search to the other and for terminating the overall optimization procedure are very similar to those required for the structured problem (1).

Another type of difficulty, not amenable to random sampling of starting points, may also arise in using the methods of this section. A very poor linear approximation to one or more of the nonlinear constraints $g(x)$, $h(x)$ may produce local optima even though the linearized problem has a concave objective and convex constraints. In such cases, use the best suboptimal successor point x^{k+1} obtained with this procedure as the starting point of one of the algorithms to be discussed in the next section. Computational efficiency in the nonlinear procedure should consequently be appreciably improved.

3.2. *Nonlinear Constraints* $g(x)$, $h(x)$

Faced with a problem of type (2), that is if there are many highly nonlinear constraints $g(x)$, $h(x)$ in problem (1) with $f(x)$ and all $g(x)$, $h(x)$ continuous, it is more efficient to start with the algorithm of this section than with the ones discussed above. Such an algorithm, and several variants, based on a suggestion by Zoutendijk [3] for optimizing unconstrained problems with large-step gradient methods are presented in this section.

These methods are members of the class of interior point methods described by Huard [13, 14]. As with all interior point methods, the main idea here is to place all of the nonlinear constraints into the objective. The new objective† is thus of the form

$$F(x^{k+1},x^k) = \{[f(x^k)-f(x^{k+1})] \prod_{j \in J_1} g_j(x) \prod_{j \in J_3} h_j(x) : \forall j \in J_1, J_3\} \quad (17)$$

for feasible x^k. Although the function defined by (17) is nonconcave, Huard shows [14] that all local optima are also global. Thus, a gradient procedure can be applied to (17) to obtain an optimum. An alternate form of the objective is

$$F(x^{k+1},x^k) = \min \{f(x^k)-f(x^{k+1}), g_j(x), h_j(x) : \forall j \in J_1, J_3\}. \quad (18)$$

Although (18) is concave, a gradient method is not directly applicable.

As an example of a method of centers algorithm of the class represented

---

† These methods are, however, valid if the constraints $g(x)$, $h(x)$ are nonconvex. Moreover, it is unnecessary to assume that $f(x)$ is concave; see footnote on p. 100.

by Eqn (17), consider a form suggested by Fiacco and McCormick [27]. In this case special attention must be extended to nonlinear equalities $h(x)$

$$F(x^{k+1}, x^k) = \frac{1}{f(x^k) - f(x^{k+1})} + \sum_{j \in J_1} \left\{ \frac{1}{g_j(x^{k+1})} \right\} - \sum_{j \in J_3} \{h_j(x)\}^2. \quad (19)$$

Other choices of penalty function, for example a logarithmic form, are possible.

Start with a point $x^0 \in S_0$, where

$$S_0 = \{x : x^* \in S, \ f(x^*) \leqslant f(x)\}. \quad (20)$$

If S_0 is empty, construct a modified problem [3]. Calculate

$$\nabla F(x^{k+1}, x^k) = \left\{ \frac{1}{[f(x^k) - f(x^{k+1})]^2} \right\} \nabla f(x^{k+1})$$

$$- \sum_{j \in J_1} \left\{ \frac{1}{g_j^2(x^{k+1})} \right\} \nabla g_j(x^{k+1}) - 2 \sum_{j \in J_3} \{\nabla h_j(x^{k+1})\}. \quad (21)$$

Minimization of the form (21) can proceed with any of the first-order gradient methods discussed in the last section. Unlike the algorithm described for the linearized constraint case, however, it is never necessary to start with a first-order procedure and then shift over to a second-order one near the optimum. The well-known tendency for gradient methods to become inefficient when the optimum is approached can be countered in another way. Huard proves [14] that it is not necessary to optimize the function $F(x^{k+1}, x^k)$ completely. Since a fairly large tolerance is possible in the solutions obtained with this algorithm, the procedure should be stopped when the euclidean norm $\|\nabla F(x^{k+1}, x^k)\|$ is reduced to some preassigned percentage of its initial value.

Since the form (19) is highly nonlinear, it is natural to use a modified Newton's method exclusively. Convergence of direction or step size sub-problems can be accelerated when a second-order procedure is used from the start by an extrapolation procedure. A method such as that used by Fiacco and McCormick in the SUMT algorithm can be applied here [28], since the parameter r appears implicitly in (21). It will be necessary, as above, to use Wolfe's algorithm. With more sophisticated gradient methods, linearize the concave objective with adjoined cutting planes after placing the objective into the constraints or by taking numerical derivatives in the neighborhood of the current trial solution x^k so that the structured linear programming algorithm of Sakarovitch and Saigal [8] can be applied. The constrained Newton's method [4] or [22] and the method of Davidon [23] may be used here in the choice of a sequence $\{d^k\}$ or $\{t^k\}$, as in the last section. Accelerating devices are also possible, for example a search along the line $x^k + t^k d^k$ for a better value for each component of t than the usual 0·5.

4. Convergence Proofs

The proof of convergence for each of the algorithms developed above for the case of "near-linear" constraints is the same as that for any other large step gradient algorithm. Adopting the point of view of Topkis and Veinott [4], as generalized by Zangwill [29, 30], consider the map A to be a feasible direction algorithm which maps any current trial solution x^k into a successor solution point x^{k+1}, generating an infinite sequence $\{x^k\}$ of solution points. A is a point to set map, that is the algorithm A defines a set of possible successor points, x^{k+1} [30, 31].

The map A is also a composition, that is $A = MD$. Composition A is defined as $A(x) = \bigcup_i \{M_i(w): w \in A(x)\}$ where the map $M(x, d)$ is the set of $w = x + td$ and $f(x + td) = \min_t \{f(x + td): t \in T\}$, T a compact set, say. Given x^k, the map D specifies a direction d^k or a step size t^k. The existence of such direction and step size functions is discussed in [4] and [29]. Given x^k and d^k, or t^k, the map M yields a point x^{k+1} which minimizes $f(x)$ on the feasible set S along the ray $x^k + t^k d^k$.

The strategy to be followed involves expressing A^1 as a composition of several algorithm subparts, each of which can be proved to have the closure property. Because the composition of closed maps is closed [31], map A can then be proved to be closed: two theorems due to Zangwill [30], the second of which is slightly generalized, are convenient in this effort.

THEOREM 1. *Let the point to set map* $A: S \to 2^S$, *a subset of* S, *define one of the modified large step gradient algorithms developed above for solving problem* (1). *Suppose that:* (a) S *is compact;* (b) $f(x)$ *is continuous;* (c) *if* x *is not a solution* $f(w) < f(x)$ *for any* $w \in A(x)$ *and if* x *is a solution* $f(w) = f(x)$, *any* $w \in A(x)$; (d) *the map* A *is closed at* x *if* x *is not a solution. Then, the accumulation point of any convergent subsequence* $\{x^k\}_{k \in K} \to \bar{x}$ *is a solution.*

Proof. Consult Zangwill [30].

THEOREM 2. *If the subsequence* x^k *converges to* $\bar{x}$, *that is* $\{x^k\}_{k \in K} \to \bar{x}$, *and the subsequence* d^k *converges to* $\bar{d}$, *that is* $\{d^k\}_{k \in K} \to \bar{d}$, *where* $\bar{d} \neq 0$, *or the sequence*

$\{t^k\}_{k\in K}\to \bar{t}$, $\bar{t}\neq 0$, points $\bar{d}$, $\bar{x}$, or $\bar{t}$, $\bar{x}$, that is $\{x^k\}_{k\in K}\to\bar{x}$, $\{d^k\}_{k\in K}\to\bar{d}$, and $d^k\in A(x^k)$ for $k\in K$ imply $\bar{d}\in A(\bar{x})$ or $\{x^k\}_{k\in K}\to\bar{x}$, $\{t^k\}_{k\in K}\to\bar{t}$, and $t^k\in A(x^k)_{k\in K}$ imply $\bar{t}\in A(\bar{x})$.

Proof. Consult Zangwill [30]. The proof for t is analogous.

4.1. *Algorithms for "Nearly-Linear" Constraints*

It is clear from the assumptions of problem (1) that conditions (*a*) and (*b*) of Theorem 1 are satisfied: (*a*) because S was assumed compact and (*b*) since $f(x)$ was assumed continuous. Property (*c*) obviously holds, for $\nabla f(x^k)\neq 0$ implies that

$$f(w)<f(x^k) \tag{22}$$

where $w = x^{k+1}$.

All that remains is to show for Theorem 2 that condition (*d*) of Theorem 1 also holds. To do this a proof that the map A^1 is closed is required; that is it is necessary to show that $x^k\to\bar{x}$ and that $d^k\to\bar{d}$ or $t^k\to\bar{t}$ when linear or quadratic programming are used to obtain optimal direction vectors d^k or step size vectors t^k, given a particular algorithm which evidences antizigzagging behavior such as Zoutendijk's procedure *P2* or Topkis and Veinott's modified Zoutendijk *P1*, or McCormick's bent gradient method, as discussed above. Zoutendijk proves [2] that $x^k\to\bar{x}$, a constrained stationary point if his first-order *P2* procedure described above is used. It is obvious, reversing the "if and only if" in Topkis and Veinott's definition† of a " usable" direction, that $\bar{x}$ a constrained stationary point implies that $\bar{d}$ is the corresponding limit point of the sequence $\{d^k\}$. Hence $d^k\to\bar{d}$, and the algorithms for linearized constraints discussed above are convergent. Similar arguments can be used for the step size t^k of McCormick's algorithm, as discussed in Section 3.1.

4.2. *Algorithms for Highly Nonlinear Constraints*

The proof of convergence of the modified interior point method for problems with highly nonlinear constraints proceeds in a slightly different fashion. Properties (*a*) and (*b*) of Theorem 1 hold for the reasons stated above; property (*c*) of Theorem 1 must also be established. To accomplish this here, notice that a point is a "solution" if $w\in C(x)$ implies $f(w) = f(x)$. Under some reasonable assumption, for example that the closure of the feasible

---

† If $f(x)$ is continuously differentiable on an open set S and d is feasible at $x \in S$, then d is usable for $f(\cdot)$ at $x \in S$ if and only if $\nabla f(x)d < 0$.

region is identical with the closure of the interior of the feasible region, it can be easily proved that a solution corresponds to an optimal point of problem (1) with nonlinear constraints, establishing property (c).

It is next necessary to prove that the method of centers, and a particular example such as (19) due to Fiacco and McCormick, as used in the algorithms, is also closed. The strategy here is to show that since the interior point method is a composition, $A^2 = A^1 C$, it too is closed. In this composition, C is the interior point method of Eqn (19) and A^1 is the feasible direction composition used to optimize the method of centers objective. In fact the map A^1 here is identical with the composition A^1, which was proved closed above. Once x^k is given, the algorithm $C(x^k)$ produces a point w such that $f(w, x^k) = \min f(x^{k+1}, x^k)$. The feasibility of w is implied by the fact that $w \in C(x^k)$.

The form of the proof of the closure of $C(x)$ is due to Zangwill [30]. Huard [14] in the general case and Fiacco and McCormick [27] in the case of the example of Eqn (19), show that $x^k \to \bar{x}$ and $w^k \to \bar{w}$ for all $k \in K$ where $w \in C(x^k)$. By the definition of w^k for any x

$$f(w^k, x^k) \leqslant f(x^{k+1}, x^k). \tag{23}$$

Using the continuity assumption, take limits so that

$$f(\bar{w}, \bar{x}) \leqslant f(x^{k+1}, \bar{x}). \tag{24}$$

Because Eqn (24) holds for any x, the value $\bar{w}$ must minimize $f(x^{k+1}, x^k)$. Therefore, $\bar{w} \in A(\bar{x})$ implying C is closed. Consequently, the convergence of interior point algorithms for highly nonlinear constraints discussed in the preceding section is proved.

5. Computational Experience

Preliminary tests on relatively low dimension problems have been made with a feasible direction algorithm using Zoutendijk's procedure P2 and a cutting plane linearization of the $g(x)$. In addition, tests with an interior point method, using the same procedure to minimize the form (21), have been made. These experiments indicate convergence in a reasonable amount of time to the correct solution of convex constraint, concave objective problems; however, most of the variant algorithms have not yet been investigated at all. So far the data generated from the algorithms is not sufficient to support firm judgements on the efficiency of any of them relative to competitive procedures such as Rosen's partition programming [9–11].

These procedures are relatively simple, however, and should therefore be more efficient than Rosen's algorithm. In the first place, the methods proposed above do not require the solution of two classes of problems: many separate linear subproblems and one problem in the variables linking the otherwise separate blocks together. Further, it can be expected that more steps will be required with the generalized inverses of the gradient projection method than with the gradient methods proposed here. Finally, the methods of this paper can each incorporate the relatively efficient basis triangularization methods of Dantzig [17] to fully exploit the special structure of the constraint matrix.

When the several algorithms of both types discussed above are more fully developed, computational experience will be published.

6. Conclusions

Powerful new algorithms for structured nonlinear programming problems with either linearized or highly nonlinear constraints have been developed in preceding sections. A number of variants have been offered. Convergence of both type 1 and type 2 have been proved by application of the concept of a closed map.

Available computational experience indicates that these new methods will be powerful tools for solving large decomposable nonlinear programming problems. On the available evidence, they appear to warrant extensive further investigation and additional development.

References

1. G. B. Dantzig (1963). "Linear Programming and Extensions", Chapter 23. Princeton University Press, Princeton, New Jersey.
2. G. Zoutendijk (1960). "Methods of Feasible Directions", Elsevier, New York.
3. G. Zoutendijk (1966). Nonlinear programming: a numerical survey. *J. Soc. ind. appl. Math. Control* 4, 194-209.
4. D. M. Topkis and A. F. Veinott Jr. (1967). On the convergence of some feasible direction algorithms for nonlinear programming. *J. Soc. ind. appl. Math. Control* 5, 268-279.
5. G. P. McCormick (1968). "Anti-Zigzagging by Bending". Advanced Research Department, Research Analysis Corporation, McLean, Virginia (Unpublished working paper).
6. G. B. Dantzig and R. M. Van Slyke, (1965). "Generalized Upper Bounded Techniques for Linear Programming-I", Operations Research Center, University of California, Berkeley, ORC 64-17 (RR).
7. G. B. Dantzig and R. M. Van Slyke (1965). "Generalized Upper Bounded Techniques for Linear Programming-II", Operations Research Center, University of California, Berkeley, ORC 64-18.

8. M. Sakarovitch and R. Saigal (1967). An extension of generalized upper bounding techniques for structured linear programs, *J. Soc. ind. appl. Math.* **15**, 906-914.

9. J. B. Rosen (1963). Convex partition programming. *In* "Recent Advances in Mathematical Programming", (R. Graves and P. Wolfe, eds.) pp. 159-176. McGraw-Hill, New York.

10. J. B. Rosen (1964). Primal partition programming for block diagonal matrices. *Num. Math.* **6**, 250-260.

11. J. B. Rosen and J. C. Ornea (1963). Solution of nonlinear programming problems by partitioning. *Mgmt. Sci.* **10**, 160-173.

12. R. E. Griffith and R. A. Stewart (1961). A nonlinear programming technique for the optimization of continuous processing systems. *Mgmt Sci.* **7**, 379-392.

13. P. Huard (1963). "Programmes Mathematiques a Constraintes Nonlineaires", Note EDF No. HX 0/868.

14. P. Huard (1967). Resolution of mathematical programming with nonlinear constraints by the method of centres. *In* "Nonlinear Programming". (J. Abadie, ed.) pp. 209-219. North-Holland Publishing Company, Amsterdam.

15. B. T. Lieu and P. Huard (1966). La methode des centres dans un espace topologique. *Num. Math.* **8**, 56-67.

16. F. B. Hildebrand (1956). "Introduction to Numerical Analysis", McGraw-Hill, New York.

17. G. B. Dantzig (1963). Compact basis triangularization for the simplex method. *In* "Recent Advances in Mathematical Programming". (R. L. Graves and P. Wolfe eds.), pp. 125-132. McGraw-Hill, New York.

18. H. Levenberg (1944). A method for the solution of certain nonlinear problems in least squares, *Q. app. Math.* **2**, 164-168.

19. P. Wolfe (1967). "On the Convergence of Gradient Methods Under Constraint", IBM Watson Research Center, Yorktown Heights, New York, RC-1752.

20. P. Wolfe (1967). Methods of nonlinear programming. *In* "Nonlinear Programming". (J. Abadie, ed.), pp. 99-131. North-Holland Publishing Company, Amsterdam.

21. B. V. Shah, R. J. Buehler and O. Kempthorne (1964). Some algorithms for minimizing a function of several variables, *Soc. ind. appl. Math.* **12**, 74-92.

21a. P. Wolfe (1959). The simplex method for quadratic programming. *Econometrica* **27**, 382-398.

22. H. D. Mills (1968). "Extending Newton's Method to Systems of Inequalities", *In* The proceeding of the Sixth International Symposium on Mathematical Programming, Princeton University, August 1967, to be published.

23. W. C. Davidon (1959). "Variable Metric Method for Minimization". AEC Research and Development Report ANL-5990.

24. E. M. L. Beale (1967). Nonlinear programming. *In* "Digital Computer User's Handbook". (M. Klerer and G. A. Korn, eds), pp. 4-152. McGraw-Hill, New York.

25. S. H. Brooks (1958). A discussion on random methods for seeking maxima. *Ops. Res.* **6**, 244-251.

26. L. E. Schwartz (1969). Application of Nonlinear Programming and Bayesian Statistics to the Theory of the Firm *In* The Proceedings of the Sixth International Symposium on Mathematical Programming, Princeton University, August 1967, to be published.

27. A. V. Fiacco and G. P. McCormick (1967). The sequential unconstrained minimization technique (SUMT) without parameters. *Ops. Res.* **15**, 820-827.
28. G. P. McCormick, Private communication.
29. W. I. Zangwill (1966). "Convergence Conditions for Nonlinear Programming Algorithms", Center for Research in Management Science, University of California, Berkeley, California, Working Paper 197.
30. W. I. Zangwill (1967). "Applications of the Convergence Conditions", Center for Research in Management Science, University of California, Berkeley, Working Paper 231.
31. C. Berge (1963). "Topological Spaces", Macmillan, New York.

Disscusion

SMART. (Gas Council). I would like to ask Dr Schwartz if he has encountered any difficulties in programming the system which he outlines.

SCHWARTZ. So far no programs have been written and all the calculations have been done by hand. However I would expect the linear part of the procedure to be the most difficult and the nonlinear part fairly straightforward.

9. The Application of Nonlinear Programming to the Automated Minimum Weight Design of Rotating Discs

B. M. E. DE SILVA

Mechanical Engineering Laboratory, English Electric Company, Leicester, England

1. Introduction

The object of the research described in this paper is to investigate the feasibility of using nonlinear programming procedures to solve a class of minimum weight structural optimization problems with nonanalytic constraints. The structural configuration of the system is completely specified by the design parameters of which some are fixed and others are permitted to vary within a prescribed range, thus making it possible to optimize the system for minimum weight. The constraints on the design variables ensure physically reasonable designs and may be expressed in the form

$$l_i \leqslant x_i \leqslant u_i \quad \text{for } i = 1, \ldots, n \tag{1}$$

where the n real variables $x_1, \ldots, x_n$ are the design variables for the system. The bounds l_i, u_i are constants or functions of the other design variables.

The behaviour or response of the system is governed by the behaviour variables (that is stresses, deflection, vibrational frequencies, and so on), which are also constrained to vary within a prescribed range to prevent failure of the system under the design loads. For instance, the behavioural constraints may include statical constraints which prevent the stresses exceeding the yield stress, instability constraints which prevent failure of the structure by buckling, dynamical constraints which restrict the natural frequencies of vibration to lie within prescribed frequency bands, and so on. The behavioural constraints may therefore be expressed in the form

$$L_j \leqslant y_j(x_1, \ldots, x_n) \leqslant U_j \quad \text{for } j = 1, \ldots, m. \tag{2}$$

115

The weight of the structure is assumed to be a single valued differentiable function of the design variables

$$W = W(x_1, ..., x_n).$$ (3)

The minimum weight solutions are obtained by minimizing Eqn (3) subject to the constraint conditions (1), (2). The functions W, y_j in general are nonlinear and the solutions are given by a nonlinear programming formulation.

The minimum weight problems considered in this paper are restricted to problems for which the behaviour variables cannot be expressed analytically as functions of the design variables. Therefore it is not possible to use closed form analytical procedures for determining the minimum weight solutions and recourse must be made to approximate or numerical procedures. The behaviour variables are functions only in the sense that they are computer oriented rules for determining the behaviour associated with a given design and are not given in a closed analytical form in terms of the design variables. Thus the behaviour variables may be regarded as a "black box" into which are put the design variables representing a given design and out of which comes the behaviour variables for that design. The box contains such devices as differential equations, finite difference procedures, a digital computer, and so on.

Consider for instance the problem of minimizing the weight of a steam turbine disc subject to specified geometrical and behavioural constraints. For purposes of simplicity, the turbine disc is idealized as a rotating circular disc (Fig. 1) of variable thickness. The behavioural constraints have been restricted to a consideration that the stresses in the disc should be below the yield stress, while the geometrical constraints impose restrictions on the dimensions and tolerances of the disc.

The weight is given by the functional expression

$$W[h] = \int_{a_1}^{a_m} 2\pi \rho r h(r) dr$$ (4)

where a_1, a_m are the inner and outer radii respectively, ρ is the density and $h(r)$ is the thickness at a radial distance r from the axis of rotation, $h(r)$ being measured parallel to the axis of rotation. The equilibrium equation for the disc is given by [1]

$$\frac{d}{dr}(h\sigma_r) + \frac{h}{r}(\sigma_r - \sigma_\theta) + \rho\omega^2 r h = 0$$ (5)

where σ_r, σ_θ are the radial and tangential stresses respectively and ω is the angular velocity of rotation of the disc. This equation has been derived on the assumption of radially symmetric plane stress. The stresses may be expressed in terms of the radial displacement $u(r)$ by the following compatibility relations

$$\sigma_r = \frac{E}{1-v^2}(e_r + ve_\theta), \qquad \sigma_\theta = \frac{E}{1-v^2}(ve_r + e_\theta), \tag{5a}$$

$$e_r = \frac{du}{dr}, \qquad e_\theta = \frac{u}{r}, \tag{5b}$$

where e_r, e_θ are the radial and tangential strains, E is Young's modulus and v is Poisson's ratio.

Therefore substituting Eqns (5a), (5b) in (5) gives the following differential equation for $u(r)$

$$\frac{d^2u}{dr^2} + \left(\frac{1}{r} + \frac{1}{h}\frac{dh}{dr}\right)\frac{du}{dr} - \left(\frac{1}{r} - \frac{v}{h}\frac{dh}{dr}\right)\frac{u}{r} + \frac{\rho\omega^2(1-v^2)}{E}r = 0. \tag{6}$$

Therefore in order to determine $u(r)$ explicitly it is necessary to specify

$$h = h(r) \qquad (a_1 \leqslant r \leqslant a_m) \tag{6a}$$

as a function of r. Then for prescribed boundary conditions on σ_r, σ_θ given by

$$[\sigma_r]_{r=a_1} = s_1; \qquad [\sigma_r]_{r=a_m} = s_m \tag{6b}$$

Eqn (6) uniquely determines $u(r)$ as a function of r. Therefore from Eqns (5a), (5b) the stresses σ_r, σ_θ may be determined as functions of r. The stresses are functionals of $h(r)$ and correspond to black box type behaviour variables.

The material of the disc is assumed to obey a yield condition of the form

$$F(\sigma_r, \sigma_\theta) \leqslant \sigma_0 \tag{7}$$

where σ_0 is the yield stress. The yield condition used in this investigation is the yield condition of Tresca defined by [2]

$$F(\sigma_r, \sigma_\theta) \equiv \max \left\{ \tfrac{1}{2}|\sigma_r - \sigma_\theta|, \tfrac{1}{2}|\sigma_r|, \tfrac{1}{2}|\sigma_\theta| \right\}. \tag{7a}$$

The variation of $h(r)$ is defined by

$$h(r) \geqslant \varepsilon \qquad (8)$$

where ε is a specified tolerance which ensures that $h(r)$ is never negative. The problem then consists of determining an optimal $h(r)$ which minimizes Eqn (4) subject to the constraint conditions (6)–(8) and is essentially a Bolza type problem in the calculus of variations [3] for which the discretized nonlinear programming approximation is characterized by nonanalytic constraints on the behaviour variables.

This paper includes: (1) reformulating the disc problem as a problem in nonlinear programming, and (2) developing minimization procedures for solving problems with nonanalytic constraints by extending existing methods and formulating new ones. Methods currently applicable are the "steepest descent–alternate step" mode of travel in design space proposed by Schmit et al. [4]–[12] for the automated weight minimization of trusses and waffle plates with instability constraints. Modifications are introduced to improve their computational efficiency and convergence rates. Generalizations lead to new methods; (3) applying these methods to obtain numerical solutions to the disc problem on an English Electric KDF9 computer for purposes of comparative evaluation.

Before discussing these topics, some preliminary design concepts are introduced which contain the framework for formulating the minimization problem.

2. Design Concepts

The design variables define a point

$$x = (x_1, ..., x_n) \qquad (9)$$

in an n-dimensional real euclidean space E_n, called the design space. Consider the functions $g_k(x)$ for $k = 1, ..., 2(n+m)$ defined by

$$
\begin{aligned}
g_k(x) &= l_k - x_k & &\text{for } k = 1, ..., n \\
&= x_{k-n} - u_{k-n} & &\text{for } k = n+1, ..., 2n \\
&= L_{k-2n} - y_{k-2n}(x) & &\text{for } k = 2n+1, ..., 2n+m \\
&= y_{k-2n-m}(x) - U_{k-2n-m} & &\text{for } k = 2n+m+1, ..., 2(n+m).
\end{aligned}
\qquad (10)
$$

Therefore the constraint conditions (1), (2) become

$$g_k(x) \leqslant 0 \text{ for } k = 1, ..., 2(n+m). \qquad (11)$$

The feasible region R is a subspace of E_n and consists of points $x \in E_n$ which satisfy the constraint conditions (1), (2) or (11), so that

$$R \equiv \left\{ x; \ g_k(x) \leqslant 0 \ \text{ for } k = 1, ..., 2(n+m) \right\}. \tag{11a}$$

Design points which belong to R are called feasible points.

There is associated with each constraint function $g_k(x)$ a hyper-surface defined by

$$G_k \equiv \left\{ x; \ g_k(x) = 0 \ \text{ for } k = 1, ..., 2(n+m) \right\}. \tag{11b}$$

The hypersurfaces for nonanalytic functions correspond to unknown surfaces in E_n.

The composite constraint surface is given by

$$G \equiv R \cap (G_1 \cup G_2 ..., \cup G_{2(n+m)}) \tag{11c}$$

and defines the boundary of R and points which belong to G are called boundary points. The weight contours

$$W(x) = c \tag{11d}$$

define a family of hypersurfaces in E_n. The minimization procedures generate a sequence of feasible designs of decreasing weight which converge to the least weight contour in R. A feasible initial design is established and is systematically improved by an alternating iterative process of analysis and design modifications. These automated design cycles correspond to motion in the design space along paths which the weight decreases. Therefore the minimization process consists in the proper selection of the directions and distances of travel in design space.

3. Illustrative Problem

The steam turbine disc to be optimized is shown in Fig. 1. The width of the hub and the rim shape have been specified to allow for the attachment of the discs and the spacing of the blades in the turbine while the depth of the hub is variable to permit adjoining discs to be shrunk onto a common shaft. The thickness distribution for the remainder of the disc is variable but symmetrically distributed about the midplane. The thickness $h(r)$ is defined by

$$h(r) = b_1 \quad \text{for } a_1 \leqslant r \leqslant a_2$$
$$= h(r) \quad \text{for } a_2 \leqslant r \leqslant a_{m-1}$$
$$= b_m \quad \text{for } a_{m-1} \leqslant r \leqslant a_m$$

B. M. E. DE SILVA

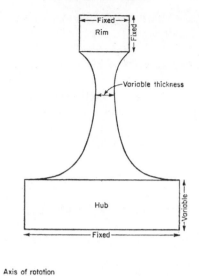

FIG. 1. Cross section of typical turbine disc.

where b_1 = width of hub (fixed), b_m = width of rim (fixed), and a_1, a_m, a_{m-1} are fixed radii while a_2 is variable. Therefore Eqn (4) becomes

$$W = \pi\rho b_1 (a_2{}^2 - a_1{}^2) + \pi\rho b_m (a_m{}^2 - a_{m-1}^2) + \int_{a_2}^{a_{m-1}} 2\pi\rho r h(r)dr$$

$$= \pi\rho b_1 (a_2{}^2 - a_1{}^2) + \pi\rho b_m (a_m{}^2 - a_{m-1}) + 2\pi\rho \sum_{j=3}^{m-1} \int_{a_{j-1}}^{a_j} r h(r)dr \quad (12)$$

where $a_1 < a_2 < a_3 < ... < a_{m-2} < a_{m-1} < a_m$. The function $h(r)$ is approximated by a sequence of linear functions $h_j (r)$ for $j = 3, ..., (m-1)$ defined by (Fig. 2).

$$h(r) \simeq h_j (r) \quad \text{for } a_{j-1} \leqslant r \leqslant a_j; j = 3, ..., (m-1) \quad (12a)$$

where

$$h_j (r) = b_{j-1} + \left(\frac{b_j - b_{j-1}}{a_j - a_{j-1}} \right) (r - a_{j-1}) \quad \text{for } a_{j-1} \leqslant r \leqslant a_j; \quad j = 3, ..., (m-1)$$

$$(12b)$$

$$h(a_j) = b_j \quad \text{for } j = 1, ..., m. \quad (12c)$$

Therefore Eqn (12) gives

$$W \simeq \pi \rho b_1 (a_2{}^2 - a_1{}^2) + \pi \rho b_m (a_m{}^2 - a_{m-1}^2) + 2\pi\rho \sum_{j=3}^{m-1} \int_{a_{j-1}}^{a_j} r h_j(r) dr$$

$$= \tfrac{1}{3}\pi\rho \sum_{j=3}^{m-2} (a_{j+1} - a_{j-1})(a_{j+1} + a_j + a_{j-1})b_j$$

$$+ \tfrac{1}{3}\pi\rho\, b_1 (-3a_1{}^2 + a_2{}^2 + a_3{}^2 + a_2 a_3)$$

$$+ \tfrac{1}{3}\pi\rho\, b_m (3a_m{}^2 - a_{m-1}^2 - a_{m-2}^2 - a_{m-1}\, a_{m-2}). \qquad (12d)$$

The integral formulation (4) has been transformed into a finite difference form (12d) by linearizing the disc. $b_1, ..., b_m$ are the thicknesses parallel to

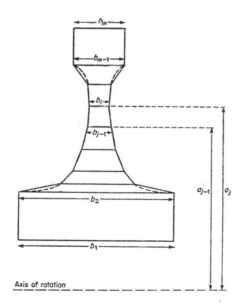

FIG. 2. Discretized nonlinear programming model.

the axis of rotation at specified radii $a_1, ..., a_m$ respectively. The disc profile is then obtained by joining adjacent thicknesses $\{b_{j-1}, b_j$ for $j = 2, ..., m\}$ by straight lines.

4. Geometrical Constraints

The following geometrical 'constraints are imposed on the disc dimensions

(1) $a_1 < a_2 < a_3 < \ldots < a_{m-2} < a_{m-1} < a_m$

(2) $b_1 = b_2$ (fixed)

(3) $b_m = b_{m-1}$ (fixed)

(4) $a_1, a_3, \ldots, a_{m-1}, a_m$ are all fixed

(5) a_2 is variable

(6) b_j is variable for $j = 3, \ldots, (m-2)$

(7) $b_j \geqslant \varepsilon_1$ for $j = 3, \ldots, (m-2)$

(8) $a_1 + \varepsilon_3 \leqslant a_2 \leqslant a_3 - \varepsilon_2$

where $\varepsilon_1, \varepsilon_2, \varepsilon_3 =$ tolerances on the design variables. Conditions (2), (3) mean that the width of the hub and rim are fixed while (4), (5) mean that the depth of the rim is fixed but the depth of the hub is variable. The tolerance ε_1 ensures non-negative b_j, while $\varepsilon_2, \varepsilon_3$ restrict a_2 to lie within specified tolerances of a_1, a_3.

Therefore the design variables for the problem are given by

$$x = (b_3, \ldots, b_{m-2}, a_2). \tag{13}$$

This corresponds to an $(m-3)$ dimensional design space. The geometrical constraints are given by

$$l \leqslant x \leqslant u \tag{13a}$$

where

$$l = (\varepsilon_1, \ldots, \varepsilon_1, a_1 + \varepsilon_3), \quad u = (\infty, \ldots, \infty, a_3 - \varepsilon_2). \tag{13b}$$

These are linear constraints and correspond to hyperplanes parallel to the coordinate planes.

5. Behavioural Constraints

The disc is symmetrical with respect to both its axis of rotation and its midplane and is in dynamic equilibrium under the action of the centrifugal and thermal loadings. The stress calculations are based on Donath's method [13, 14] which consists essentially in replacing the disc by a series of annular rings of constant width. The stresses at the outer edge of a ring are determined in terms of the stresses at the inner edge. Continuity is ensured by equating the radial displacement and the radial load at the interface of adjacent rings. The stress equations are summarized below for ready reference.

Within each ring the thickness $h(r)$ is constant so that Eqn (6) reduces to

$$\frac{d^2u}{dr^2} + \frac{1}{r}\frac{du}{dr} - \frac{u}{r^2} + \frac{\rho\omega^2(1-v^2)}{E}r = 0, \tag{14}$$

that is
$$u = C_1 r + \frac{C_2}{r} - \frac{\rho\omega^2(1-v^2)}{8E}, \tag{14a}$$

where C_1, C_2 are constants of integration. Therefore from Eqns (5a), (5b) the rotational stresses are given by

$$\left.\begin{array}{l} \sigma_r = \alpha - \dfrac{\beta}{r^2} - \dfrac{\rho\omega^2(3+v)}{8}r^2 \\[4mm] \sigma_\theta = \alpha + \dfrac{\beta}{r^2} - \dfrac{\rho\omega^2(1+3v)}{8}r^2 \end{array}\right\} \tag{14b}$$

where α, β are constants within each ring.

Similarly the thermal stresses are given by

$$\frac{d}{dr}(h\sigma_r) + \frac{h}{r}(\sigma_r - \sigma_\theta) = 0, \tag{14c}$$

where
$$\left.\begin{array}{l} \sigma_r = \dfrac{E}{1-v^2}[(e_r - \alpha\phi) + (e_\theta - \alpha\phi)] \\[4mm] \sigma_\theta = \dfrac{E}{1-v^2}[v(e_r - \alpha\phi) + (e_\theta - \alpha\phi)] \end{array}\right\} \tag{14d}$$

α is the coefficient of linear expansion and ϕ is the temperature. Substituting (5b), (14d) in (14c) gives

$$\frac{d^2u}{dr^2} + \left(\frac{1}{r} + \frac{1}{h}\frac{dh}{dr}\right)\frac{du}{dr} - \left(\frac{1}{r} - \frac{v}{h}\frac{dh}{dr}\right)\frac{u}{r}$$

$$-(1+v)\left(\frac{1}{h}\frac{dh}{dr} + \frac{d\phi}{dr}\right) = 0. \qquad (14e)$$

Therefore within each ring

$$\frac{d^2u}{dr^2} + \frac{1}{r}\frac{du}{dr} - \frac{u}{r^2} - (1+v)\alpha\frac{d\phi}{dr} = 0$$

that is

$$u = A_1 r + \frac{A_2}{r} + (1+v)\frac{\alpha}{r}\int^r r\phi\, dr$$

where A_1, A_2 are constants of integration. Thus the thermal stresses are given by (14d), (5b).

$$\left.\begin{aligned}
\sigma_r &= -\frac{\alpha E}{r^2}\int r\phi\, dr + \gamma - \frac{\sigma}{r^2} \\[2mm]
\sigma_\theta &= \frac{\alpha E}{r^2}\int r\phi\, dr - \alpha E\phi + \gamma + \frac{\delta}{r^2}
\end{aligned}\right\} \qquad (14f)$$

where γ, δ are constants. The temperature $\phi(r)$ is a prescribed function of r. The resultant stresses are then given by

$$\left.\begin{aligned}
\sigma_r &= \sigma_r\,(\text{rot}) + \sigma_r\,(\text{thermal}) \\[2mm]
\sigma_\theta &= \sigma_\theta\,(\text{rot}) + \sigma_\theta\,(\text{thermal})
\end{aligned}\right\} \qquad (14g)$$

In general, the analysis phase of the redesign cycles consists of a series of black boxes into which are fed the design variables and out of which comes the behaviour variables. The contents of the boxes which include structural models and mathematical procedures for determining the behaviour variables do not play a significant role in the subsequent design modification iterations and may be ignored. So that what is essential is the output from the black boxes which enables the behaviour variables to be checked against the behavioural constraints to ensure designs that do not violate the behavioural requirements for the problem. A more sophisticated analysis procedure merely means more accurate values for the behaviour variables associated with a given design and does not necessarily provide any new information on the minimization procedures. Therefore from this standpoint, Donath's method is a very acceptable form of analysis. It is relatively simple and was already available at the time this investigation was started.

At each stress calculation the computer program subdivides the intervals $[a_{j-1}, a_j]$ for $j = 3, ..., (m-1)$ into further subintervals by points $r_2, r_3, ...,$ r_{n-1} where

$$a_2 = r_2 < r_3 < ... < r_{n-1} = a_{m-1}. \left. \begin{array}{c} \\ \\ \end{array} \right\}$$

In addition $r_1 = a_1; \quad r_n = a_m.$ (15)

The criterion for subdividing the interval $[a_{j-1}, a_j]$ is

$$|b_{j-1} - b_j| > \tfrac{1}{2}\varepsilon (b_{j-1} + b_j) \qquad (15a)$$

where ε is a positive tolerance. If this criterion is satisfied $[a_{j-1}, a_j]$ is subdivided into u equal parts by points $q_0, q_1, ..., q_u$

$$a_{j-1} = q_0 < q_1 < ... < q_u = a_j. \qquad (15b)$$

The corresponding thicknesses at these points are given by

$$p_i = h(q_i) \quad \text{for } i = 0, ..., u \qquad (15c)$$

so that

$$|b_j - b_{j-1}| = |p_u - p_0|$$
$$= |(p_u - p_{u-1}) + (p_{u-1} - p_{u-2}) + ... + (p_2 - p_1) + (p_1 - p_0)|$$
$$\leqslant \tfrac{1}{2}\varepsilon[(p_u - p_{u-1}) + (p_{u-1} + p_{u-2}) + ... + (p_2 + p_1) + (p_1 + p_0)]$$
$$\leqslant \varepsilon u K_j$$

where $K_j = \max (b_j, b_{j-1})$

so that $u = 1 + \left\langle \dfrac{|b_j - b_{j-1}|}{K_j} \right\rangle$ (15d) where $\langle x \rangle$ is the largest integer not exceeding x. The total number of points of subdivision for each of the intervals $[a_{j-1}, a_j]$ is n, the points being labelled $r_1, r_2, ..., r_n$ with thickness $h_1, h_2, ... h_n$ respectively. The reason for this subdivision is to obtain a better estimate for the stress distribution. The number n varies from design to design.

For each design the stresses σ_r, σ_θ at $r_1, ..., r_n$ are calculated. Therefore the principal shearing stresses at these radii are given by [2]

$$\tau_1 = \tfrac{1}{2}|\sigma_r - \sigma_\theta|, \qquad \tau_2 = \tfrac{1}{2}|\sigma_r|, \qquad \tau_3 = \tfrac{1}{2}|\sigma_\theta|. \tag{16}$$

The stress constraints are defined by the Tresca yield condition

$$\tau \leqslant \tau_0 \tag{16a}$$

where τ_0 is the critical stress and τ is the maximum principal shearing stress.

$$\tau = \max(\tau_1, \tau_2, \tau_3). \tag{16b}$$

Therefore the behaviour variables are given by

$$y(x) = (\tau_{r_1}, \tau_{r_2}, ..., \tau_{r_n}) \tag{17}$$

while the behavioural constraints are given by

$$L \leqslant y(x) \leqslant U \tag{17a}$$

where

$$L = (0, 0, ..., 0), \qquad U = (\tau_0, \tau_0, ..., \tau_0). \tag{17b}$$

Due to the black box nature of the stresses the behavioural constraints correspond to unknown surfaces in design space.

6. Weight Function

The weight $W = W(b_3, ..., b_{m-2}, a_2)$ given by Eqn (12d) is a quadratic in a_2 but linear in b_j. The function W and the feasible region R are in general nonconvex and the problem may possess relative minima.

7. Nonlinear Programming Formulation

The disc problem may be formulated mathematically as a nonlinear programming problem as follows.

Given l, u, L, U determine a design x which satisfies the conditions

(1) $l \leqslant x \leqslant u$

(2) $L \leqslant y(x) \leqslant U$

and minimizes the weight $W(x)$.

8. Nonlinear Programming Procedures

Nonlinear programming procedures applicable to structural problems with analytic constraints include:
(1) Cutting plane method [15, 16] for transforming a nonlinear problem to a series of linear programming problems.
(2) Rosen's gradient projection method [17–19].
(3) Penalty function methods for transforming a constrained problem to a series of unconstrained minimization problems [20–22] each of which can be solved using any of several well-known methods on unconstrained minimization [23–25].
(4) Lagrangian methods [26, 27] using the properties of the saddle point of the Lagrangian function.
(5) Methods for leaving the boundary of the feasible region along the constant weight surface [28], the direction for the "bounce" being given by a quadratic programming problem.
(6) Steepest descent procedures [29–31] for automated weight minimization using matrix methods of structural analysis.
Equations (6), (14e), (14g) applied to (16a) may be written in the form

$$\tau[h] \equiv \int^r \phi\left(r, h(r), \frac{d}{dr} h(r)\right) dr \leqslant \tau_0. \tag{18}$$

The above methods do not apply to constraints of the type (18). A "steepest descent – alternate step" procedure developed by Schmit et al. [4–12] may however be readily adapted to describe these problems; they started from an initial feasible point and moved in the direction of steepest descent to a better design some finite distance away. This procedure is repeated until a constraint is encountered which prevents further moves in the gradient direction. Then an alternate step is taken which is a move along the constant weight surface. After the alternate step a feasible point should have been obtained from which a steep descent can be made. The process is continued until no move can be made by either mode—at which time an optimum is said to be achieved. The reasoning behind this technique is that since the gradient direction points in the direction of greatest change it is the best direction to move in to improve the design. If a move cannot be made in the best direction, then a move is made which at least does not increase the weight of the design.

A fixed incremental step length is used in conjunction with steepest descent motion, the step length being doubled at each feasible iteration. This doubling process is repeated until a design is reached which violates on a main constraint (geometrical constraints are ignored at this stage); the total distance of travel back to an already feasible point is then halved, and the direction reversed. In all subsequent iterations, the distance is always halved and the direction reversed after each transition between a violated and non-violated condition. Thus, this halving and doubling process is directed to and converges upon the constraint surface. A random number generator was then used to propagate the directions of search along the constant weight surface. A sequence of proposed new designs was generated which was tested in turn against the geometrical and behavioural constraints. If any one of these designs was found to be feasible steepest-descent motion was continued as before. This method for leaving the boundary of R is called the method of alternate base planes [8] and will be described in the following section. The methods described in this paper use an accelerated steepest-descent mode of travel in the feasible region, the step length being estimated to the nearest constraint. The step length decreases as a constraint is approached and this enables a constraint to be encountered more rapidly than a straightforward doubling process. When a design violates a constraint, a linear interpolation technique is used to converge to the constraint surface, the interpolations being always between a violated and non-violated design. In general, this ensures a better convergence rate than a doubling and halving process.

The method of alternate base planes was applied to the disc problem and thereafter more selective methods were sought for leaving the boundary of R. A direction of search was generated whereby the sections of the disc not at yield stress were thinned in proportion to their stress levels relative to the yield stress, while the section at yield was thickened by a predetermined factor. The step length was then calculated using the equal weight condition, which gave a quadratic equation for the step length. A major difficulty was the possibility of obtaining complex roots and even if real roots were forthcoming there was no guarantee that the geometrical and behavioural constraints were not violated. Therefore a method was devised which always guaranteed non-violation of the geometrical constraints.

In this method the proposed design need only be tested against the yield criterion. The linearity of the geometrical constraints enables a step length to be easily calculated which ensures an alternate step within the design variable bounds. The direction is then determined from the conditions of equal weight and normalization. To obtain real determinate solutions the number of unknowns is reduced to two by assigning zero values to the remaining variables. This corresponds to changing two design variables and leaving the rest unaltered. The section at yield stress is thickened, while the section furthest from yield is thinned so as to leave the weight unchanged. If the design violates the yield criterion the step length is progressively halved a specified number of times, and if no feasible design is forthcoming a different combination of direction cosines is set to zero, generating a different direction of search. If the yield condition is still violated this method is scrapped and the random method is used to determine an alternate step design.

The nonconvexity of W and R in general gives rise to pockets of relative minima. There is no known method yet of establishing whether a proposed solution is in fact a global solution or not. However, it is possible to establish a reasonable degree of confidence by searching a fairly wide region of design space. It is also possible to select two different initial points and run the minimization procedures along distinct paths. If the solution is the same (to within a reasonable tolerance) in the two cases, it is reasonable to assume that the proposed solution is a global one.

9. Minimization Procedures

The disc optimization problem [32, 33] is characterized by:

(1)   Multi-dimensional design space
(2)   Nonlinear weight function
(3)   Relative minima
(4)   Linear geometrical constraints
(5)   Stresses "black box" type functions

while the optimization procedure is characterized by (Fig. 3):

(1) Accelerated steepest descent motion in the feasible region until a constraint is encountered.
(2) Constrained steepest descent motion from a geometrical constraint. Since a move in the direction of steepest descent cannot generally be made without piercing through the constraint, the method moves in the next best direction, the projection of the direction of steepest descent on the constraint surface.
(3) Equal weight redesign from a behavioural constraint surface. Constrained steepest descent motion cannot take place as the surfaces are unknown. A move is therefore made which at least does not increase the weight of the design.

10. Steepest Descent Motion

The computer program starts from an initial feasible design and enters
steepest motion defined by the following iterative equation

$$x^{(q+1)} = x^{(q)} + t^{(q)}\psi^{(q)} \tag{19}$$

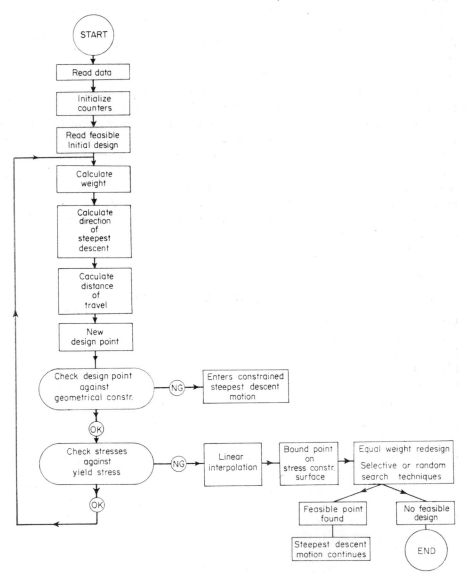

FIG. 3.   Flow diagram for structural synthesis based on a stress constant.

where

$$
\left.\begin{aligned}
x^{(q)} &= (b_3^{(q)}, ..., b_{m-2}^{(q)}, a_2^{(q)}), \quad \psi^{(q)} = -\nabla W(x^{(q)})/|\nabla W(x^{(q)})|, \\
\nabla &= \left(\frac{\partial}{\partial b_3}, ..., \frac{\partial}{\partial b_{m-2}}, \frac{\partial}{\partial a_2} \right), \\
t^{(q)} &= \text{step length}, \\
q &= \text{design cycle counter.}
\end{aligned}\right\} \quad (19a)
$$

Therefore from Eqn (12d)

$$
\left.\begin{aligned}
\frac{\partial W}{\partial b_j} &= \frac{\pi \rho}{3} (a_{j+1} - a_{j-1})(a_{j+1} + a_j + a_{j-1}) \text{ for } j = 3, ..., (m-2) \\
\frac{\partial W}{\partial a_2} &= \frac{\pi \rho}{3} (2a_2 + a_3)(b_1 - b_3).
\end{aligned}\right\} \quad (19b)
$$

Equation (19) therefore reduces to

$$
\left.\begin{aligned}
b_j^{(q+1)} &= b_j^{(q)} - \frac{\pi \rho}{3} (a_{j+1} - a_{j-1})(a_{j+1} + a_j + a_{j-1}) t^{(q)}/N^{(q)} \\
&\qquad \text{for } j = 3, ..., (m-2) \\
a_2^{(q+1)} &= a_2^{(q)} - \frac{\pi \rho}{3} (2a_2 + a_3)(b_1 - b_3) t^{(q)}/N^{(q)}
\end{aligned}\right\} \quad (19c)
$$

where the normalization factor $N^{(q)}$ is given by

$$
\begin{aligned}
N^{(q)} = \frac{\pi \rho}{3} \Bigg[&\sum_{j=3}^{m-2} (a_{j+1} - a_{j-1})^2 (a_{j+1} + a_j + a_{j-1})^2 \\
&+ (2a_2 + a_3)^2 (b_1 - b_3)^2 \Bigg]^{\frac{1}{2}}.
\end{aligned} \quad (19d)
$$

The distance to a behavioural constraint cannot be determined exactly as the surfaces are unknown. Therefore the step length is estimated as follows. Let

$$ h_i^{(q)} = \text{thickness at radius } r_i; $$

$$ \tau_{r_i}^{(q)} = \text{maximum principal shearing stress at } r_i. $$

For purposes of this calculation, it is assumed that each $h_i^{(q)}$ can be changed independently without affecting the stress distribution elsewhere. Therefore to bring $h_i^{(q)}$ to yield stress it must be changed to $\bar{h}_i^{(q)}$ given by

$$(2\pi r_i \bar{h}_i^{(q)})\tau_0 \simeq (2\pi r_i h_i^{(q)})\tau_{r_i}$$

$$\bar{h}_i^{(q)} \simeq \frac{\tau_{r_i}}{\tau_0} h_i^{(q)}. \tag{20}$$

This relation is derived on the assumption that the load remains unchanged. Therefore the distance $t_i^{(q)}$ to the constraint surface at r_i is given by

$$\bar{h}_i^{(q)} = h_i^{(q)} - t_i^{(q)}\phi_i^{(q)} \qquad (0 \leqslant \phi_i \leqslant 1)$$

so that

$$t_i^{(q)} = \left(\frac{\tau_0 - \tau_{r_i}}{\tau_0}\right) \frac{h_i^{(q)}}{\phi_i^{(q)}}$$

where

$$\left. \begin{aligned} \phi_i^{(q)} &= \frac{\psi_j^{(q)}(r_i - a_{j-1}) + \psi_{j-1}^{(q)}(a_j - r_i)}{a_j - a_{j-1}} \\ a_{j-1} &\leqslant r_i \leqslant a_j \quad \text{for } j = 3, \ldots, (m-2) \end{aligned} \right\} \tag{20a}$$

and

then

$$t^{(q)} = \min_{3 \leqslant i \leqslant n-2} t_i^{(q)}. \tag{20b}$$

Thus $t^{(q)}$ decreases as a behavioural constraint surface is approached. At each iteration the design is checked against the geometrical and behavioural constraints. The design is first checked against the geometrical constraints and if the geometrical constraints are not violated, the corresponding stress distribution is calculated and then checked against the yield criterion. If the stresses are below the yield stress, the design is feasible and steepest descent motion continues until a non-feasible design is encountered. A non-feasible design corresponds to a region of constraint violation, that is violation of either the geometrical or the stress constraints.

11. Geometrical Constraint Violation

The design lies outside the geometrical bounds. The distances from the last feasible design to the geometrical constraints are calculated and the least positive distance is taken, giving a point lying on the nearest constraint. Let $x^{(q+1)}$, $x^{(q)}$ be the non-feasible and feasible designs respectively. Therefore from Eqns (19c), (13a), (13b), the distances to the geometrical constraints are given by

$$
\left.
\begin{aligned}
t_j &= \frac{3(b_j{}^{(q)} - \varepsilon_1)N^{(q)}}{\pi\rho(a_{j+1} - a_{j-1})(a_{j+1} + a_j + a_{j-1})} \quad \text{for } j = 3, ..., (m-2) \\[2mm]
t_1 &= \frac{3(a_2{}^{(q)} - a_1 - \varepsilon_1)N^{(q)}}{\pi\rho(2a_2 + a_3)(b_1 - b_3)} \\[2mm]
t_2 &= \frac{3(a_3 - \varepsilon_2 - a_2{}^{(q)})N^{(q)}}{-\pi\rho(2a_2 + a_3)(b_1 - b_3)}.
\end{aligned}
\right\}
\tag{21}
$$

Therefore the required design is given by

$$
x^* = x^{(q)} + t^*\psi^{(q)}
\tag{21a}
$$

where
$$
t^* = \min_{1 \leqslant j \leqslant m-2}(t_j; t_j > 0).
\tag{21b}
$$

The point x^* is checked against the behavioural constraints and, if satisfactory, the program enters constrained steepest descent motion.

12. Behavioural Constraint Violation

A linear interpolation procedure is used to converge to a boundary point on a behaviour constraint (to within a specified tolerance). Due to their linearity the geometrical constraints are never violated during the subsequent inter-polations, which are always between a feasible and non-feasible design (violating the yield criterion). Let $x^{(q+1)}$, $x^{(q)}$ be the non-feasible and feasible designs respectively. The corresponding behaviour functions are given by

$$
\left.
\begin{aligned}
y(x^{(q+1)}) &= (\tau_{r_1}^{(q+1)}, ..., \tau_{r_n}^{(q+1)}) \\
y(x^{(q)}) &= (\tau_{R_1}^{(q)}, ..., \tau_{R_N}^{(q)})
\end{aligned}
\right\}
\tag{22}
$$

where the stresses are evaluated at radii $(r_1, ..., r_n)$ $(R_1, ..., R_N)$ respectively. Suppose the yield stress is exceeded at a section of the disc at a radial distance r_k.

Let $\qquad \tau_v = \tau_{r_k}^{(q+1)} > \tau_0 \qquad (1 \leqslant k \leqslant n)$

$$\tau_f = (\tau^{(q)})_{r=r_k}$$

$$= \frac{\tau_{R_{t+1}}^{(q)}(r_k - R_t) + \tau_{R_t}^{(q)}(R_{t+1} - r_k)}{(R_{t+1} - R_t)} < \tau_0 \qquad (22a)$$

where $\qquad R_t \leqslant r_k \leqslant R_{t+1} \qquad (1 \leqslant t \leqslant N-1).$

τ_f is the corresponding mean stress at r_k in the feasible design $x^{(q)}$. Therefore the linear interpolations are defined by

$$\hat{x}^{(1)} = x^{(q)} \qquad \Delta^{(1)} = x^{(q)}$$

$$\bar{x}^{(r)} = \hat{x}^{(r)} + \delta^{(r)} \psi^{(q)} \qquad \text{for } r = 1, 2, \dots$$

$$\delta^{(r)} = \frac{\tau_0 - \tau_f}{\tau_r - \tau_f} \Delta^{(r)}$$

$$\Delta^{(r+1)} = \Delta^{(r)} - \delta^{(r)} \qquad \text{if } \hat{x}^{(r)} \text{ is feasible}$$

$$= \delta^{(r)} \qquad \qquad \text{otherwise}$$

where $\delta^{(r)}$ = step length at rth interpolation; $\Delta^{(r)}$ = distance between current feasible and non-feasible designs; $\hat{x}^{(r)}$ = current feasible design, when yield criterion is violated at several radial points r_s

$$\delta^{(r)} = \min_s \delta s^{(r)}.$$

These interpolations continue until $x^{(r)}$ converge to a constraint surface (that is when the design lies on the constraint surface to within 99·2 per cent yield stress or when the incremental distance $\Delta^{(r)} \leqslant 0·01$).

13. Equal Weight Redesign

Let x = boundary point on a behavioural constraint surface, $\bar{x}$ = proposed alternate step design, that is

$$\bar{x} = x + t\lambda$$

where $\qquad \lambda = (\lambda_1, \dots, \lambda_{m-3}); \qquad t = \text{step length}.$

The proposed new design lies on the constant weight surface, so that

$$W(x) = W(x+t\lambda). \tag{23}$$

Substituting Eqn (23) in Eqn (12d) and simplifying

$$\lambda_1 \lambda_{m-3}^2 t^3 - \lambda_{m-3}[(b_1-b_3)\lambda_{m-3}-(a_3+2a_2)\lambda_1]t^2$$

$$-\left[\sum_{j=3}^{m-2}(a_{j+1}-a_{j-1})(a_{j+1}+a_j+u_{j-1})\lambda_{j-2}+(b_1-b_3)(a_3+2a_2)\lambda_{m-3}\right]t = 0.$$

There is a common factor of t, indicating a zero root, which is to be expected since $t = 0$ satisfies Eqn (23). Therefore

$$\lambda_1 \lambda_{m-3}^2 t^2 - \lambda_{m-3}[(b_1-b_3)\lambda_{m-3}-(a_3+2a_2)\lambda_1]t - \left[\sum_{j=3}^{m-2}(a_{j+1}-a_{j-1})\right.$$

$$\left. \times (a_{j+1}+a_j+a_{j-1})\lambda_{j-2}+(b_1-b_3)(a_3+2a_2)\right] = 0. \tag{23a}$$

14. Method of Alternate Base Planes

The direction of search [8] is defined by

$$\lambda_i^{(i)} = 0 \qquad \text{for } i = 1, ..., (m-3) \tag{24}$$

$$\lambda_j^{(i)} = \frac{R_j}{N} \qquad \text{for } j = 1, ..., (m-3); j \neq i \tag{24a}$$

where R_j are random numbers and N is the normalization factor defined by

$$N = \left(\sum_{j=1}^{m-3} R_j\right)^{\frac{1}{2}}.$$

Therefore the distances to the geometrical constraints are given by

$$t_j^{(i)} = \frac{b_j - \varepsilon_1}{-\lambda_{j-2}^{(i)}} \qquad \text{for } j = 3, ..., (m-2)$$

$$t_1^{(i)} = \frac{a_2 - a_1 - \varepsilon_3}{-\lambda_{m-3}^{(i)}}; \qquad t_2^{(i)} = \frac{a_3 - \varepsilon_2 - a_2}{\lambda_{m-3}^{(i)}}.$$

Let
$$\Delta_1^{(i)} = \min_{1 \le j \le m-2} (t_j; t_j > 0)$$

$$\Delta_2^{(i)} = \max_{1 \le j \le m-2} (t_j; t_j < 0).$$

Define

$$\left.\begin{aligned}
\Delta_r^{(i)} &= R_r \, \Delta_1^{(i)} \qquad \text{for } r = 1, 2, 3 \\[2mm]
&= R_r \, \Delta_2^{(i)} \qquad \text{for } r = 4, 5, 6
\end{aligned}\right\} \tag{24b}$$

where $\qquad 0 < R_r < 1 \qquad$ for $r = 1, ..., 6.$

Therefore the step length for equal weight redesign is given by

$$t = \Delta_r^{(i)} \tag{25}$$

and Eqn (23a) becomes

$$\lambda_1 \lambda_{m-3}^2 \, \Delta_r^{(i)2} - \lambda_{m-3} \left[(b_1 - b_3)\lambda_{m-3} - (a_3 + 2a_2)\lambda_1\right] \Delta_r^{(i)}$$

$$- \left[\sum_{j=3}^{m-2}(a_{j+1} - a_{j-1})\,(a_{j+1} + a_j + a_{j-1})\lambda_{j-2} + (b_1 - b_3)\,(a_3 + 2a_2)\right] = 0.$$

This equation is used to redetermine $\lambda_i^{(i)}$ where $\lambda_j^{(i)}$ for $j \ne i$ are given by Eqn (24a) and $\Delta_r^{(i)}$ by Eqn (24b).

Consider the designs

$$\bar{x}_i^{((r)} = x + \Delta_r^{(i)}\lambda \qquad \text{for } r = 1, ..., 6 \tag{25a}$$

where
$$W(b_3, ..., b_i, ..., b_{m-2}, a_2)$$

$$= W(b_3 + \Delta_r^{(i)}\lambda_1, ..., \bar{b}_i^{(r)}, ..., b_{m-2} + \Delta_r^{(i)}\lambda_{m-4}, a_2 + \Delta_r^{(i)}\lambda_{m-3}).$$

The designs are tested against the design requirements and if any one of these is feasible, steepest descent motion proceeds until a constraint is encountered. If none of these designs is feasible, the base plane is changed $(i \to i+1)$ and a new set of proposed designs is generated. This process is continued until a feasible design is obtained or the current boundary design is accepted as the proposed optimum.

15. Selective I

This was the first attempt at using the physics of the problem to move away from a behavioural constraint. For a given direction λ, Eqn (23a) is a quadratic in the step length. Let the behaviour variables for the boundary point be given by

$$y(x) = (\tau_{r_1}, \ldots, \tau_{r_n})$$

where $\qquad \tau_{r_q} = \tau_{0_i} \quad \text{for } a_{l-1} \leqslant i_q \leqslant u_l; \quad 1 \leqslant q \leqslant n; \quad 2 \leqslant l \leqslant m.$

Define $\qquad\qquad \tau_{a_k} = \max(\tau_{a_{l-1}}, \tau_{a_l})$

where $\qquad\qquad\qquad k = (l-1) \text{ or } 1.$ $\qquad\qquad\qquad\qquad$ (26)

Therefore the direction of search is given by

$$\lambda_j = (\tau_{a_j} - \tau_0)/N < 0 \qquad\qquad \text{for } j \neq k$$

$$= \frac{\partial W}{\partial b_k}(x^{(q)})/|\nabla W(x^{(q)})| > 0 \qquad \text{for } j = k \qquad (26a)$$

where the normalization factor N is given by

$$N^2 = \left(\sum_{j \neq k}(\tau_{a_j} - \tau_0)^2\right)\bigg/(1 - \lambda_k^2).$$

The method of alternate base planes consumed computer time in searching through the random directions to find a line which would give a feasible point on the same weight contour. Selective I reduces the degree of randomness by examining only those directions which on physical considerations move away from a behavioural constraint. The disadvantages are, (1) possibility of complex roots, (2) even if real roots are forthcoming, the step length may be negative, and (3) geometrical constraints may be violated.

16. Selective II

This is a more intelligent version designed to overcome the above difficulties. From Eqns (13a), (13b) a step length defined by

$$t = \min_i (x_i - l_i, u_i - x_i) \qquad\qquad\qquad (27)$$

gives an alternate step within the design variable bounds. Therefore

$$l \leqslant \bar{x} \leqslant u. \tag{27a}$$

The direction is then determined from the equal weight condition (23a) and the normalization condition

$$\sum_{i=1}^{m-3} \lambda_i^2 = 1. \tag{27b}$$

Equations (23a), (27b) are indeterminate. To obtain determinate solutions the number of unknowns is reduced to two by assigning predetermined values to $(m-5)$ cosines. These are made zero to obtain real solutions. The following designs are considered:

$$\bar{x}^{(r)} = x + (t/2^r) \lambda^{(r)} \qquad \text{for } r = 0, ..., 3 \tag{28}$$

The designs are tested against the behavioural constraints and if any one of these is feasible, steepest descent motion continues as before. If no feasible design is forthcoming, a different direction of search is generated corresponding to a different combination of direction cosines being assigned the value zero.

Define
$$\tau_{a_s} = \min_{2 \leqslant j \leqslant (m-2)} \tau_{a_j}.$$

The following cases are considered

Case 1. $s \neq 2$.

$$\begin{aligned}
\bar{b}_s &= b_s + t\lambda_s, & \lambda_s &< 0 \\
\bar{b}_k &= b_k + t\lambda_k, & \lambda_k &> 0 \\
\bar{b}_j &= b_j & &\text{for } j = 3, ..., (m-2); \ j \neq k, s \\
\bar{a}_2 &= a_2 & t &= b_s - \varepsilon_1
\end{aligned}$$

Therefore from Eqns (23a), (27b)

$$\lambda_s = -\frac{\mu}{\sqrt{(1+\mu^2)}}, \qquad \lambda_k = \frac{1}{\sqrt{(1+\mu^2)}} \tag{29}$$

where
$$\mu = \frac{(a_{k+1}-a_{k-1})(a_{k+1}+a_k+a_{k-1})}{(a_{l+1}-a_{l-1})(a_{l+1}+a_{l-1})} > 0.$$

Case 2. $s = 2, k \neq 3$.

$$\bar{b}_k = b_k + t\lambda_k, \qquad \lambda_k > 0$$

$$\bar{b}_j = b_j \qquad \text{for } j = 3, ..., (m-2); \quad j \neq k$$

$$\bar{a}_2 = a_2 + t\lambda_{m-3}, \qquad \lambda_{m-3} < 0$$

$$t = a_2 - (a_1 + \varepsilon_3).$$

Therefore from Eqns (23a), (27b)

$$\alpha t^2 \lambda_{m-3}^4 + 2\alpha(a_3 + 2a_2)t\lambda_{m-3}^3 + [1 + \alpha(a_3 + 2a_2)^2]\lambda_{m-3}^2 - 1 = 0 \qquad (29a)$$

where
$$\alpha = \frac{(b_1 - b_3)^2}{(a_{k+1} - a_{k-1})^2(a_{k+1} + a_k + a_{k-1})^2}.$$

Equation (29a) has a real root in $[1, 0]$ and is determined using linear interpolations being always between function values of opposite sign. Therefore from (27b), λ_k is given by, $\lambda_k = \sqrt{(1 - \lambda_{m-3}^2)}$.

Case 3. $s = 2, \quad k = 3$.

$$\bar{b}_3 = b_3 + t\lambda_1 \qquad \lambda_1 > 0$$

$$\bar{b}_j = b_j \qquad j = 4, ..., (m-2);$$

$$\bar{a}_2 = a_2 + t\lambda_{m-3} \qquad \lambda_{m-3} < 0$$

$$t = a_2 - (a_1 + \varepsilon_3).$$

Equations (23a), (27b) reduce to

$$t^4\lambda_{m-3}^6 + 2\gamma t^3\lambda_{m-3}^5 + (\gamma^2 - 2\beta^2 + \alpha^2 - t^2)t^2\lambda_{m-3}^4 + 2(\alpha^2 - \beta^2 - t^2)\gamma t\lambda_{m-3}^3$$
$$+ (\beta^4 + \alpha^2\gamma^2 - \gamma^2 t^2 + 2\beta^2 t^2)\lambda_{m-3}^2 + 2\beta^2\gamma t\lambda_{m-3} - \beta^4 = 0. \qquad (29b)$$

when
$$\alpha = (b_1 - b_3), \qquad \beta^2 = (a_4 - a_2)(a_4 + a_3 + a_2), \qquad \gamma = (a_3 + 2a_2).$$

As before Eqn (29b) has a real root in $[-1, 0]$.

17. Constrained Steepest Descent Motion

Constrained steepest descent motion is defined by

$$x^{(q+1)} = x^{(q)} + t^{(q)}\psi^{(q)} \tag{30}$$

where $t^{(q)}$ is given by Eqn (20b) and $\psi^{(q)}$ is determined as follows.

Case 1. $x^{(q)}$ lies on $b_k = \varepsilon_1$, $3 \leqslant k \leqslant (m-2)$.

Then

$$\psi_j^{(q)} = -\frac{1}{N}\left(\frac{\partial W}{\partial b_{j+2}}\right) \qquad \text{for } j = 1, ..., (m-4);\ \ j \neq k$$

$$= 0 \qquad\qquad \text{for } j = k$$

$$= -\frac{1}{N}\left(\frac{\partial W}{\partial a_2}\right) \qquad \text{for } j = (m-3)$$

where

$$N = \left[\sum_{j \neq k}\left(\frac{\partial W}{\partial b_{j+2}}\right)^2 + \left(\frac{\partial W}{\partial a_2}\right)^2\right]^{\frac{1}{2}}.$$

Case 2. $x^{(q)}$ lies on $a_2 = a_1 + \varepsilon_3$ or $a_2 = a_3 - \varepsilon_2$.

Then

$$\psi_j^{(q)} = -\frac{1}{N}\left(\frac{\partial W}{\partial b_{j+2}}\right) \qquad \text{for } j = 1, ..., (m-4)$$

$$= 0 \qquad\qquad\qquad \text{for } j = (m-3)$$

where

$$N = \left[\sum_{j=1}^{m-4}\left(\frac{\partial W}{\partial b_{j+2}}\right)^2\right]^{\frac{1}{2}}.$$

This is a simplified form of Rosen's gradient projection method for linear constraints.

18. Numerical Results

The following cases were considered.

Case 1. A standard steam turbine disc with seven points of division.

Case 2. An arbitrary shaped disc with the same number of divisions.

TABLE I

Case	Dimension of design space $(m-3)$	Initial weight (lbs)	Final Weight (lb)		Number of iterations		Run time (mins)	
			Case (a)	Case (b)	Case (a)	Case (b)	Case (a)	Case (b)
1	4	$3 \cdot 58934 \times 10^3$	$1 \cdot 66187 \times 10^3$	$2 \cdot 25877 \times 10^3$	62	80	5	7.833
2	4	$3 \cdot 60248 \times 10^3$	$1 \cdot 64547 \times 10^3$	$2 \cdot 32714 \times 10^3$	74	40	5	4·9
3	11	$3 \cdot 58973 \times 10^3$	$1 \cdot 61401 \times 10^3$	$2 \cdot 4537 \times 10^3$	186	408	30	60
4	11	$1 \cdot 65165 \times 10^3$	$1 \cdot 03400 \times 10^3$	—	188	—	30	—

Case 3. A standard disc with fourteen points of division.

Case 4. Final design for case (1) but with a finer division.

These cases were run using the Selective II and the method of alternate base planes in turn and are labelled cases (a), (b), respectively. They are shown in Figs 4–14, and are also summarized in Table I above for ready reference.

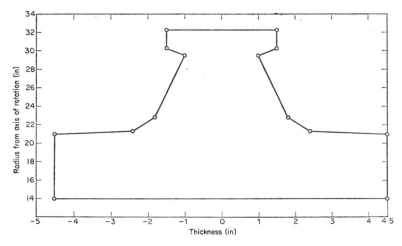

FIG. 4.   Cases 1a, 1b. Initial design. Weight = $3 \cdot 58934 \times 10^3$ lb.

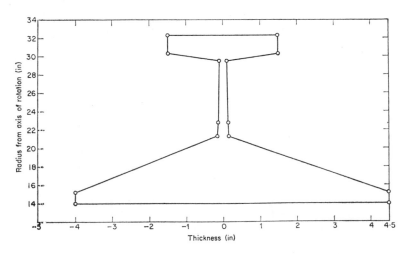

FIG. 5.   Case 1a; 62 cycles. Final design. Weight = $1 \cdot 66187 \times 10^3$ lb.

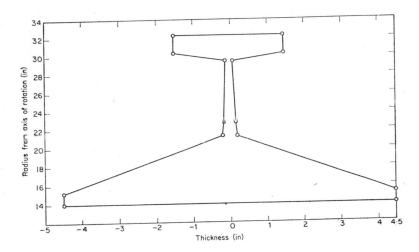

Fig. 6.   Case 1b; 80 cycles. Final design. Weight $= 2 \cdot 25877 \times 10^3$ lb.

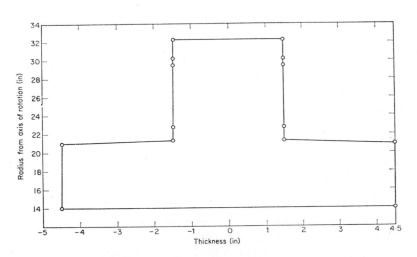

Fig. 7.   Cases 2a, 2b. Initial design. Weight $= 3 \cdot 60248 \times 10^3$ lb.

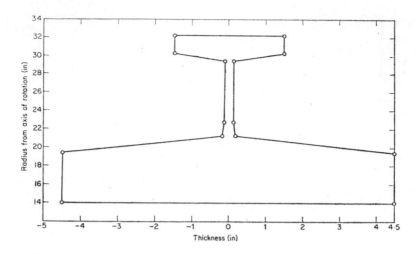

FIG. 8.   Case 2a; 74 cycles. Final design. Weight $= 1 \cdot 64547 \times 10^3$ lb.

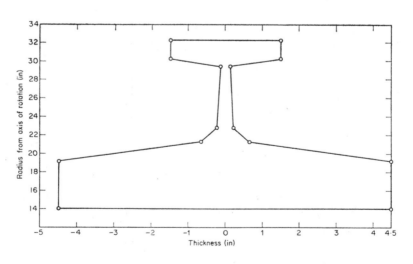

FIG. 9.   Case 2b; 40 cycles. Final design. Weight $= 2 \cdot 32714 \times 10^3$ lb.

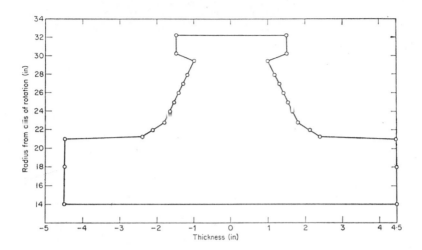

FIG. 10.   Cases 3a, 3b. Initial design. Weight $= 3 \cdot 58973 \times 10^3$ lb.

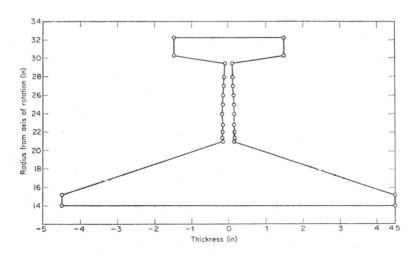

FIG. 11.   Case 3a; 186 cycles. Final design. Weight $= 1 \cdot 61401 \times 10^3$ lb·

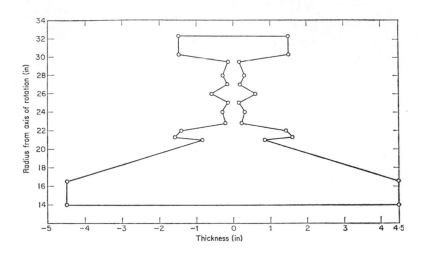

FIG. 12. Case 3b. Final design. Weight = 2·14537 × 10³ lb.

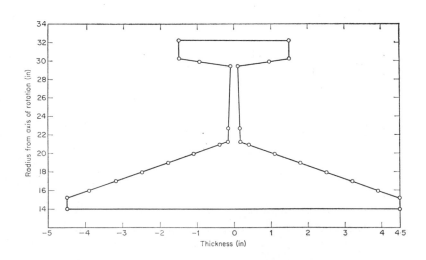

FIG. 13. Case 4. Initial design. Weight = 1·65165 × 10³ lb.

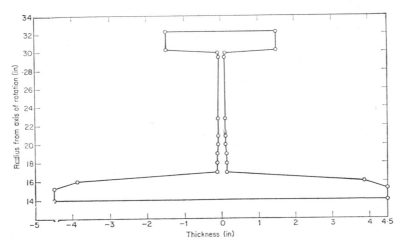

FIG. 14.   Case 4; 188 cycles. Final design. Weight $= 1.03400 \times 10^3$ lb.

The discs were made of mild steel for which the density and elastic properties were assumed constant. The numerical work was carried out on an English Electric KDF9 computer using Algol compiler language.

19. Discussion

Although the initial designs for cases (1), (2) differ in weight by less than 0.005 per cent they are radically different in configuration; but the resulting designs tend to have approximately the same weight and configuration. Case (2) was run primarily to test for relative minima to establish whether the starting design influenced the final outcome. Cases (3), (4) were run to investigate the stability of the minimization paths. Initially the weight reductions were relatively rapid (Figs 15–16) but tended to slow down as the optimum was reached. As the iteration progressed equal weight redesign tended to give design points lying close to the behavioural constraints thereby slowing down the weight reductions. The random method consumed considerable computer time in searching through the random directions to locate a feasible design. However, Selective II was always able to locate a feasible design after one or two trials. Selective I never worked since almost always complex roots were generated for the quadratic equation for determining the step length; in the few occasions when real roots were forthcoming, the geometrical constraints were violated giving negative thicknesses. The number of iterations to obtain a specified weight reduction depends primarily on the dimension of the design space.

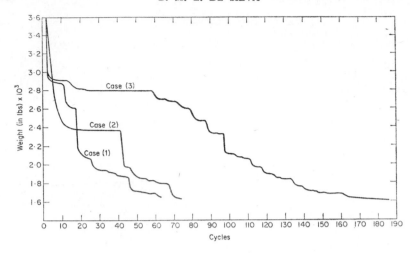

FIG. 15. Weight *versus* total redesign attempts. Based on selective search techniques for moving away from a bound point.

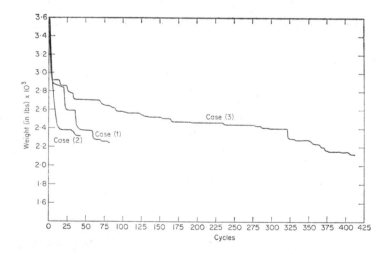

FIG. 16. Weight *versus* total redesign attempts. Based on random search.

The estimated step length used in the steepest descent mode of travel enabled a behavioural constraint to be encountered after about two or three iterations and thereafter the linear interpolation technique gave rapid convergence onto the constraint.

Acknowledgements

The research described in this paper was sponsored and supported by the Mechanical Engineering Laboratories of the English Electric Company, by whose permission this paper is published. Acknowledgement is also made to Dr W. A. Green of the Department of Theoretical Mechanics, University of Nottingham, for many valuable suggestions and discussions.

References

1. L. Cavallaro (1965). Stress analysis of rotating discs. *Nucl. Struct. Engng*, **2**, 271-281.
2. W. Prager and P. G. Hodge. "Theory of Perfectly Plastic Solids", Chapter 1. John Wiley, New York.
3. B. M. E. de Silva (1968). "Non-classical variational Problems in Minimum Weight Structural Optimization Theory". English Electric Company. Rpt. No. W/M(4B)p.1382.
4. L. A. Schmit (1960). "Structual Design by Systematic Synthesis". Proceedings of the 2nd National Conference of Electronic Computation, Structural Division A.S.C.E., 105-132.
5. L. A. Schmit and T. P. Kicher (1962). "Synthesis of Material and Configuration Selection". Structural Division, A.S.C.E., **88**, No. ST3, 79-102.
6. L. A. Schmit and W. M. Morrow (1963). "Structural Synthesis with Buckling Constraints". Structural Division, A.S.C.E., **89**, No. ST2, 107-126.
7. L. A. Schmit and R. H. Mallett (1962). "Design Synthesis in a Multi-dimensional Design Space with Automated Material Selection". E.D.C. Rpt. 2-62-2, Case Inst. Tech.
8. L. A. Schmit, T. P. Kicher and W. M. Morrow (1963). "Structural Synthesis Capability for Integrally Stiffened Waffle Plates". A.I.A.A., **1**, No. 12, 2820-2836.
9. L. A. Schmit and T. P. Kicher (1962). "Structural Synthesis of Symmetric Waffle Plates". N.A.S.A. Tech. Note D-1691.
10. L. A. Schmit and T. P. Kicher (1961). "Structural Synthesis of Symmetric Waffle Plates with Integral Orthogonal Stiffeners". E.D.C. Rpt. 2-61-1, Case Inst. Tech.
11. L. A. Schmit and R. L. Fox (1963). "Synthesis of a Simple Shock Isolator". E.D.C. Rpt. 2-63-4, Case Inst. Tech.
12. L. A. Schmit (1966). "Automated Redesign". International Scientific Tech. 63-78.
13. E. D'Sylva. "Disc Stresses". English Electric Company Rpt. W/M(4A) p.88.
14. L. G. Roberts. "Amendments to Program 00011 which Calculates Disc Stresses". English Electric Company Rpt. No. W/M(6A).1010.
15. J. E. Kelley Jr. (1960). The cutting plane method for solving convex programs. *J. Soc. ind. appl. Math.*, **8**, (4) 703-712.
16. F. Moses (1964). "Optimum Structural Design using Linear Programming". Structural Division, A.S.C.E., **90**, No. ST6, 89-104.
17. J. B. Rosen (1960). The gradient projection method, Part I. *J. Soc. ind. appl. Math.*, **8**, (1) 181-217.

18. J. B. Rosen (1961). The gradient projection method for nonlinear programming, Part II. *J. Soc. ind. appl. Math.*, **9**, (4), 514-532.
19. D. M. Brown and A. Ang. (1965). "A Nonlinear Programming Approach to the Elastic Minimum Weight Design of Steel Structures". University of Illinois, Urbana, Illinois.
20. A. V. Fiacco and G. P. McCormick (1964). The sequential unconstrained minimization technique for nonlinear programming; a primal-dual method. *Mgmt. Sci.*, **10**, (2), 360-365.
21. L. A. Schmit and R. L. Fox (1964). An integrated approach to structural synthesis and analysis. *Am. Inst. Aerom. Aseran.*, **3**, (6), 1104-1112.
22. L. A. Schmit and R. L. Fox (1964). An Integrated Approach to Structural Synthesis and Analysis. 5th Annual Conference on Structural Materials. New York, 294-315.
23. J. B. Crockett and H. Chernoff (1955). Gradient methods of maximization. *Pacif. J. Math.*, **5**, (1), 33-50.
24. D. J. Wilde (1964). "Optimum Seeking Methods". Prentice-Hall Englewood, New Jersey.
25. J. Kowalik (1966). Nonlinear programming procedures and design optimization. *Acta. polytech. scand.*, series d, **13**,.
26. B. Klein (1955). Direct uses of extremal principles in solving certain types of optimization problems involving inequalities. *J. Ops. Res.*, **3**, (2), 168-175.
27. H. W. Kuhn and A. W. Tucker (1951). "Nonlinear Programming". Proceedings 2nd Berkeley Symposium on Mathematical Statistics and Probability. (J. Neumann, ed.), pp. 481-492. University California Press, Berkeley.
28. J. L. Greenstadt (1966). A ricocheting gradient method for nonlinear programming. *Soc. ind. appl. Math.*, **14**, (3), 429-445.
29. R. A. Gellatly and R. H. Gallagher (1966). A procedure for automated minimum weight structural design—Part I. *Aeronaut. Q.*, **17**, (3), 216-230.
30. R. A. Gellatly and R. H. Gallagher (1966). A procedure for minimum weight structural design—Part II. *Aeronaut. Q.*, **17**, (4), 332-342.
31. G. C. Best (1964). Completely automatic weight minimization method for high speed digital computers. *J. Aircraft*, **1**, (3), 129-133.
32. B. M. E. de Silva (1965). "On the Application of Nonlinear Programming to the Automatic Structural Synthesis of Turbine Discs". English Electric Company, Rpt. No. W/M(6B) 1022.
33. B. M. E. de Silva (1967). "The Minimum Weight Design of Steam Turbine Discs". English Electric Company, Rpt. No. W/M(4B) 1323.

10. An Integral Equation Approach to Second Variation Techniques for Optimal Control Problems

G. S. Tracz* and B. Bernholtz

*Department of Industrial Engineering, University of Toronto,
Toronto, Canada*

1. Introduction

While the theoretical development of optimal control theory has advanced at a rapid pace in a few short years, particularly in certain areas [1, 2, 3, 4], it is probably fair to say that numerical procedures for solving such problems have lagged somewhat behind, especially for non-separable problems with given terminal conditions. In this paper, a new method for solving the following class of optimal control problems will be considered.

Determine control functions $u(t)$ in $t_0 \leqslant t \leqslant t_1$ so as to minimize a performance index of the form

$$J(u) = K[x(t_1)] + \int_{t_0}^{t_1} L[x(t), u(t), t]dt \tag{1}$$

with
$$\frac{dx(t)}{dt} = \dot{x}(t) = f[x(t), u(t), t], \quad x(t_0) = a \tag{2}$$

and
$$G[x(t_1)] = 0. \tag{3}$$

The state variable $x(t)$ is a column vector with n components, while the control variable $u(t)$ is a column vector with m components, $0 < m \leqslant n$. K and L are real-valued scalar functions which possess continuous partial derivatives in the required variables up to the second order for all t. The vector a of initial conditions is given, and $G[x(t_1)] = 0$ is a column vector of final conditions on the state vector with q components, $q \leqslant n$.

This class of fixed-time optimal control problems includes the fixed endpoint problem $x(t_1) = b$ as a special case.

*Present address: Department of Educational Planning, The Ontario Institute for Studies in Education, Toronto 181, Ontario, Canada.

The method of solution described in this paper is iterative in nature and each iteration proceeds as follows. Let $\bar{J}(u, v)$ denote an augmented performance index given by

$$\bar{J}(u, v) = J(u) + \int_{t_0}^{t_1} p^T(t)\{f[x(t), u(t), t] - \dot{x}(t)\}dt + v^T G[x(t_1)] \qquad (4)$$

where v is a q-component column vector of Lagrange multipliers, and $p(t)$ is an n-component column vector of Lagrange multiplier functions introduced to take into account constraints (2) and (3), respectively. The transpose of an expression ( ) is denoted by ( )^T. Suppose u and v are now interpreted as the current control function vector and Lagrange multiplier vector; let δu and dv be proposed perturbations in these quantities. The change $\bar{J}(u, v) - \bar{J}(u + \delta u, v + dv)$ is approximated by a quadratic function $-\delta \bar{J}$ in δu and dv. The variations δu and dv are then chosen so as to maximize $-\delta \bar{J}$ subject to constraints imposed to ensure improved compliance with the terminal condition (3).

Section 2 of the paper is devoted to a derivation of the quadratic approximation, while Section 3 shows that the solution to the subsidiary quadratic maximization problem can be obtained by solving simultaneously a Fredholm linear matrix integral equation and a system of linear algebraic equations in δu and dv.

2. Derivation of the Quadratic Approximation to the Variation in the Augmented Performance Index

Written more fully, the augmented performance index $\bar{J}$ is

$$\bar{J}(u,v) = K[x(t_1)] + v^T G[x(t_1)]$$
$$+ \int_{t_0}^{t_1} \left\{ L[x(t), u(t), t] + p^T(t)\{f[x(t), u(t), t] - \dot{x}(t)\} \right\} dt. \qquad (5)$$

Equation (5) can be rewritten as

$$\bar{J}(u, v) = \phi[v, x(t_1)] + \int_{t_0}^{t_1} [H(x, p, u, t) - p^T \dot{x}]dt \qquad (6)$$

where the Hamiltonian

$$H(x, p, u, t) = L(x, u, t) + p^T f \qquad (7)$$

and

$$\phi[v, x(t_1)] = K[x(t_1)] + v^T G[x(t_1)]. \qquad (8)$$

If the current $u(t)$ and v are perturbed by variations $\delta u(t)$ and dv, variations $\delta x(t)$ and $\delta p(t)$ are produced. Retaining terms up to the second order, the change $\bar{J}(u, v) - \bar{J}(u+\delta u, v+dv)$ may be approximated by $-\delta\bar{J}$ where $\delta\bar{J}$ is given by ($H_{pp} = 0$ due to Eqn 7):

$$\delta\bar{J} = [\delta x^T\phi_x + dv^T G + dv^T G_x \delta x + \tfrac{1}{2}\delta x^T\phi_{xx}\delta x]_{t=t_1}$$

$$+ \int_{t_0}^{t_1} [\delta x^T H_x + \delta u^T H_u + \delta p^T H_p + \tfrac{1}{2}\delta x^T H_{xx}\delta x + \tfrac{1}{2}\delta u^T H_{uu}\delta u$$

$$+ \tfrac{1}{2}\delta x^T H_{xu}\delta u + \tfrac{1}{2}\delta u^T H_{ux}\delta x + \tfrac{1}{2}\delta u^T H_{up}\delta p + \tfrac{1}{2}\delta p^T H_{pu}\delta u$$

$$+ \tfrac{1}{2}\delta x^T H_{xp}\delta p + \tfrac{1}{2}\delta p^T H_{px}\delta x - \delta p^T\dot{x} - p^T\delta\dot{x} - \delta p^T\delta\dot{x}]dt. \qquad (9)$$

On integrating

$$-\int_{t_0}^{t_1} p^T\delta\dot{x}\,dt$$

by parts, Eqn (9) can be rearranged as follows [5]:

$$\delta\bar{J} = [\delta x^T(\phi_x - p) + dv^T G + dv^T G_x\delta x + \tfrac{1}{2}\delta x^T\phi_{xx}\delta x]_{t=t_1}$$

$$+ p(t_0)\delta x(t_0) + \int_{t_0}^{t_1} \{\delta p^T(H_p - \dot{x}) + \delta x^T(H_x + \dot{p})$$

$$+ \delta u^T H_u + \tfrac{1}{2}\delta u^T H_{uu}\delta u + \tfrac{1}{2}\delta u^T H_{ux}\,dx + \tfrac{1}{2}\delta x^T H_{xu}\delta u$$

$$+ \tfrac{1}{2}\delta x^T H_{xx}\delta x - \delta p^T[\delta\dot{x} - H_{px}\delta x - H_{pu}\delta u]\}dt. \qquad (10)$$

If the given system of Eqns (2) is linearized about the current trajectory $x(t)$, then

$$\delta\dot{x}(t) = f_x(t)\delta x(t) + f_u(t)\delta u(t)$$

$$= H_{px}(t)\delta x(t) + H_{pu}(t)\delta u(t), \quad \delta x(t_0) = 0. \qquad (11)$$

In addition, the following condition which is a necessary condition for an extremal path [6] will be imposed on $p(t)$ at each iteration:

$$\dot{p}(t) = -H_x \qquad (12)$$

with

$$p(t_1) = \phi_x[v, x(t_1)] = [K_x + v^T G_x]_{t=t_1}. \qquad (13)$$

Furthermore, in terms of the Hamiltonian H, Eqn (2) can be expressed as

$$\dot{x}(t) = H_p, \quad x(t_0) = a. \qquad (14)$$

Substitution of Eqns (11–14) in Eqn (10) yields

$$\delta \bar{J} = [dv^T G + dv^T G_x \, \delta x + \tfrac{1}{2} \delta x^T \phi_{xx} \, \delta x]_{t=t_1}$$

$$+ \int_{t_0}^{t_1} \delta u^T H_u \, dt + \tfrac{1}{2} \int_{t_0}^{t_1} [\delta x^T \delta u^T] \begin{bmatrix} H_{xx} & H_{xu} \\ H_{ux} & H_{uu} \end{bmatrix} \begin{bmatrix} \delta x \\ \delta u \end{bmatrix} dt. \tag{15}$$

Up to this point, the procedure is very similar to that taken by Kelley *et al.* [7], Breakwell *et al.* [5], McReynolds and Bryson [8], and Mitter [9]. Now, however, instead of eliminating the control increment $\delta u(t)$ and solving a coupled system of differential equations in $\delta x(t)$ and $\delta p(t)$, the quantity $\delta \bar{J}$ will be expressed explicitly as a quadratic functional in $\delta u(t)$ and dv over the entire interval $[t_0, t_1]$.

The matrices $f_x(t)$ and $f_u(t)$ in Eqn (11) are matrix functions of $x(t)$ and $u(t)$ so that (11) represents a system of time-varying but linear differential equations. For this system

$$\delta x(t) = \int_{t_0}^{t} \Phi(t, s) f_u(s) \delta u(s) \, ds \tag{16}$$

where the fundamental or transition matrix $\Phi(t, s)$ satisfies the matrix differential equation

$$(d/dt) \, \Phi(t, s) = f_x(t) \Phi(t, s), \quad \Phi(s, s) = I. \tag{17}$$

Equivalently Eqn (16) can be expressed as

$$\delta x(t) = \int_{t_0}^{t} S(t, s) \delta u(s) \, ds + \int_{t}^{t_1} S(t, s) \delta u(s) \, ds = \int_{t_0}^{t_1} S(t, s) \delta u(s) \, ds \tag{18}$$

where, by definition,

$$S(t, s) = \begin{cases} \Phi(t, s) f_u(s) & t \geqslant s \\ 0 & t < s \end{cases} \tag{19}$$

Substituting Eqn (18) in (15), the quantity $\delta \bar{J}$ is then expressed explicitly as a quadratic functional in $\delta u(t)$ and dv:

$$\delta \bar{J} = dv^T G[x(t_1)] + dv^T G_x(t_1) \int_{t_0}^{t_1} S(t_1, t) \delta u(t) \, dt$$

$$+ \int_{t_0}^{t_1} \delta u^T H_u \, dt + \tfrac{1}{2} \int_{t_0}^{t_1} \int_{t_0}^{t_1} \delta u^T(t) W(t, s) \delta u(s) \, dt \, ds \tag{20}$$

where

$$W(t, s) = S^T(t_1, t)\phi_{xx}(t_1)S(t_1, s) + H_{uu}(t)\delta(t-s)$$

$$+ H_{ux}(t)S(t, s) + S^T(s, t)H_{xu}(s)$$

$$+ \int_{\theta = \max\{t, s\}}^{t_1} S^T(\theta, t)H_{xx}(\theta)S(\theta, s)d\theta. \tag{21}$$

In Eqn (21), $\delta(t-s)$ is the Dirac delta function [10],

$$\int_{t_0}^{t_1} \delta(t-s)\delta u(s)ds = \delta u(t), \qquad t \in [t_0, t_1]. \tag{22}$$

The $m \times m$ matrix $W(t, s)$ will be referred to as the *second-order weighting matrix*.

The motivation for this approach is as follows. Gradient methods or first-order methods in control function space are numerous—the approach was initially developed and applied to aeronautical problems by Kelley [11] and Bryson and Denham [12]. However, to the authors' knowledge, no second-order method in control function space has been proposed.

The structure of the second-order weighting matrix $W(t, s)$ is particularly interesting. In it, the contribution of $H_{uu}(t)$ is seen to be an impulse occurring at the time $t = s$. The values of H_{ux} and H_{xu} are modified by the quantity $S(t, s)$ in which

$$S(t, s) = \begin{cases} \Phi(t, s)f_u(s) & t \geqslant s \\ 0 & t < s \end{cases} \tag{23}$$

so that the transition matrix which evolves as a function of time, and $f_u(s)$ which relates $x(s)$ and $u(s)$, are brought into play. $W(t, s)$ also includes H_{xx}, in the form of an integral expression involving a variable lower limit of integration.

Note that if $H_{ux} = H_{xu}^T$ and H_{xx} are omitted, then the weighting matrix (assuming $\phi_{xx} = 0$) is simply H_{uu}; this would result in a suboptimal scheme discussed by Merriam, and which was found inadequate [13].

Finally it may be noted that the term in Eqn (20) involving both δu and dv is an indication of the interaction due to making changes simultaneously in u and v. This term plays an essential role in the procedure given in Section 3 for determining δu and dv so as to maximize $-\delta \bar{J}$, the approximation to the change $\bar{J}(u, v) - \bar{J}(u+\delta u, v+dv)$, while at the same time obtaining improved compliance with the given terminal conditions.

3. Derivation of the Control Algorithm

A procedure will now be developed for selecting the variations $\delta u(t)$ in the control vector to yield the maximum value of $-\delta \bar{J}$ or the minimum value of $\delta \bar{J}$ where $\delta \bar{J}$ is given by Eqn (20), while at the same time obtaining improved compliance with the given terminal conditions. Both requirements can be met by introducing a correction $\delta G(t_1) = G_x(t_1)\delta x(t_1)$ and determining the $\delta u(t)$ which yield a minimum value for $\delta \bar{J}$ under the restriction that $\delta G(t_1) = G_x(t_1)\delta x(t_1)$, where $\delta G(t_1)$ is a given variation.

Note that the quantities $\delta u(t)$ and dv^T can be varied independently. Further, if dv^T is kept fixed, the quantity $dv^T G[x(t_1)]$ in Eqn (20) is a fixed quantity.

The minimum value of $\delta \bar{J}$ will occur when $\delta(\delta \bar{J}) = \delta^2 \bar{J} = 0$ for an arbitrary $\delta(\delta u(t)) = \delta^2 u(t)$. Now,

$$\varrho^2 \bar{J} = \int_{t_0}^{t_1} \delta^2 u^T(t)[H_u(t) + S^T(t_1, t)G_x{}^T(t_1)dv]dt$$

$$+ \tfrac{1}{2}\int_{t_0}^{t_1}\int_{t_0}^{t_1} \delta^2 u^T(t)W(t, s)\delta u(s)dsdt + \tfrac{1}{2}\int_{t_0}^{t_1}\int_{t_0}^{t_1} \delta u^T(t)W(t, s)\delta^2 u(s)dsdt \quad (24)$$

$$= \int_{t_0}^{t_1} \delta^2 u^T(t)[H_u(t) + S^T(t_1, t)G_x{}^T(t_1)dv]dt$$

$$+ \tfrac{1}{2}\int_{t_0}^{t_1}\int_{t_0}^{t_1} \delta^2 u^T(t)W(t, s)\delta u(s)dsdt + \tfrac{1}{2}\int_{t_0}^{t_1}\int_{t_0}^{t_1} \delta^2 u^T(s)W^T(t,s)\delta u(t)dsdt. \quad (25)$$

Since s, t are dummy variables of integration, and $W(t, s) = W^T(s, t)$, then

$$\delta^2 \bar{J} = \int_{t_0}^{t_1} \delta^2 u^T(t)[H_u(t) + S^T(t_1, t)G_x{}^T(t_1)dv]dt$$

$$+ \int_{t_0}^{t_1}\int_{t_0}^{t_1} \delta^2 u^T(t)W(t, s)\delta u(s)dsdt \quad (26)$$

$$= \int_{t_0}^{t_1} \delta^2 u^T(t)[H_u(t) + S^T(t_1, t)G_x{}^T(t_1)dv + \int_{t_0}^{t_1} W(t, s)\delta u(s)ds]dt. \quad (27)$$

As the coefficient of $\delta^2 u^T$ must vanish for $\delta^2 \bar{J} = 0$, then the following linear Fredholm matrix integral equation with kernel $W(t, s)$ is obtained:

$$H_u(t) + S^T(t_1, t)G_x^T(t_1)dv + \int_{t_0}^{t_1} W(t, s)\delta u(s)ds = 0. \tag{28}$$

The integral Eqn (28) takes into account the dynamics $\dot{x}(t) = f(x, u, t)$ of the control optimization problem in which a choice of $\delta u(t)$ for $t = \tau$ produces effects for all subsequent time $t > \tau$, as outlined by Merriam [13]. However, steepest-descent or first-order methods in control function s pace neglect this factor, so that the algorithm becomes a static minimization of the Hamiltonian with respect to the control variable at a finite number of time intervals.

Note that the term involving both δu and dv introduces the term $S^T(t_1, t)G_x(t_1)dv$ in Eqn (28), so that the solution $\delta u(t)$ of the integral equation depends on dv.

Returning to Eqn (28), it can be written as

$$H_u(t) + S^T(t_1, t)G_x^T(t_1)dv + \mathscr{E}\{\delta u\} = 0 \tag{29}$$

where the operator $\mathscr{E}$ is assumed to be invertible. Therefore

$$\delta u(t) = -[\mathscr{E}]^{-1}\{H_u(t) + S^T(t_1, t)G_x^T(t_1)dv\}. \tag{30}$$

From Eqn (18)

$$\delta G(t_1) = G_x(t_1)\delta x(t_1) = G_x(t_1)\int_{t_0}^{t_1} S(t_1, t)\delta u(t)dt. \tag{31}$$

Now $G[x(t_1)]$ may or may not satisfy the constraint $G[x(t_1)] = 0$. Substitution of Eqn (30) in (31) allows dv to be determined as a function of $\delta G(t_1)$, where $\delta G(t_1)$ is chosen so that the trajectory corresponding to $u(t) + \delta u(t)$ comes closer to satisfying the constraint at $t = t_1$ than did the trajectory corresponding to $u(t)$. Note that $\delta G(t_1)$ must be chosen small enough so that the linearization by which $\delta x(t_1)$ is obtained remains valid.

Substitution of Eqn (30) into (31) yields

$$-G_x(t_1)\int_{t_0}^{t_1} S(t_1, t)[\mathscr{E}]^{-1}\{H_u(t) + S^T(t_1, t)G_x^T(t_1)dv\}dt = \delta G(t_1). \tag{32}$$

The dv can be calculated from Eqn (32), since they are the only unknown quantities in that equation. Having calculated the dv, the control increments

$\delta u(t)$ can be evaluated by using Eqn (30). The values of $p(t_1)$ are updated as follows:

$$v_{new} = v_{old} + dv \tag{33}$$

and
$$p_{new}(t_1) = K_x(t_1) + v_{new}^T G_x(t_1). \tag{34}$$

Note that in Eqn (32), if $\delta G(t_1) = 0$, that is if the constraint $G[x(t_1)] = 0$ is satisfied at some iteration, the value of dv will be zero only if $H_u(t)$ is zero for all t. For if $H_u(t)$ is not equal to zero, then the vector $\delta u(t)$ itself is not zero, as seen from Eqn (30). Thus, at the next iteration, $G[x(t_1)]$ may be different from zero. Only when both $H_u(t)$ and $\delta G(t_1)$ are zero will dv be zero, with the result that the optimal $u^*(t)$ and $p^*(t_1)$ have been determined.

Summary of the Iterative Scheme

(1) Select an initial control vector $u(t)$ and an initial vector v of Lagrange multipliers.

(2) Integrate the system differential equations $\dot{x}(t) = f[x(t), u(t), t]$ forward in time with the control vector $u(t)$ and given values of $x(t_0) = a$. Record the functions $x(t), u(t)$.

(3) Linearize the system differential equations about the current $x(t), u(t)$ to obtain the matrices $f_x(t)$ and $f_u(t)$.

(4) Integrate the adjoint system $\dot{p}(t) = -H_x(x, p, u, t)$ backward in time with $p(t_1) = K_x(t_1) + v^T G_x(t_1)$ as starting conditions.

(5) Calculate the Hamiltonian function $H = L + p^T f$ and its derivatives $H_u, H_{uu}, H_{ux}, H_{xu}$ and H_{xx}.

(6) Evaluate the transition matrix $\Phi(t, s)$ for the linearized system in (3) and then calculate Eqn (23)

(7) The second variation weighting matrix $W(t, s)$ can now be calculated using Eqn (21).

(8) Express the increment $\delta u(t)$ in the control vector as a function of dv by solving the linear matrix integral Eqn (29).

(9) Choose an appropriate value of the correction $\delta G(t_1) = G_x(t_1)\delta x(t_1)$; if $G[x(t_1)] = 0$, then $\delta G(t_1)$ is set to zero.

(10) Evaluate the Lagrange multipliers dv by solving Eqn (32).

(11) With the dv known, the increments $\delta u(t)$ can be calculated from Eqn (30), and the new control functions become $u(t) + \delta u(t)$; let the Lagrange multipliers be $v + dv$.

(12) Return to step (2) and repeat the sequence of steps (2)–(11) until both $H_u(t) = 0$ and $\delta G(t_1) = 0$, that is $G[x(t_1)] = 0$.

4. Fixed End Point, Linear Systems with a Quadratic Performance Index

Consider the fixed end point problem with the system being described by a set of linear time-varying differential equations

$$\dot{x}(t) = A(t)x(t) + B(t)u(t), \qquad (x(t_0) = a, \quad x(t_1) = b) \qquad (35)$$

and a quadratic performance index

$$J = \int_{t_0}^{t_1} \tfrac{1}{2}[x^T(t)Q(t)x(t) + u^T(t)R(t)u(t)]dt. \qquad (36)$$

$Q(t)$ and $R(t)$ are, respectively, $n \times n$ and $m \times m$ positive definite symmetric matrices for all t. The matrices $A(t)$ and $B(t)$ are of order $n \times n$ and $n \times m$, respectively. All four matrices are assumed to be continuous functions of t.

If $u^0(t)$ is some initial control function and v^0 an initial choice of the unknown Lagrangian multiplier, then the values of $\delta u^0(t)$ and dv^0 are obtained by solving

$$H_{u^0}(t) + S^T(t_1, t)dv^0 + \int_{t_0}^{t_1} W(t, s)\delta u^0(s)ds = 0 \qquad (37)$$

while satisfying the constraint

$$x(t_1) = b \qquad (38)$$

should yield one-step convergence to the minimum. To verify that this is indeed so, it will be shown that $H_u(t) = 0$, where $u(t) = u^0(t) + \delta u^0(t)$, $x(t_1) = b$, and $x(t_1) = x^0(t_1) + \delta x^0(t_1)$.

For this class of problems, the Hamiltonian H is

$$H = \tfrac{1}{2}[x^T(t)Q(t)x(t) + u^T(t)R(t)u(t)] + p^T(t)[A(t)x(t) + B(t)u(t)] \qquad (39)$$

so that

$$H_u = Ru + B^T p, \quad H_{uu} = R, \quad H_{ux} = 0 = H_{xu}, \quad H_{xx} = Q. \qquad (40)$$

Further, since

$$\delta\dot{x}(t) = A(t)\delta x(t) + B(t)\delta u(t), \quad \delta x(t_0) = 0 \qquad (41)$$

then

$$\delta x(t) = \int_{t_0}^{t_1} S(t, s)\delta u(s)ds \qquad (42)$$

where

$$S(t, s) = \begin{cases} \Phi(t, s)B(t) & t \geqslant s \\ 0 & t < s \end{cases} \qquad (43)$$

and $$(d/dt)\Phi(t,s) = A(t)\Phi(t,s), \quad \Phi(s,s) = I. \qquad (44)$$

The matrix integral equation (28) then takes the following form:

$$R(t)u^0(t) + B^T(t)p^0(t) + S^T(t_1,t)dv^0 + R(t)\delta u^0(t)$$

$$+ \int_{t_0}^{t_1} \mathbf{W}(t,s)\delta u^0(s)ds = 0 \qquad (45)$$

where $p^0(t)$ is the adjoint variable corresponding to $u^0(t)$ and v^0, and

$$\mathbf{W}(t,s) = R(t)\delta(t-s) + \mathbf{W}(t,s)$$

with $$\mathbf{W}(t,s) = \int_{t_0}^{t_1} S^T(\theta,t)H_{xx}(\theta)S(\theta,s)d\theta. \qquad (46)$$

Note that it is now assumed that dv^0 has been determined such that $x(t_1) = x^0(t_1) + \delta x^0(t_1) = b$ by using Eqn (42). This is always possible since only linear systems are being considered. By definition,

$$H_u(t) = R(t)u(t) + B^T(t)p(t)$$

$$= R(t)[u^0(t) + \delta u^0(t)] + B^T(t)p(t)$$

$$= R(t)u^0(t) + R(t)R^{-1}(t)[-R(t)u^0(t) - B^T(t)p^0(t)$$

$$- S^T(t_1,t)dv^0 - \int_{t_0}^{t_1} \mathbf{W}(t,s)\delta u^0(s)ds] + B^T(t)p(t) \qquad (47)$$

$$= B^T(t)\delta p^0(t) - S^T(t_1,t)dv^0 - \int_{t_0}^{t_1} \mathbf{W}(t,s)\delta u^0(s)ds \qquad (48)$$

where $$\delta p^0(t) = p(t) - p^0(t). \qquad (49)$$

Therefore one-step convergence is obtained if

$$B^T(t)\delta p^0(t) = S^T(t_1,t)dv^0 + \int_{t_0}^{t_1} \mathbf{W}(t,s)\delta u^0(s)ds \qquad (50)$$

that is

$$B^T(t)\delta p^0(t) = S^T(t_1,t)dv^0 + \int_{t_0}^{t_1}\int_{t_0}^{t_1} S^T(\theta,t)Q(\theta)S(\theta,s)\delta u^0(s)dsd\theta. \qquad (51)$$

It will now be shown that Eqn (51) does in fact hold. From the definition of the adjoint vector $p(t)$ in Eqns (12), (13)

$$\dot{p}(t) = -A^T(t)p(t) - Q(t)x(t), \qquad p(t_1) = v \tag{52}$$

where $v = v^0 + dv^0$. Solving for $p(t)$,

$$p(t) = \Phi^T(t_1, t)v - \int_{t_1}^{t} \Phi^T(\theta, t)Q(\theta)x(\theta)d\theta \tag{53}$$

so that

$$B^T(t)p(t) = S^T(t_1, t)v + \int_{t_0}^{t_1} S^T(\theta, t)Q(\theta)x(\theta)d\theta. \tag{54}$$

Note that

$$S^T(\theta, t) = \begin{cases} B^T(t)\Phi^T(\theta, t) & \theta \geqslant t \\ 0 & \theta < t \end{cases} \tag{55}$$

Similarly,

$$B^T(t)p^0(t) = S^T(t_1, t)v^0 + \int_{t_0}^{t_1} S^T(\theta, t)Q(\theta)x^0(\theta)d\theta. \tag{56}$$

Subtracting Eqn (56) from (54) and using

$$\delta p^0(t) = p(t) - p^0(t), \quad \delta x^0(\theta) = x(\theta) - x^0(\theta), \quad dv^0 = v - v^0 \tag{57}$$

yields

$$B^T\delta p^0(t) = S^T(t_1, t)dv^0 + \int_{t_0}^{t_1} S^T(\theta, t)Q(\theta)\delta x^0(\theta)d\theta$$

$$= S^T(t_1, t)dv^0 + \int_{t_0}^{t_1}\int_{t_0}^{t_1} S^T(\theta, t)Q(\theta)S(\theta, s)\delta u^0(s)dsd\theta. \tag{58}$$

Therefore Eqn (51) is valid and $H_u(t) = 0$ for all t, $t_0 \leqslant t \leqslant t_1$, with $x(t_1) = b$ as required. This shows that the incorporation of the second-order weighting matrix $W(t, s)$ allows one-step convergence when the performance index J is quadratic, and the system equations are linear time-varying with a terminal constraint of the form $x(t_1) = b$.

5. Numerical Results

To illustrate the theory presented in this paper, numerical results for two examples are presented in this section.

First, given the third-order linear system

$$\begin{aligned}
\dot{x}_1(t) &= x_2(t), & x_1(0) &= 1\cdot0 \\
\dot{x}_2(t) &= x_3(t) - x_2(t), & x_2(0) &= 0 \\
\dot{x}_3(t) &= u(t) - x_3(t), & x_3(0) &= 0
\end{aligned} \right\} \tag{59}$$

it is required to minimize the quadratic performance index

$$J = \int_0^{2.5} [x^2_1(t) + u^2(t)]dt \tag{60}$$

subject to

$$x_1(2\cdot5) = 0, \quad x_2(2\cdot5) = 0, \quad x_3(2\cdot5) = 0. \tag{61}$$

The system being linear with a quadratic performance index and given terminal conditions, only one iteration should be needed for convergence from any starting point.

In the computer solution, the duration time of the process, $t_1 - t_0 = 2\cdot5$, is subdivided into N equal intervals Δt, where the choice of N is left to the discretion of the optimizer. The integral

$$\int_0^{2.5} W(t, s)\delta u(s)ds \tag{62}$$

is approximated by a sum; setting

$$u(i\Delta t) = u(i) \quad \text{and} \quad W(i\Delta t, j\Delta t) = W(i, j)$$

the integral equation law (28) can be reduced to the solution of the following set of $(N+1)$ linear equations,

$$H_u(i) + S^T(N, i)dv + \sum_{j=0}^{N} W(i, j)h(j)\delta u(j)\Delta t = 0 \tag{63}$$

$(i = 0, 1, 2, ..., N)$ where $h(0) = h(N) = \frac{1}{2}$, and $h(j) = 1$, $j = 1, 2, ..., N-1$.

In the example solved, N was chosen to be 20, so that $\Delta t = 0\cdot125$; the initial control function $u^0(i)$ is 0, $i = 0, 1, 2, ..., 20$, and

$$p_1{}^0(20) = v_1{}^0 = 0, \quad p_2{}^0(20) = v_2{}^0 = 0, \quad p_3{}^0(20) = v_3{}^0 = 0. \tag{64}$$

The results following a single iteration are tabulated in Table I. Effectively one-step convergence is obtained within the allowable limits of the expected error due to the discretization of Eqn (28) and its approximation to the form (63).

TABLE I

t	$x_1(t)$	$x_2(t)$	$x_3(t)$	$u(t)$	$H_u(t)$
0·00	1·00	0·00	0·00	−4·19	0·01
0·25	0·99	−0·10	−0·81	−3·18	0·01
0·50	0·94	−0·31	−1·21	−2·11	0·00
0·75	0·83	−0·52	−1·29	−1·02	0·00
1·00	0·68	−0·68	−1·10	0·06	0·00
1·25	0·51	−0·73	−0·73	1·02	0·00
1·50	0·33	−0·68	−0·26	1·73	0·00
1·75	0·17	−0·53	0·22	1·99	0·00
2·00	0·07	−0·32	0·56	1·47	0·00
2·25	0·01	−0·11	0·58	0·27	0·00
2·50	0·002	−0·011	0·009	−3·89	0·00

$v_1 = 19·15$, $v_2 = -30·97$, $v_3 = 7·79$, $J = 10·60$. IBM 7094 computer execution time: 1 sec.

The error due to discretization can be minimized by either using a larger value of N or by additional iterations. For example, after a total of three iterations, the value of J became 11·08 (analytical value of $J = 11·09$), while the values of the vector v were (19·65, −32·08, 8·18) with $|H_u(i)| < 0·001$ for all i.

Consider now the nonlinear system,

$$\dot{x}_1(t) = [1 - x_2^2(t)]x_1(t) - x_2(t) + u(t), \qquad x_1(0) = 0$$
$$\dot{x}_2(t) = x_1(t), \qquad x_2(0) = 1·0 \tag{65}$$

with

$$J = \int_0^{5.0} [x_1^2(t) + x_2^2(t) + u^2(t)]dt. \tag{66}$$

This problem was solved by Merriam [13]. (The initial values of the control function $u^0(t)$ that he used are not included in his results.)

In the example here, N was chosen to be 10, so that $\Delta t = 0·50$. The initial control function $u^0(i)$ selected was based on Merriam's optimum solution, plotted as Fig. 10.9, p. 226 in [13]. Since this is a free-end point problem, the dv in Eqn (63) were set to zero with $p_1(N) = p_2(N) = 0$. The problem was also solved by using a first-order method (steepest descent of the Hamiltonian.) Table II illustrates the results.

TABLE II

| Initial Control Function | | Optimum Control Function $u^*(t)$ | | |
| | | | Integral | Steepest |
t	$u^0(t)$	Merriam	equation	descent
0·0	−0·18	−0·40	−0·38	−0·30
0·5	0·62	0·57	0·60	0·59
1·0	0·98	1·01	1·02	1·00
1·5	0·98	0·99	1·00	1·00
2·0	0·72	0·76	0·75	0·74
2·5	0·48	0·46	0·46	0·47
3·0	0·32	0·23	0·23	0·25
3·5	0·26	0·09	0·08	0·14
4·0	0·17	0·00	−0·002	0·06
4·5	0·05	−0·02	−0·02	−0·001
5·0	0·00	0·00	0·00	0·00
No. of iterations		—	2	46
IBM 7094 computation time (sec)		—	1·30	7·05
Value of J		2·8669	2·8678	2·8744
$\sum_{i=0}^{10} |H_u(i)|$		—	0·40	0·97

The most obvious point to observe is that the steepest descent procedure became intolerably slow in the vicinity of the optimum solution. In this example, forty-six iterations were required to produce an acceptable solution. With the integral equation formulation, however, only two iterations were needed; moreover, the value of J obtained is 2·8678 compared to 2·8744 with the first-order procedure (Merriam's value is 2·8669). In addition, the quantity

$$\sum_{i=0}^{10} |H_u(i)|$$

is calculated to be 0·40 for the second-order method, and 0·97 for steepest descent. This indicates that the integral equation solution is more accurate. The time per iteration for the integral equation was 0·65 sec (IBM 7094) compared to 0·18 sec for steepest descent.

Table III provides another way of assessing the results obtained for Merriam's example. This is done by introducing an equivalent diagonal matrix Y_{EQ} where any diagonal element i is by definition

$$Y_{EQ}(i, i) = -\delta u(i)/kH_u(i) \quad (i = 0, 1, 2, ..., N). \tag{67}$$

If Table III is studied, it is seen that the entries of the matrix Y_{EQ} provide a good indication of the performance that results from the application of these two methods.

In the case of steepest descent, the control law used is

$$\delta u(i) = -kH_u(i) \qquad (i = 0, 1, 2, ..., N)$$

and hence $Y_{EQ} = I$, the unit matrix for *every* iteration, at every instant of time. Consequently, the value of k itself must be such that the solution converges successfully. This leads to some guessing procedure for a suitable value of k.

In the integral equation formulation, the value of any ith element of Y_{EQ} is the ratio $\{-\delta u(i)/H_u(i)\}$ $(k = 1)$, where $\delta u(i)$ is obtained by solving Eqn (63) with $dv = 0$. It is seen that some entries of Y_{EQ} are negative (although $H_{uu}(i) = 2\cdot0$ for all i).

TABLE III
$Y_{EQ}(i, i)$

i	0	1	2	3	4	5	6	7	8	9	10
Steepest descent	1·0	1·0	1·0	1·0	1·0	1·0	1·0	1·0	1·0	1·0	1·0
Integral equation	0·41	0·13	−0·43	−0·08	−0·22	0·08	0·19	0·26	0·27	0·25	0·0

The actual values of Y_{EQ} are indeed indicative of the advantages gained by using not only the second partial derivatives of the Hamiltonian with respect to x and u, but especially by incorporating the integral equation into the control scheme. This allows the choice for the value of δu to be made not only with respect to present time, but more important with respect to *future* time, so that the dynamic interplay between x and u which affects the Hamiltonian (which, in turn, includes the performance index) is not neglected.

The matrix $W(i, j)$ must be positive definite for the application of the integral equation scheme of this paper to be valid. Before applying the method, a check on the necessary condition of positive definiteness can be made by evaluating the eigenvalues of $W(i, j)$.

6. Summary and Recommendations for Further Study

A new approach to the solution of optimal control problems with specified terminal conditions has been presented. In the notation of this paper, the change in the augmented performance index $\bar{J}$ due to perturbations in the control function $u(t)$, and in the Lagrange multiplier vector v, has been approximated by a quadratic functional expansion in $\delta u(t)$ and dv. The control law is expressed by a linear matrix Fredholm integral equation and a system of linear algebraic equations which reflect the dynamics of the optimal control problem and the conditions imposed by the terminal conditions. The kernel $W(t, s)$ of this integral equation is called a second-order weighting matrix; it incorporates in a systematic manner the second partial derivatives of the Hamiltonian function H with respect to the state and control variables. While questions of convergence have not been discussed, the numerical examples demonstrate that the method works as expected—at least, for small problems. Further areas of investigation will now be outlined briefly.

Let $u(t)$, $v(t)$ be any two control vectors. The real-valued function

$$(u, v) = \int_{t_0}^{t_1} \int_{t_0}^{t_1} u^T(t) W(t, s) v(s) \, ds \, dt$$

is clearly bilinear in u and v. Suppose (u, u) is positive definite, that is $(u, u) \geqslant 0$, with equality holding if and only if $u = 0$. Then $||u|| = (u, u)^{1/2}$ defines a norm in control space and the length of the vector u can be defined to be $||u||$. By means of this norm, a "natural" metric $D(u, v)$ can be introduced in control space. The distance $D(u, v)$ between the points u and v is defined to be the length of the vector $u-v$, that is $D(u, v) = ||u-v||$, so that

$$[D(u, v)]^2 = \int_{t_0}^{t_1} \int_{t_0}^{t_1} [u(t)-v(t)]^T W(t, s) [u(s)-v(s)] \, ds \, dt. \qquad (68)$$

While the components of a control vector may have different physical dimensions, it can be seen from the structure of $W(t, s)$ that the individual terms under the integral sign in Eqn (68) make sense physically and that $D(u, v)$ has the dimensions of J. This property justifies the use of the term "natural".

The positive definiteness of (u, u) and the invertibility of the operator $\mathscr{E}$ defined in Eqn (29) are clearly related properties. With a metric defined in control space, it is possible to take a different although related point of view to that taken in this paper, as will now be illustrated by means of the free-end point problem defined by Eqns (1) and (2).

Rather than maximizing the quadratic approximation to the change in the augmented performance index at each iteration, one can determine the direction δu in control space in which $J(u)$ is decreasing most rapidly with respect to the natural metric (Eqn 68). This direction is obtained by solving the following problem.

Minimize

$$\int_{t_0}^{t_1} \delta u^T H_u \, dt$$

subject to the constraint

$$[D(u, u+\delta u)]^2 = \int_{t_0}^{t_1} \int_{t_0}^{t_1} \delta u^T(t) W(t, s) \delta u(s) ds dt = 1.$$

The solution δu must satisfy the matrix Fredholm integral equation

$$H_u(t) + \int_{t_0}^{t_1} W(t, s) \delta u(s) ds = 0$$

that is the δu obtained using the steepest descent procedure is exactly the same as the δu obtained using the quadratic approximation method of this paper. However there is a difference in interpretation. In the quadratic approximation approach, if u is the current control vector, then $u+\delta u$ is the new control vector. In the steepest descent approach the new control vector is $u+\alpha\delta u$, where α is a positive scalar. The step parameter α may be chosen in various ways, for example, so as to minimize $J(u+\alpha\delta u)$. If a fixed end-point problem is being considered, it would be desirable as well to impose conditions on α that would ensure that the trajectory corresponding to $u+\alpha\delta u$ comes closer to satisfying the end-point constraint than did the trajectory corresponding to u.

First-order gradient methods in control space imply the choice of a metric, namely the "euclidean" metric where $w_{ij}(t, s)$ the elements of $W(t, s)$, have the following properties for $i = 1, 2, ..., m$, and $j = i, 2, ..., m$:

$$w_{ij}(t, s) = 0 \quad \text{for all } (t, s) \text{ if } i \neq j$$

and
$$w_{ii}(t, s) = \begin{cases} 1 & \text{if } t = s \\ 0 & \text{if } t \neq s \end{cases}$$

This metric does not allow for the interactions between the components of the control vector and indeed physically it may not always make sense except with a *scalar* control function. For many physical problems the natural metric is non-euclidean [11, 14]. One of the results of this paper is that a natural metric for the control space of the class of optimal control problems considered is given by Eqn (68).

Crockett and Chernoff have already investigated steepest descent procedures for calculus minimization problems assuming a general metric in n-space [15]. They have shown that the "best" metric is that in which the metric tensor is the matrix of second derivatives evaluated at the optimum point which is, of course, not known. This suggests using a variable metric in which the metric tensor is the matrix of second derivatives. Applied to descent procedures, this implies that the current metric tensor for determining the next direction of descent should be taken as the matrix of second derivatives evaluated at the point just reached in the descent procedure. Continually updating and inverting the matrix of second derivatives is, of course, a computational problem of some magnitude. The various algorithms derived by Davidon [16] and Fletcher and Powell [17] use this as their point of departure.

A useful computational investigation, then, would be to examine the problems of updating the weighting matrix $W(t, s)$ and of inverting $\mathscr{E}$—possibly in some manner analogous to that used by Fletcher and Powell in their modification of Davidon's work. For example, it is well known that initially the greatest rate of reduction in the performance index is obtained by steepest-descent methods. Thus, at first, $W(t, s)$ can be taken as the euclidean metric defined above. As the solution approaches its optimum value, and second-order effects dominate, this matrix is updated by adding terms H_{uu}, H_{ux}, H_{xx}, and so on, in the form suggested by $W(t, s)$. Such a procedure becomes a *true* combination of first-order and second-order descent in control function space. Limited numerical experience with this hybrid method by the authors has shown that it combines both the stability afforded by initial use of first-order descent and the quadratic terminal convergence given by second-order descent. Furthermore, information regarding the maximum possible rate of convergence can be obtained by studying the eigenvalues of $W(t, s)$; Crockett and Chernoff have derived results for the classical calculus minimization problem [15].

Returning now to the procedures of this paper, the expansion of the augmented performance index $\bar{J}$ as a quadratic functional explicitly in $\delta u(t)$ and dv permits the method to be modified to take into account inequality constraints on the control and state variables. At present, a common weakness in all other second variation techniques [5, 7–9, 13] is their inability to incorporate directly inequality constraints on the control functions as these are eliminated by substitution. It is hoped that in a subsequent paper the control law, in the presence of inequality control constraints, can be shown to be expressed in the form of a nonlinear Hammerstein integral equation. Note that in the absence of control constraints, the integral equation defining the control law is linear.

Another possibility for handling constraints is to proceed directly from Eqn (20), and, with the addition of the appropriate constraining equations, to apply known quadratic programming algorithms. These suggestions deserve further investigation as they will combine the ideas of nonlinear integral equations, nonlinear programming techniques and optimal control theory [18].

In closing, some limitations of the method described in this paper will be discussed. Firstly, it may be numerically tedious to derive the transition matrix $\Phi(t, s)$, which is necessary in order to obtain the kernel $W(t, s)$ in the linear integral equation. To reduce some of the numerical computations, it might be useful to apply a simplified transition matrix evaluation proposed by White [19].

Secondly, the scheme relies on a linearized perturbation theory. The $u(t)$ program is changed by an amount $\delta u(t)$ which is sufficiently small for the linearization of the equations describing small perturbations about the nominal trajectory to have acceptable validity. There is a danger then that $\delta u(t)$ may be too large for the linearization to be valid, and the iterations may diverge. It may be possible to prevent this by specifying an upper bound on the quantity

$$\int_{t_0}^{t_1} \delta u^T(t) A \delta u(t) dt$$

where A is some arbitrary positive definite quantity. This drawback, however, is common to all optimization procedures when applied to solving nonlinear problems.

Acknowledgements

The authors would like to acknowledge the financial support by the National Research Council of Canada during the course of this work. George S. Tracz was the grateful recipient of a National Research Council Studentship during the years 1963-66; and he is indebted to Dr W. F. Denham, Analytical Mechanics Associates, New York, for an enlightening discussion and helpful comments. Both authors also wish to thank Mr G. Horne for obtaining the numerical results in Section 5.

References

1. L. S. Pontryagin, V. G. Boltyanskii, R. V. Gamkrelidze and E. F. Mishchenko (1962). "The Mathematical Theory of Optimal Processes". John Wiley, New York.
2. R. Bellman and S. E. Dreyfus (1962). "Applied Dynamic Programming". Princeton University Press, New Jersey.
3. G. M. Kranc and P. E. Sarachik (1963). An application of functional analysis to the optimal control problem. *J. bas. Engng* **85**, 143-150.
4. B. Paiewonsky (1965). Optimal control: a review of theory and practice. *Am. Inst. Aeron. Astron.* **3**, (11), 1985-2006.
5. J. V. Breakwell, J. L. Speyer and A. E. Bryson (1963). Optimization and control of nonlinear systems using the second variation. *J. Soc. ind. appl. Math. Control* **1**, (2), 193-223.
6. J. V. Breakwell (1959). The optimization of trajectories. *J. Soc. ind. appl. Math.* **7**, (2), 215-247.

7. H. J. Kelley, R. E. Kopp and H. G. Moyer (1963). "A Trajectory Optimization Technique Based upon the Theory of the Second Variation". AIAA Astrodynamics Conference, Yale University, Paper No. 63-415.

8. S. R. McReynolds and A. E. Bryson (1965). "A Successive Sweep Method for Solving Optimal Programming Problems". Proceedings Joint Automatic Control Conference, pp. 551-555, Troy, New York.

9. S. K. Mitter (1966). Successive approximation methods for the solution of optimal control problems. *Automatica* **3**, 135-149.

10. A. Erdelyi (1961). From delta functions to distributions. *In* "Modern Mathematics for the Engineer". (E. F. Beckenbach, ed.), McGraw Hill, New York.

11. H. J. Kelley (1962). Method of gradients. *In* "Optimization Techniques". (G. Leitmann, ed.), pp. 205-254. Academic Press, New York.

12. A. E. Bryson and W. F. Denham (1962). A steepest-ascent method for solving optimum programming problems. *J. appl. Mech.* **29**, 247-257.

13. C. W. Merriam III (1965). Direct computational methods for feedback control optimization. *Inf. Control* **8**, (2), 215-232.

14. M. M. Denn and R. Aris (1965). Green's functions and optimal systems: the gradient direction in decision space. *Ind. Engng. Chem. Fundam.*, **4**, (2), 213-222.

15. J. B. Crockett and H. Chernoff (1955). Gradient methods of maximization. *Pacif. J. Math.* **5**, 33-50.

16. W. C. Davidon (1959). "Variable Metric Method for Minimization". Argonne National Lab., Rept. No. ANL-5990, (Available from Office of Technical Services).

17. R. Fletcher and M. J. D. Powell (1963). A rapidly convergent descent method for minimization. *Comput. J.* 165-168.

18. Proceedings of the First International Conference on Programming and Control, U.S. Air Force Academy, Colorado, April 15-16, 1965. *J. Soc. ind. appl. Math. Control* **4**, (1), 1-245.

19. J. White (1966). "Simplified Calculation of Transition Matrices for Optimal Navigation". NASA Ames Research Centre, Moffett Field, California, Report TN D-3446, p. 30.

Discussion

FLETCHER. I would like to ask Dr Tracz a question. You mention at the end of your paper that some people had tried using conjugate gradients to try to solve control problems rather than the metric method you describe. Would there be an improvement in using your method as against the conjugate gradient method?

TRACZ. I would expect that timewise there would not be much improvement but iterationwise there would be. The metric method requires more time because of the need to invert a matrix: one would only do this say every five iterations because it is too costly to invert.

FLETCHER. Am I correct in assuming that the order of the problem is the number of intervals into which one derives the time scale?

TRACZ. Yes, this might typically be a hundred points. You may note that the method lends itself to quadratic programming. In the δJ expression the double integration is replaced by a double summation, thus translating the problem from being one of optimal control to one of nonlinear programming.

11. Computerized Hill Climbing Game for Teaching and Research

L. P. HYVARINEN AND G. M. WEINBERG

IBM European Systems Research Institute, Geneva, Switzerland

1. Introduction

The systematic trial and error method, Hill Climbing, is best applicable for optimization in two main types of problems.

Type 1. When the analytic relationship of the independent (manipulable) variables and the objective function is not known but the value of the objective function (response) can be evaluated at individual points by experiments. This is often the case in the optimization of industrial processes.

Type 2. When the analytic form of the objective function is known but there is no finite algorithm for obtaining the extreme value(s) in a closed form. Most nonlinear programming problems fall in this category.

The main shortcomings of the method are that the number of points (experiments) increases very rapidly with the number of independent variables, and that there is no criterion to tell whether the obtained extremum is a global optimum.

The efficiency of a Hill Climbing stategy, relative to another, is defined in terms of a suitable cost function such as the number of points or experiments required for localizing the optimum within a specified range.

2. Hill Climbing Program

The computer program described in this paper was designed primarily as an educational tool for teaching Hill Climbing techniques; it gives the student an opportunity of applying and comparing different methods. It is now being used in connection with a course on industrial process modeling.

The principle of the program is as follows. A suitable objective function $y(x, z)$ of two independent variables is chosen by the instructor, unknown to the students. A student can interrogate the program by introducing the coordinates (x_i, z_i) of one or more points on punched cards, then he receives

171

as an answer the respective values $y_i = y(x_i, z_i)$ of the objective function. After analysing the outcomes of this and previous series he enters new point(s) according to the applied technique until it seems evident that the optimum has been reached.

The program records the cumulative cost status of each student (team), and this is printed in the output together with x_i, z_i and y_i.

3. Contest

In order to give an added flavor to this exercise it was introduced in the form of a Hill Climbing Contest. The formal rules of the contest are given in Appendix A. Appendix B contains a few sample outputs of the program. The first line gives, in addition to the team identification, the run number, that is the sequence number of the series of points run simultaneously. This is followed by x_i, z_i, y_i, sequence number i of the point and the cumulative cost or penalty. A row of asterisks separates a point from the next. Note: x, z and y are given in the floating point format, for example $7{\cdot}99999E-01 = 0{\cdot}8$.

4. Penalties

The penalty P_i per point (see Appendix A) in the present form of the program is not constant. It has been made proportional to $(Y-y_i)$, where Y is a constant greater than the optimum response y^*.

$$P_i = b(Y-y_i) = a-by_i$$

b is the constant coefficient of proportionality and $a = bY$. With this penalty structure, points in the neighborhood of the optimum are less expensive than far from it. The student is given at the outset of the contest the values of a and b. Thus, $Y = a/b$ gives him an upper bound of the optimum y^*.

A student terminates the climb by introducing his terminal coordinates (x_T, z_T) assumed to be closest to the optimum. The deviation from the true optimum generates a terminal penalty proportional to the square of the deviation (y^*-y_T) where $y_T = y(x_T, z_T)$. The total cost is evaluated as the cumulative penalty ΣP_i plus the terminal penalty. The winner of the game is the team with the lowest total cost.

5. Secrecy

To protect against unauthorized use of the program (obtaining points outside the penalty record) each team has a secret pass word punched on the first card to identify itself to the program. Respective files are opened at the outset of the contest by the instructor. An attempt to use the program with a wrong pass word is reported by the program to the instructor. When a team terminates the climb its file is closed automatically and the team has no longer an access to the program if not re-opened by the instructor.

6. Contour Map

The program includes, in addition, a routine for printing the contour map of the objective function (see Appendix D). This gives the student an opportunity to trace afterwards his climb on the map and evaluate his progress and errors. Figures 4-7 have been drawn from computer printed contour maps.

7. Modifications

Modifications and extensions are easy to make within the present framework of the program. The penalty structure is easily modified by changing the form of the function P_i and the same applies to the terminal penalty.

A possible change suggested for the P_i is to introduce a penalty contribution which is a monotonically increasing function of the distance of successive points x_i, z_i to x_{i+1}, z_{i+1}. This would be more appropriate in the case of optimization of a physical process, where great changes in the manipulable variables upset the process and require long transient periods to settle in a steady state.

Another obvious modification is to increase the number of independent variables into a multidimensional Hill Climbing problem.

A modification already implemented in the program is a random noise component superimposed on the objective function. That is, the obtained responses y_i are no longer accurate but subject to a random error of a given magnitude. The noise can be suppressed by defining the amplitude equal to zero.

8. Evaluation of Climbing Strategies

The present program can be used, in addition to the educational purposes, for testing different techniques of the search for the optimum. In general all such techniques can be divided into three separate stages or strategies.

1. Initial exploration.
2. Climbing.
3. Final adjustment.

The first stage consists of a thin coverage of the accessible (constrained) domain to locate an advantageous starting point for the climb and to get an overall survey of the terrain. The final adjustment is normally done by a polynomial fitting in the neighborhood of the optimum.

The greatest variety of different strategies have been developed for the middle game, climbing such as the gradient ascent, the contour tangent, and the partan methods.

Climbing strategies are based on sequential experiments so that the choice of the next point, say (x_{i+1}, z_{i+1}), is based on the outcomes $y_1, y_2, \dots y_i$ of the preceding experiments and their coordinates. Hence, if we are using all the information available in the previous points for the choice of the next point, the strategy cannot be periodic in nature, that is it cannot be based on the recursive application of a program loop or routine.

It will be understood, however, that the most recent points have the strongest influence in this decision and, therefore, a sufficient strategy can be based on not more than, say, the n most recent experiments, where n, the order of the strategy, is a finite number. For example, a two-dimensional gradient ascent method uses, in addition to the fixed step size, only the values of the three most recent points being, consequently, of order three.

It is conceivable that a strategy of a higher order can be designed to be more efficient than one of a lower order since it makes more use of the available information. In order to design an algorithm for an efficient optimum seeking method it would be desirable to find logical rules of decision of a high enough order n for the coordinates of the next experiment.

9. Triangulation

The strategy, called triangulation, described in this section is an effort towards such an algorithm. The strategy is of the order seven utilizing a minimum of four points; it is based on a hexagonal system of coordinates in a two-dimensional space.

The decisions are not based on the exact magnitudes of the y_i but only on their relative **magnitudes**. This approach also tends to make the method insensitive to noise to a certain extent, that is inaccuracies of y_i. The logical rules of decision in this method are given by the block diagram of Appendix C.

The triangular basic pattern of four points y_0, y_1, y_2, y_3 is more symmetrical than, for example, the three points placed at right angles in the gradient method, and they are thus surveying all directions more efficiently.

The method is best understood by following the case examples in Appendix D step by step observing the sequence numbers and the logic of the diagram in Appendix C. The examples show that relatively few points are needed in areas where the slope is approximately linear. At points of irregularities such as ridges the basic pattern develops itself into a complete hexagon.

Note that also the step size is automatically decreased when a hexagonal pattern is encountered with the highest point at the center, which gives a strong indication of an optimum inside the hexagon. How to determine the initial step size and when to terminate the climb are not included in the logic.

Preliminary experiments run on the present program for a number of different types of objective functions indicate that the algorithm will converge more reliably to the optimum than the gradient ascent or the contour tangent method, even with relatively complex surfaces.

The same basic approach could evidently be extended to the three-dimensional case based on a tetragonal basic pattern; however, the decision rules for this case have not been worked out.

APPENDIX A

1. The Hill Climbing Contest

1.1. *Objective*

The objective of the contest is to find the *highest point* y^* of the response surface $y = y(x, z)$ in the two-dimensional domain $0 \leqslant x \leqslant 1, 0 \leqslant z \leqslant 1$, at the lowest possible cost.

1.2. *Rules*

(1) The optimum is sought by performing experiments on the system. Denote the outcome of the ith experiment at point $x_i z_i$ by $y_i = y(x_i, z_i)$.

(2) Simultaneous and sequential experiments are allowed.

(3) The participants are obliged to keep their own cost status, the number of experiments performed, and the outcomes of their experiments to themselves. Pooling of information between teams is considered cheating.

(4) The participant has full freedom in selecting the points (x_i, z_i) within the accessible domain using a maximum of d decimal places.

(5) The response y_i is free of error up to f significant digits.

(6) All conceivable mathematical and other methods are allowed in the search for the optimum (except cheating). A reasonable amount of computer time is permitted for the analysis of the gathered information.

(7) Each experiment i is associated with a penalty (cost) P_i determined by $P_i = a - b y_i$; $a > b y^*$.

(8) The participant may terminate the contest after any experiment i by submitting the terminal coordinates (x_T, z_T), which need not necessarily be any of the points of previous experiments. After termination the participant no longer has an access to the program.

(9) The total number of experiments is limited to a maximum of N $(i \leqslant N)$.

(10) The terminal cost is computed by the formula

$$C_T = c[y^* - y_T]^2$$

where $y_T = y(x_T, z_T)$ and y^* is the true optimum.

(11) At the start of the contest each participant will be given the values of the parameters a, b, c, d, f and N, and possibly some information on the general properties of the response surface.

(12) The winner (the Hill Climbing Champion of the class) will be the participant or the team with the lowest total cost

$$C = C_T + \sum_i P_i.$$

APPENDIX B

```
TEAM NUMBER   1          RUN NUMBER   1
         HYVARINEN'S DOWNFALL


X= 9.99999E-02          Z= 2.99999E-01          Y= 3.97186E+00;
TEAM.POINTS=          1  TEAM.PENALTY=              7.028;
* * * * * * * * * * * * * * * * * * * * * * * * * * * * * * * *

X= 3.99999E-01          Z= 7.99999E-01          Y= 4.59339E-01;
TEAM.POINTS=          2  TEAM.PENALTY=             17.568;
* * * * * * * * * * * * * * * * * * * * * * * * * * * * * * * *

X= 7.99999E-01          Z= 5.49999E-01          Y= 5.06437E-01;
TEAM.POINTS=          3  TEAM.PENALTY=             28.061;
* * * * * * * * * * * * * * * * * * * * * * * * * * * * * * * *

X= 7.50000E-01          Z= 1.49999E-01          Y= 4.44264E-01;
TEAM.POINTS=          4  TEAM.PENALTY=             38.616;
* * * * * * * * * * * * * * * * * * * * * * * * * * * * * * * *

X= 5.77999E-01          Z= 4.70999E-01          Y= 1.63000E+00;
TEAM.POINTS=          5  TEAM.PENALTY=             47.985;
* * * * * * * * * * * * * * * * * * * * * * * * * * * * * * * *

END OF JOB


TEAM NUMBER   1          RUN NUMBER   2
         HYVARINEN'S DOWNFALL


X= 3.49999E-01          Z= 2.41999E-01          Y= 3.14783E+00;
TEAM.POINTS=          6  TEAM.PENALTY=             55.837;
* * * * * * * * * * * * * * * * * * * * * * * * * * * * * * * *

X= 2.30999E-01          Z= 5.20999E-01          Y= 1.84022E+00;
TEAM.POINTS=          7  TEAM.PENALTY=             64.996;
* * * * * * * * * * * * * * * * * * * * * * * * * * * * * * * *

END OF JOB



         TEAM NUMBER   1          RUN NUMBER   9
                  HYVARINEN'S DOWNFALL


    X= 2.19999E-01          Z= 3.09999E-01          Y= 9.60518E+00
    TEAM.POINTS=         19  TEAM.PENALTY=          118.489;
    * * * * * * * * * * * * * * * * * * * * * * * * * * * * * *
    TEAM.PENALTY=          118.886;
    THIS IS THE FINAL WORD

         END OF JOB
```

APPENDIX C

Hill Climbing by Triangulation

(1) Define an appropriate step size r and the starting point (x_0, z_0) with $y_0 = y(x_0, z_0)$.

(2) Place the next three points along the circle of radius r about the point y_0 symmetrically at angles of 120°; the "phase" of these points is arbitrary, in principle. If no other preferences, place the first along axis x. This pattern of points is called the triangulation.

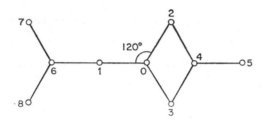

(3) The four points in a triangulation are referred to as y_0, y_1, y_2, y_3 so that apart from the center $y_0, y_1 \geqslant y_2 \geqslant y_3 \geqslant y_4$.

(4) According to the rules given in the diagram, a triangulation can be followed by a new triangulation of the two basic types above. The triangulation with points y_4, y_2, y_5, y_3 is said to be "about point y_4, using points y_2 and y_3 (of the starting triangulation)". The triangulation with points y_6, y_1, y_7, y_8 is said to be "about point y_6, using point y_1". This is the terminology to define a triangulation uniquely.

(5) The decision rules for placing the experiments are given in the block diagrams on the next pages (Figs. 1-3.)

Comments (refer to numbers on flow charts):

(1) A maximum, a saddle point or a ridge in the range r, of exploration.

(2) A maximum in the range of exploration.

(3) If center point is lowest, a minimum in range. Response surface can be bimodal.

(4) Possibility of a saddle point and a bimodal response surface.

(5) A curved ridge in range of exploration.

(6) A maximum, a ridge or a saddle point in range of exploration.

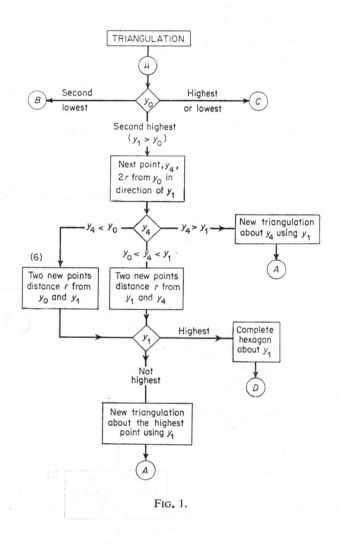

FIG. 1.

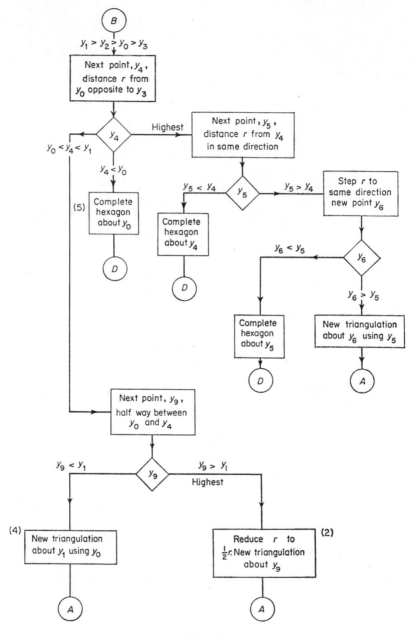

FIG. 2.

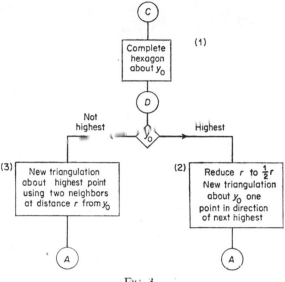

FIG. 3.

APPENDIX D

Figures 4–7 present case examples of optimum search using the triangulation, gradient ascent, and contour tangent methods.

The objective function (unknown to the experimenter) is in Figs 4 and 5.

$$y(x,z) = \frac{9 \cdot 72}{1 + f_1(x,z) + f_2(x,z)}$$

where

$$f_1(x,z) = |8 \cdot 3x_1{}^2 - 8 \cdot 9x_1 + 2 \cdot 3 - z_1|$$

$$f_2(x,z) = \ln\{1 + |x_1 + 1 \cdot 45z_1{}^2 + 0 \cdot 268z_1 - 0 \cdot 85|\}$$

and

$$x_1 = 0 \cdot 8x \cos \alpha + (z - 0 \cdot 2) \sin \alpha$$

$$z_1 = -0 \cdot 8x \sin \alpha + (z - 0 \cdot 2) \cos \alpha$$

$$\tan \alpha = 0 \cdot 7.$$

On Figs 6 and 7 the response is

$$y(x,z) = [g_1(x,z) + g_2(x,z)](1 + 0 \cdot 1 R)$$

where

$$g_1(x,z) = \frac{0 \cdot 091}{0 \cdot 01 + (x - 0 \cdot 22)^2 + (z - 0 \cdot 31)^2}$$

$$g_2(x,z) = \frac{0.07}{0.01 + |x - 0.6| + |z - 0.5|}$$

R = pseudorandom number uniformly distributed between $-\frac{1}{2}$ and $+\frac{1}{2}$.

The origin $x = 0$, $z = 0$, is the upper left-hand corner. The vertical axis is x and the horizontal axis z.

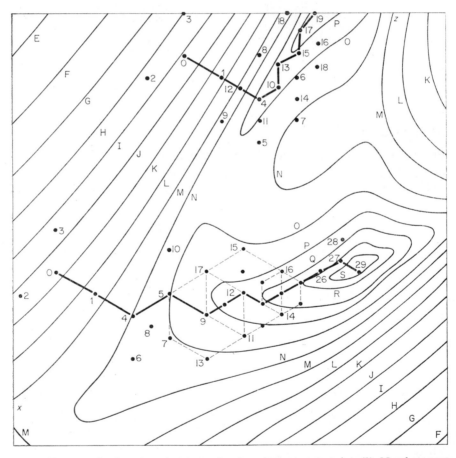

FIG. 4. Two searches by triangulation starting from two separate points (0). Numbers next to points indicate sequence of search. Ascending sequence of letters indicates ascending responses.

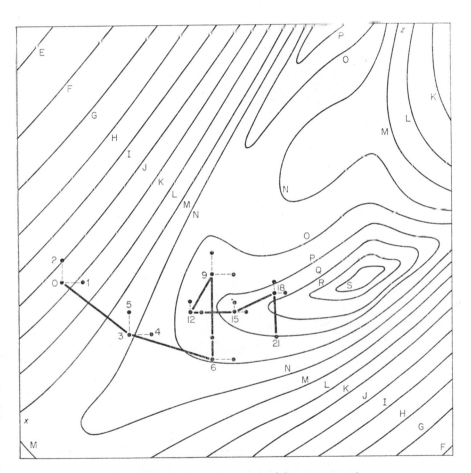

FIG. 5. Search by gradient method (steepest ascent).

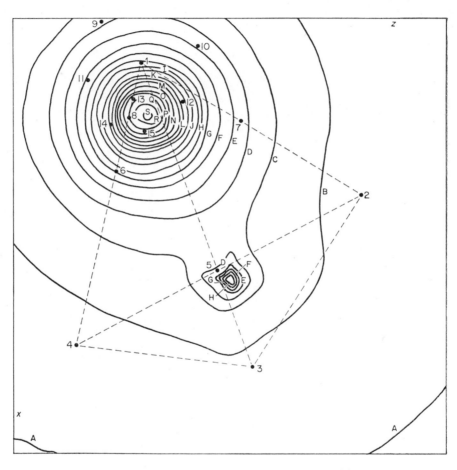

FIG. 6. Search by triangulation. Seven first points represent initial exploration.

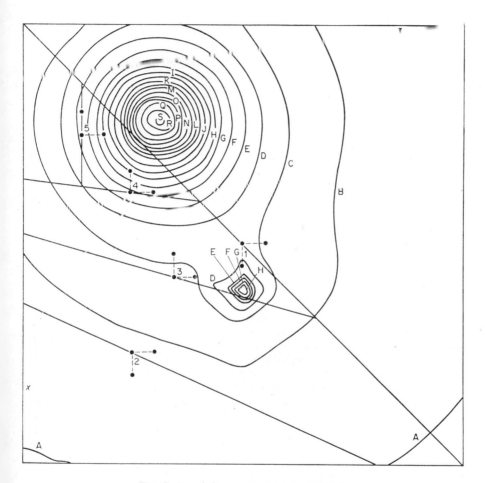

FIG. 7. Search by contour tangent method.

Discussion

VAJDA. Does each player use the same function and if so is any player allowed to see how the other is getting on?

HYVARINEN. Yes, each player or team uses the same function, but has to keep the information to himself: to do otherwise would be considered as cheating.

VAJDA. Does any player know when he has reached the maximum?

HYVARINEN. No, he is only given an upper bound on this.

VAJDA. I cannot see how this device could be used to test methods which assume that more is known about the function at each approximation, for instance the directions of steepest ascent and descent.

HYVARINEN. It could be arranged that this information be given to the players.

BEALE. Is the game restricted to functions of two variables?

HYVARINEN. At the moment, yes. However it would be possible to extend the technique of triangularization to three dimensions.

BEALE. I asked the question because it is often possible to solve two-dimensional optimization problems by methods based on *ad hoc* rules about the way things look. These do work fairly well in two dimensions: it is only when we get up to higher dimensional problems that the real difficulties set in and these methods do not work too well.

ANON. I would like to ask about step sizes in triangularization. Are these initially arbitrary and are they ever increased?

HYVARINEN. The step sizes are initially arbitrary and although there are cases in which the step size is decreased there are none in which it would be increased.

ANON. That is, if you entered a region in which smaller steps were required and then left it again, the smaller step length would continue?

HYVARINEN. Yes. Incidentally I would like to make a comment about the angles in this method. I said these were placed symmetrically at 120°, but in many cases these angles were not unique because of different possible scalings of the variables. Similarly the performance of the method will be affected by change of scale, or by any affine transformation. I wonder whether this fact has been considered sufficiently in many of the other optimizing schemes discussed here.

ANON. This triangularization scheme seems very similar to the simplex method designed for optimizing on a response surface, which is based on a triangle of three points and which can be generalized very readily in many dimensions. It seems that the former scheme will not generalize nearly as readily because of the complicated order relationships involved.

HYVARINEN. Yes, to some extent I agree. However one might argue that a method will be more successful, according to the number of points it takes into account when determining the next choice. For instance to obtain the gradient would require three points; in triangularization, up to seven points can be used in determining the next trial. Because of this I would expect the method to be more efficient. Of course you are only comparing function values and not looking at their exact magnitudes, but the effect of this is to make the method less sensitive to inaccuracies in the function.

12. Review of Constrained Optimization

D. Davies and W. H. Swann

I.C.I. Ltd, Wilmslow, England

1. Introduction

A paper by Fletcher presented at the conference (Chapter 1, this Volume) has considered methods for optimizing an unconstrained function $f(x_1, x_2, x_3, ..., x_n)$, and this paper extends his treatment of the problem to cover the constrained case where the function is subject to constraints of the form

$$c_i(\mathbf{x}) \geqslant 0 \qquad (i = 1, 2, ..., m)$$

$$e_j(\mathbf{x}) = 0 \qquad (j = 1, 2, ..., p).$$

In general, the function $f(\mathbf{x})$, $c_i(\mathbf{x})$ and $e_j(\mathbf{x})$ will all be nonlinear and since maximization and minimization are equivalent problems, only the latter need be considered.

Other papers in the conference have discussed methods that approximate the nonlinear functions by linear ones and solve the resulting problem by linear programming. At the optimum of the linearized function, a new approximation is made and the process repeated. Perhaps the best known and most efficient of these methods is the generalization of Wolfe's reduced gradient method due to Abadie and Carpentier [1].

In the discussion of the various approaches in this paper, the opinions given on the value of methods are mainly from personal experience and are based on reliability (that is success on a wide range of problems), reasonable computer usage and minimal entry to the model for function and constraints. This latter criterion is useful since the iteration times of different methods are usually small in comparison to the time taken to evaluate the model for a practical industrial problem.

2. Transformations

We begin by considering the rather obvious remark that if constraints can be removed, this should be done. For example, with an equality constraint $e(\mathbf{x}) = 0$, a rearrangement to give

$$x_k = E(x_1, x_2, ..., x_n) \tag{1}$$

will allow the optimization to be performed in variables $x_1, x_2, ..., x_{(k-1)}$, $x_{(k+1)} ..., x_n$ and x_k would be determined from Eqn (1) either directly or by iteration.

Box [2] has suggested that transformations be made to the variables in order to remove constraints. He argues that even though the model may be more complicated in terms of transformed variables, one does get a better defined minimum and one of the powerful unconstrained methods should obtain a solution.

Suppose that there is a constraint $x \geqslant c$; then this could be handled by defining a new variable y by

$$x = c+y^2 \text{ or } x = c+e^y. \tag{2}$$

Similarly, $l \leqslant x \leqslant u$ can be represented by a variable y defined by

$$x = l+(u-l) \sin^2 y \tag{3}$$

and in each case, the optimization is performed with respect to the new variable, y.

Let us consider a simple example taken from Box and co-workers [3] with constraints $x_1 \geqslant 0$, $x_2 \geqslant 0$, $x_1+x_2 \leqslant 1$. The feasible region is illustrated in Fig. 1. In polar coordinates, $x_1 = r \cos \theta$, $x_2 = r \sin \theta$ and the constraints

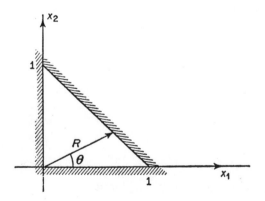

FIG. 1.

can be written as

$$0 \leqslant r \leqslant R, \qquad 0 \leqslant \theta \leqslant \tfrac{1}{2}\pi$$

where $R = 1/(\cos\theta + \sin\theta)$. Using Eqn (3), new variables y_1, y_2 may be defined by

$$r = R\sin^2 y_1, \qquad \theta = \tfrac{1}{2}\pi\sin^2 y_2$$

and the resulting optimization with respect to these variables will automatically satisfy the constraints.

Many more transformations exist other than (2) and (3) and use of them will be of value if the constraints are relatively simple. However, with many constraints, the analysis to apply the transformation is often tedious and one of the remaining techniques of this paper is preferable.

3. Classification of Methods

Unconstrained optimization methods are often based on an iteration defined by

$$x^{i+1} = x^i + h^i d^i \tag{4}$$

where d^i is a direction of search, h^i the current step length and x^i, x^{i+1} are successive points of the iteration. The search direction d^i is chosen from knowledge of function and possibly partial derivative values from previous iterations, but when a constraint is present this too plays an important part in the choice of d^i.

In Fig. 2, suppose the point B had been reached by an unconstrained optimizer. Continued use of the optimizer may give a new direction pointing towards the unconstrained optimum, M, whereas the search should be

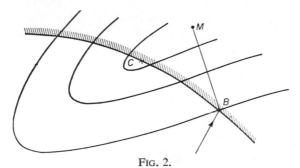

FIG. 2.

directed towards the constrained optimum, C. Thus the problem of constrained optimization is to incorporate into the search direction the appropriate information about the constraint, and on nonlinear constraints this can be very difficult and time consuming.

Many approaches have been put forward for solving the constrained problem and they mostly fall into two categories as follows.

1. *Function modification followed by unconstrained optimization.* These methods seek to define a new function that has an unconstrained optimum at the same point as the optimum of the given constrained problem. Optimization of this new function will then define the required change in the search direction.

2. *Direction modification without altering the function.* Some of these methods attempt to follow a constraint while others try to rebound from them and so continue the search in the feasible region. An example of this latter approach is given by Greenstadt [4].

The main advantage of (1) is that the constraints are virtually ignored; but there are serious disadvantages in that some methods require function values outside the feasible region (which may not be possible), the true function is not being optimized, and the modified one may exhibit behaviour such as ridges that will hinder the progress of an unconstrained technique. Thus if the optimization has to stop with an intermediate solution due to time or other considerations, the point reached may not be better than the starting value and, most likely, it will not be as good as a point obtained by a type (2) method in the same time.

Although this is an advantage of the type (2) methods, cases arise, when the constraints are highly nonlinear, which lead to many modifications of the search direction and a consequent loss of the advantage. The obvious conclusion is therefore that no one method for constrained optimization is superior to all others and care must be exercised in choosing the appropriate one for a given situation. Naturally for "one-off" problems many of the methods will give the solution and time could be unimportant, but for problems where the same function is to be optimized frequently, for example on-line computer control, the selection of the fastest optimizer could be an important consideration.

4. Lagrangian Functions

The use of Lagrangian functions and the Kuhn–Tucker theorem immediately suggest a possible method of solution, and an iterative scheme has been proposed by Arrow and Uzawa [5]. A Lagrangian function is defined as

$$\phi(\mathbf{x}, \lambda) = f(\mathbf{x}) + \sum_{i=1}^{m} \lambda_i c_i(\mathbf{x}). \tag{5}$$

This summation does not include the equality constraints but it is possible to include them by rewriting them as two inequalities to give

$$e(x_1, ..., x_n) \geqslant 0 \quad \text{and} \quad e(x_1, ..., x_n) \leqslant 0.$$

The Kuhn–Tucker theorem states that under certain conditions on the function and constraints, the problem is solved by finding a saddle point of the function $\phi(\mathbf{x}, \lambda)$ in Eqn (5), that is a minimum with respect to $\mathbf{x}$ and a maximum with respect to λ. The λ_i are taken to be zero if a constraint is satisfied and negative otherwise so that the effect on the function when a constraint is violated is to impose a penalty proportional to the amount of the violation.

The iterative scheme proposed by Arrow and Uzawa uses the following formulae:

$$\mathbf{x}^{i+1} = \mathbf{x}^i - h^i \, B_1^{-1} \, \mathbf{g}_i$$

$$\lambda^{i+1} = \min \quad [0, \lambda^i + h^i \, B_2^{-1} \, \Delta_i]$$

$$\mathbf{g}_i \quad = \left[\frac{\partial \phi}{\partial x_1}, ..., \frac{\partial \phi}{\partial x_n} \right]_{\mathbf{x} \, = \, \mathbf{x}^i}$$

$$\Delta_i \quad = [c_1(\mathbf{x}^i), ..., c_m(\mathbf{x}^i)]$$

and B_1^{-1}, B_2^{-1} are weighting matrices chosen in a manner as defined by Fletcher in his paper (Chapter 1, this Volume). Using this scheme, one possible approach would be to use Davidon's method separately for $\mathbf{x}$ and λ so that B_1^{-1} and B_2^{-1} would be estimates of the corresponding Hessian matrices.

We have tried this method but did not achieve very satisfactory results when compared to those given by some of the later methods to be described.

5. Rosenbrock's Method

One of the earliest methods in use in our Company, and a most robust one, is that due to Rosenbrock [6]. To understand the method for applying the constraints it is useful firstly to give a description of his method as applied to the unconstrained problem.

Each iteration uses a set of orthogonal directions, denoted, for iteration i, by $\xi_1^{(i)}, \xi_2^{(i)}, ..., \xi_n^{(i)}$ and associated with each of these are step-lengths to be taken along them, $viz \ d_1, d_2, ..., d_n$. The initial set of directions is usually chosen as the coordinate axes, and new directions are chosen to reflect the progress made on the last completed iteration. A "success" is defined to be a step in any of the directions such that the resulting function value is less than or equal to the previous best value. Similarly, a "failure" is a step that results in a greater function value.

An iteration consists of taking a single step along each direction in turn until a "success" followed by a "failure" has occurred on each of the directions. When a step along any direction $\xi_j^{(i)}$ succeeds, the corresponding step-length d_j is increased the next time the direction is considered. Similarly if a failure is recorded, a smaller step in the opposite direction is tried when $\xi_j^{(i)}$ is next considered. Thus a typical step on direction $\xi_j^{(i)}$ is

$$\mathbf{x}_j^{(i)} = \mathbf{x}_{j-1}^{(i)} + d_j^{(i)} \xi_j^{(i)}$$

where $\mathbf{x}_{j-1}^{(i)}$, $\mathbf{x}_j^{(i)}$ are the points obtained on iteration i along directions $\xi_{j-1}^{(i)}$ and $\xi_j^{(i)}$ respectively, and $d_j^{(i)}$ is the step-length associated with $\xi_j^{(i)}$.

If $$f(\mathbf{x}_j^{(i)}) \leqslant f(\mathbf{x}_{j-1}^{(i)})$$

then $$d_j^{(i+1)} = \alpha d_j^{(i)} \qquad \text{where } \alpha > 1$$

and $$j = \begin{cases} j+1 & j \leqslant n-1 \\ 1 & j = n \end{cases}$$

otherwise $$d_j^{(i+1)} = \beta d_j^{(i)} \qquad \text{where } \beta < 0$$

                $$\mathbf{x}_j^{(i)} = \mathbf{x}_{j-1}^{(i)}$$

and $$j = \begin{cases} j+1 & j \leqslant n-1 \\ 1 & j = n \end{cases}$$

In considering the constrained problem, it is useful to rearrange the constraints to the form

$$l_i \leqslant x_i \leqslant u_i \qquad (i = 1, 2, ..., r \geqslant n) \tag{6}$$

where the variables $x_{n+1}, ..., x_r$ are the constraints $c_i(\mathbf{x})$ suitably arranged. Where no limit u_i is given for a particular constraint, then a convenient limit should be imposed by the user. Associated with each pair of constraints as in Eqn (6), Rosenbrock defines a boundary region of "width" $10^{-4}(u_i - l_i)$ so that a variable x_i $(i = 1, 2, ..., r)$ is within its boundary region if either

$$l_i \leqslant x_i \leqslant l_i + 10^{-4}(u_i - l_i)$$

or $$u_i \geqslant x_i \geqslant u_i - 10^{-4}(u_i - l_i). \tag{7}$$

Rosenbrock's search procedure continues as described above until a variable enters its boundary region. This occurrence is taken to indicate that the minimum of the function probably lies outside the feasible region, and so the function is modified within the boundary region to create a feasible minimum. The search can then continue and further modifications are made to the function as boundary regions of other constraints are encountered.

Suppose for example that x_i lies within the boundary region as defined by Eqn. (7). Then the "depth of penetration" into the region is denoted by

$$\lambda = \frac{\text{amount of penetration}}{\text{width of boundary region}} = \frac{l_i + 10^{-4}(u_i - l_i) - x_i}{10^{-4}(u_i - l_i)}$$

and the computed function value f is replaced by f' where

$$f' = f - (f - f^*)(3\lambda - 4\lambda^2 + 2\lambda^3)$$

and f^* is the least function value obtained for which no boundary region is entered.

We can see that for $\lambda = 0, f' = f$ and at $\lambda = 1, f' = f^*$ where $f^* \geqslant f$ so that an unconstrained minimum has been created for some λ between 0 and 1.

The method has proved to be very reliable. Many variations tried by Box [7] have not given any significant improvement. Wood [8] gives a slight modification that may improve the rate of convergence and also remarks that he believes the method has received less attention in the United States than it deserves. In conclusion, it is noted that Rosenbrock's method for constrained problems works well in conjunction with his unconstrained algorithm, but with other algorithms based on linear searches (for example DSC method, Swann [9]) this has not been the case and other methods must be considered.

6. Penalty Functions

The modification to the function described by Rosenbrock is equivalent to adding a penalty term as a constraint is approached but, as mentioned above, it is not suitable for application to other methods of unconstrained optimization. Instead a more general penalty function approach may be adopted.

Consider the function

$$F(\mathbf{x}, \mathbf{k}) = f(\mathbf{x}) + \sum_{i=1}^{m} k_i [c_i(\mathbf{x})]^2$$

where the weights k_i are zero if $c_i(\mathbf{x}) \geqslant 0$ and are assigned positive values if

$c_i(\mathbf{x}) < 0$. If a constraint is violated a penalty is added to $f(\mathbf{x})$ and the penalty is larger as $c_i(\mathbf{x})$ becomes increasingly negative. If no constraints are violated, there is no penalty and the function is equivalent to $f(\mathbf{x})$. Under certain conditions, it can be shown theoretically that, as $k_i \to \infty$, an unconstrained optimum of $F(\mathbf{x}, \mathbf{k})$ will correspond to a constrained optimum of $f(\mathbf{x})$. Any unconstrained optimization method may be used, with the weights k_i being successively modified as the optimization proceeds and a convenient scheme for their computation is given by Leitmann [10].

The application of penalty functions in this way works reasonably well but there are several disadvantages with the approach. Steep valleys and discontinuous derivatives are created at the constraint boundary and these features are often difficult to overcome, particularly with gradient methods. Also the method requires the availability of function values at non-feasible points and this may not be possible in practice. Since there are other methods which do not suffer from these difficulties, this penalty function approach is not recommended as a general method.

7. Constraints with Hooke and Jeeves' Method

One of the most simple and effective direct search methods for unconstrained optimization is that due to Hooke and Jeeves [11] and many techniques, including penalty functions, have been suggested for extending it to constrained problems. In considering one or two of these, it is useful to begin with a description of the unconstrained technique.

The method uses a succession of simple moves known as *exploratory* and *pattern* moves. An exploratory move examines each variable in turn by adding an increment to it, recalculating the function value and retaining the point if the value is an improvement. If no improvement is recorded a negative step of the same size is taken from the original point, and this new point is retained if its function value is an improvement. If neither step succeeds,

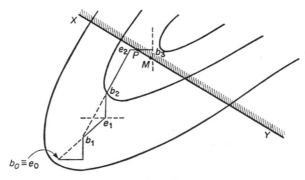

FIG. 3.

the corresponding variable is left at its original value and the next one considered. The exploratory move is terminated when all the variables have been considered. Suppose the exploratory move begins at a point e_i and ends at point b_{i+1}, then the pattern move forms a new point $e_{i+1} = 2b_{i+1} - b_i$. An illustration of the use of Hooke and Jeeves' method is given in Fig. 3 where, ignoring XY, it can be seen that the method successfully generates directions $(b_{i+1} - b_i)$ that move along the ridge formed by the function. If a constraint XY is now imposed, it will be seen that the exploratory move is unable to set up a new direction to b_3. Reduction of the step-length and new exploratory moves from e_2 would cause premature termination at some point in the neighbourhood of P whereas the constrained optimum is at point M. Hence it becomes necessary to define an alternative strategy that will select a direction pointing towards M.

One simple approach, due to Klingman and Himmelblau [12], has been termed by them the Multiple Gradient Summation Technique. The technique for one constraint is illustrated in Fig. 4 where a new search direction NSD

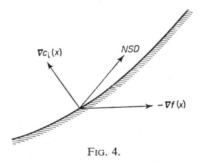

FIG. 4.

is set up by taking a linear combination of the function and constraint gradient, that is

$$NSD = \frac{\nabla c_i(\mathbf{x})}{|\nabla c_i(\mathbf{x})|} - \frac{\nabla f(\mathbf{x})}{|\nabla f(\mathbf{x})|}.$$

With several constraints a term is added for each constraint that would otherwise be violated—but the method is not very efficient in practice.

The method of Glass and Cooper [13] is a rather more sophisticated technique which calculates a new search direction from a linear programming problem obtained by linear approximation to the function and constraints. The authors have no experience of this method but Murray, in a private communication, points out that the solution of the linear program can be avoided, and that the method is slow but very reliable.

In applying Hooke and Jeeves' method, mainly with on-line computer control, the authors have found it preferable to apply the penalty function approach due to Carroll [14] and improved by Fiacco and McCormick [15]. Excellent results have been obtained using this technique with many different unconstrained optimization techniques, and use of the method is fairly widespread.

8. Carroll's Method

This method is known as the Created Response Surface Technique and is so named because it generates within the feasible region a series of surfaces whose successive unconstrained optima converge to that of the given constrained problem. To minimize $f(\mathbf{x})$ subject to m constraints $c_i(\mathbf{x}) \geqslant 0$, consider the function

$$F(\mathbf{x}, k) = f(\mathbf{x}) + k \sum_{i=1}^{m} \frac{w_i}{c_i(\mathbf{x})} \tag{8}$$

where k, w_i are positive weights. It can be seen that as any $c_i(\mathbf{x})$ tends to zero the contribution to the penalty term in Eqn (8) tends to infinity so that an unconstrained minimum has been created within the feasible region. This minimum may be obtained with any unconstrained minimization technique and the method has the advantage, unlike the previous penalty functions considered, that the function is only calculated at feasible points.

Each of the weights w_i may be chosen as zero in a well-scaled problem until the corresponding constraint is violated, in which case it is reset to unity and the minimization continued from the last feasible point. The choice of k is not so straightforward. An initial positive value k_0 is chosen and a minimum of $F(\mathbf{x}, k_0)$ is calculated at $\mathbf{x}_0$. The value of k is then reduced to $k_1 \geqslant 0$ and a minimum of $F(\mathbf{x}, k_1)$ is obtained using $\mathbf{x}_0$ as the starting approximation whence further reductions of k and repeated minimizations are performed until the required accuracy is obtained.

For the method to be efficient, successive surface optima should be good starting approximations for the next surface so that k must not be reduced too sharply; a multiplying factor between 0·02 and 0·1 has been found satisfactory by many users (for example Box *et al.* [3]). In choosing the initial value of k, this must not be too large so that the second term in $F(\mathbf{x}, k)$ dominates, and not too small so that excessive constraint violations are incurred in calculating the minimum of $F(\mathbf{x}, k_0)$. Two automatic schemes for selecting k are given by Fiacco and McCormick, and Box and co-workers give a good empirical one.

A further advantage of the method is that equality constraints may be incorporated into Eqn (8). To minimize subject to p constraints $e_j(\mathbf{x}) = 0$, the function is modified to

$$F(\mathbf{x}, k) = f(\mathbf{x}) + k \sum_{i=1}^{m} \frac{w_i}{c_i(\mathbf{x})} + k^{-\frac{1}{2}} \sum_{j=1}^{p} [e_j(\mathbf{x})]^2$$

and the technique proceeds as before. It should be noted that the initial approximation for $\mathbf{x}$ need not satisfy identically the equality constraints.

Although this form of penalty function is widely used, other authors have found it useful to consider different penalty terms. For example, Box and co-workers square $c_i(\mathbf{x})$ in Eqn (8), Frisch [16] and Parisot [17] use a logarithmic term, and Pomentale [18] has proved convergence for any general penalty term that is positive and continuous in the feasible region and becomes infinite on all constraints.

9. Linear Constraints

All methods so far considered have not made any distinction between linear constraints and nonlinear ones, but in the special case of linear constraints techniques may often be applied to handle them explicitly whilst leaving only the nonlinear constraints to be dealt with by general methods. Use of these techniques improves the performance of many optimization methods on nonlinear programming problems and we consider a modification to the DSC method (Swann [9]) for linear constraints.

The DSC method performs linear searches along n orthogonal directions which are recomputed at the end of an iteration. Suppose the current set of directions are $\boldsymbol{\xi}_1^k, \boldsymbol{\xi}_2^k, ..., \boldsymbol{\xi}_n^k$ and the iteration begins at $\mathbf{x}_0$. Searching along $\boldsymbol{\xi}_1^k$ gives $\mathbf{x}_1$, then from $\mathbf{x}_1$ along $\boldsymbol{\xi}_2^k$ gives point $\mathbf{x}_2$, and so on, and we define

$$\boldsymbol{\zeta}_1 = \mathbf{x}_n - \mathbf{x}_0$$

$$\boldsymbol{\zeta}_2 = \mathbf{x}_n - \mathbf{x}_1$$

$$\boldsymbol{\zeta}_n = \mathbf{x}_n - \mathbf{x}_{n-1}$$

Suppose also that $(i-1)$ directions $\boldsymbol{\xi}_1^{k+1}, \boldsymbol{\xi}_2^{k+1}, ..., \boldsymbol{\xi}_{i-1}^{k+1}$ have already been chosen for the $(k+1)$th iteration and that D_{i-1} is the $n \times (i-1)$ matrix formed by these vectors. Then

$$\boldsymbol{\xi}_i^{k+1} = \boldsymbol{\zeta}_i - D_{-1} \boldsymbol{\lambda}_i$$

where $\boldsymbol{\lambda}_i$ is chosen so that $\boldsymbol{\xi}_i^{k+1}$ is orthogonal to all previous directions, that is to every column of D_{i-1}. This gives

$$\lambda_i = (D'_{i-1} D_{i-1})^{-1} D'_{i-1} \zeta_i$$

so that

$$\xi_i^{k+1} = \{I - D_{i-1} (D'_{i-1} D_{i-1})^{-1} D'_{i-1}\} \xi_i = P_{i-1} \xi_i. \tag{9}$$

The matrix P_{i-1} is a projection matrix and may be obtained from a recurrence relation

$$P_i = I - D_i (D_i'D_i)^{-1} D_i' = P_{i-1} - \frac{P_{i-1} \xi_i^{k+1} \xi_i^{k+1\prime} P_{i-1}}{\xi_i^{k+1} P'_{i-1}\xi_i^{k+1}} . \tag{10}$$

In the unconstrained problem, all the columns of D_i are orthogonal and $D_i'D_i$ is the unit matrix and the calculation is the normal Gram-Schmidt orthogonalization procedure. Now suppose that p linear constraints represented by

$$N'\mathbf{x} = \mathbf{b} \tag{11}$$

are to be satisfied where N is the $(n \times p)$ matrix with columns $\mathbf{n}_1, \mathbf{n}_2, ..., \mathbf{n}_p$. In this case, only $(n-p)$ directions are to be calculated and they must be orthogonal to the columns of N. Let $\xi_i^{k+1} = \mathbf{n}_i$ for $i = 1, 2, ..., p$ and calculate P_p from Eqn (10). Then successive application of Eqn (9) will yield the required directions $\xi_{p+1}^{k+1}, \xi_{p+2}^{k+1}, ..., \xi_n^{k+1}$. Note that in this calculation, since only $(n-p)$ directions are used at each iteration, only $(n-p)$ of the vectors $\zeta_1, \zeta_2, ..., \zeta_n$ will be independent. This situation is taken care of because $P_{i-1}\zeta_i = 0$ if a dependent ζ is encountered, in which event that vector is ignored and the next one considered.

This describes a method of selecting directions so that linear equality constraints may be handled, and it is readily extended to inequality constraints by considering the appropriate constraint normal as a direction of search when the inequality has been violated. When progress is made in an iteration along a constraint normal, the constraint is no longer active and its normal is removed from the list of directions.

There are many other schemes for optimizing a nonlinear function subject to linear constraints but the basis of two more is readily derived from formula (9). We replace ξ_i by

$$\mathbf{g} = \left(-\frac{\partial f}{\partial x_1}, ..., -\frac{\partial f}{\partial x_n}\right)$$

and obtain the projection of the vector of steepest descent onto the intersection of the active constraints. This projection is used by Rosen [19] in his gradient projection method but being a steepest descent method the usual oscillations arise on a function exhibiting a sharp ridge. It will be recalled from Fletcher's paper (Chapter 1) that Davidon's method overcomes these difficulties in the unconstrained case, and an extension of his method by Goldfarb and Lapidus [20] provides an efficient method for linear constraints that converges for a quadratic objective function subject to p constraints in $(n-p)$ iterations. Some suggested improvements to the method of Goldfarb and Lapidus are given by Fletcher [21].

10. Projection Matrices and Nonlinear Constraints

The method of Goldfarb and Lapidus has been used by ourselves in conjunction with Carroll's technique for nonlinear constraints and represents a powerful optimization routine. In our experience, the performance of the routine is improved if the constraint basis is not altered on the first iteration following a surface change. It has been found that this avoids unnecessary changes of the constraint basis at later iterations.

Davies [22] proposes a method for extending the use of projection matrices to deal with nonlinear constraints and has obtained results that compare very favourably with those from Carroll's technique. Basically, the method proceeds in the same manner as Goldfarb and Lapidus' method until a

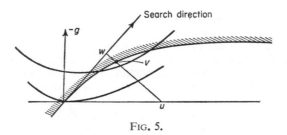

FIG. 5.

nonlinear constraint becomes active. The nonlinear constraint is then replaced by its tangent hyperplane and the direction of search projected into it (see Fig. 5). On a convex constraint, a move along this direction will give a nonfeasible point **w** and a move must be made from **w** to a feasible point **v** on the constraint. Davies, following Rosen [23], defines a return direction

$$\mathbf{r} = D_i (D_i' D_i)^{-1} \boldsymbol{\phi} \tag{12}$$

where $\boldsymbol{\phi}$ is the p-dimensional vector whose elements are the absolute values of the p active constraints at **w**. The point **u** is calculated along **r** in the normal plane to the steepest descent vector at the point x from where the iteration was begun. Interpolation between w and u gives the required point v, from where a further step is made in the search direction and the above procedure repeated.

Another procedure using projection matrices for nonlinear constraints is that of Kalfon and co-workers [24]. Their method forms a linear combination of Eqns (9) and (12) to give a feasible search direction and wherever possible uses the method of conjugate gradients (Fletcher and Reeves [25]) to generate new search directions.

11. Conclusions

Whilst Rosenbrock's method has been most successfully applied since 1960, it is a little slow by present standards and Carroll's method is to be preferred. It is also suggested that wherever a method can take account of linear constraints, this is advantageous. For constrained optimization using gradient methods, particularly Davidon's, limited experience with Davies' technique shows this to be a good alternative to Carroll's method for nonlinear constraints.

References

1. J. Abadie and J. Carpentier (1969). "Generalization of the Wolfe reduced gradient method to the case of nonlinear constraints", Chapter 4, this Volume.
2. M. J. Box (1966). A comparison of several current optimization methods, and the use of transformations in constrained problems. *Comput. J.* 9, 67.
3. M. J. Box, D. Davies and W. H. Swann (1969). "Nonlinear Optimization Techniques". ICI Monograph on Mathematics and Statistics. Oliver and Boyd, Edinburgh.
4. J. L. Greenstadt (1966). A ricocheting gradient method for nonlinear optimization. *J. Soc. ind. Appl. Math.* **14**, 3.
5. K. J. Arrow, L. Hurwicz and H. Uzawa, (1958). "Studies in Linear and Nonlinear Programming". Stanford University Press, Stanford, California.
6. H. H. Rosenbrock (1960). An automatic method for finding the greatest or least value of a function. *Comput. J.* **3**, 175.
7. M. J. Box (1965). A new method for constrained optimization and a comparison with other methods. *Comput. J.* **8**, 42.
8. C. F. Wood (1965). "Review of Design Optimization Techniques". IEEE Transactions on Systems Science and Cybernetics. Vol. SSC-1 No. 1.
9. W. H. Swann (1964). "Report on the Development of a New Direct Search Method of Optimization". ICI Ltd, Central Instrument Laboratory Research Note 64/3.
10. G. Leitmann (ed.) (1962). "Optimization Techniques with Applications to Aerospace Systems". Academic Press, New York.

11. R. Hooke and T. A. Jeeves (1961). Direct search solution of numerical and statistical problems. *J. Ass. comput. Mach.* **8**, 212.
12. W. R. Klingman and D. M. Himmelblau (1964). Nonlinear programming with the aid of a multiple-gradient summation technique. *J. Ass. comput. Mach.* **11**, 400.
13. H. Glass and L. Cooper (1964). Sequential search: a method for solving constrained optimization problems. *J. Ass. comput. Mach.* **12**, 71.
14. C. W. Carroll (1961). The created response surface technique for optimizing nonlinear restrained systems. *Op. Res.* **9**, 169.
15. A. V. Fiacco and O. P. McCormick (1964). The sequential unconstrained minimization technique for nonlinear programming, a primal dual method. *Mgmt. Sci.* **10**, 360.
16. K. R. Frisch (1955). "The Logarithmic Potential Method of Convex Programming". Memo of University Institute of Economics, Oslo.
17. G. R. Parisot (1960). "Les Programmes Logarithmiques—Application aux Calculs des Programmes Convexes". Unesco, NS, ICIP, D.1, 5. IBM France.
18. T. Pomentale (1965). A new method for solving conditioned maxima problems. *J. Math. Analysis Applic.* **10**, 216.
19. J. B. Rosen (1960). The gradient projection method for nonlinear programming, part 1—linear constraints. *J. Soc. ind. appl. Math.* **8**, 181.
20. D. Goldfarb and L. Lapidus (1968). Conjugate gradient method for nonlinear programming problems with linear constraints. *Ind. Engng. Chem. Fundam.* **1**, 142.
21. R. Fletcher (1968). "Programming under Linear Equality and Inequality Constraints". ICI Management Services Report MSDH/68/19.
22. D. Davies (1968). "The Use of Davidon's Method in Nonlinear Programming". ICI Management Services Report MSDH/68/110.
23. J. B. Rosen (1961). The gradient projection method for nonlinear programming, Part 2—Nonlinear constraints. *J. Soc. ind. appl. Math.* **9**, 514.
24. P. Kalfon, G. Ribiere and J. C. Sogno (1968). "A Method of Feasible Directions using Projection Operators". Presented at IFIPS Congress, Edinburgh.
25. R. Fletcher and C. M. Reeves (1964). Function minimization by conjugate gradients. *Comput. J.* **7**, 149.

Discussion

WOLFE. With regard to the transformation methods which you mention for converting a constrained problem into an unconstrained problem, I would like to submit a thesis that these methods are not really very good at all. Also considering the last method you described, does this not have a lot in common with Abadie's Reduced Gradient Method?

DAVIES. In certain cases transformation methods have worked, but as a general method for real problems then I would agree. With regard to Abadie's method, I would think that the method which I outlined would take rather less function evaluations.

WOLFE. I think that we could suggest tests for methods for constrained optimization similar to that of quadratic termination for the unconstrained problems. I would suggest that if a method for constrained optimization does not terminate in the linear case then it must be bad. In these cases I would say that the method should bear some relation to the Simplex Method. It should either terminate or at least get pretty close to the answer in the same number of steps. It is on these heuristic grounds that I think the transformation methods are bad.

I wonder also if you could expand a little on some of the words you were using. One word you use in the technical sense is "reliable": another word which many too many people use without trying to define it is "successful". Could you say something about that.

DAVIES. By "reliable" I mean that the method has solved most of our problems: Davidon's method with CRST of Carroll is such an example. Rosenbrocks' method too, although not very fast, usually gets there in the end. The word "successful" we would give a somewhat similar interpretation.

ABADIE. Certain methods work well on problems of up to say twenty variables but fail on larger problems of say fifty variables. Perhaps we should call these methods "almost successful" or "almost reliable".

13. Acceleration Techniques for Nonlinear Programming

R. FLETCHER* AND A. P. McCANN

Computing Laboratory, University of Leeds, Leeds, England

One of the most successful approaches to solving the general nonlinear programming problem without equality constraints is the Created Response Surface Technique (CRST) of Carroll [1]. As described in the previous paper, the function

$$P(\mathbf{x}, r) = F(\mathbf{x}) + r \sum_{i \in I} 1/c_i(\mathbf{x}) \tag{1}$$

is minimized successively for given values of $r > 0$ (surfaces). The value of r is reduced by a constant multiple (*rdec*) after minimizing on each surface and the sequence of minima so obtained tends to the solution of the nonlinear programming problem. The set I denotes the effective constraints; however, it does not affect the argument to include noneffective constraints in the sum because for these $1/c_i$ does not tend to infinity so that their contribution to P as $r \to 0$ becomes negligible.

It is also reasonable to choose the most rapidly convergent method known (Davidon's method: see [2]) to minimize each surface, especially as the approximating matrix H which occurs can be carried forward from one iteration to start the next. Consequently a library procedure for Davidon's method was taken and used as the basis for the implementation of the CRST.

The aim of the project was to analyse the behaviour of successive surfaces of the CRST and see how this could be used to accelerate convergence. The outstanding contribution here is by Fiacco and McCormick [3] who showed how the minima of the surfaces were related and how this information could be used to obtain a good starting approximation for each surface (the minimization method they used was the Newton or Taylor series method). There have been reports that the acceleration technique is not very efficient when

---

* Present address: Theoretical Physics Division, AERE, Harwell, England.

used with Davidon's method and one objective was to check whether this was so.

Another objective was to examine the behaviour of the H matrix at each surface to see whether a more accurate starting matrix could be obtained for each surface. This was done bearing in mind a paper by Murray [4] showing how the condition number of H could tend to infinity. The investigations brought out another source of inefficiency, namely that the linear search in the library Davidon procedure (based on cupic interpolation) was not appropriate to a function with singularities, and this was also examined. One factor to be taken into consideration when examining the results is that the computer used had a 48 bit (12 decimals) word length and that all scalar products were accumulated to double length. It might not be possible to reproduce some of these results on a computer with a smaller word length.

After preliminary trials an initial value of $r = 1$ and a value of $rdec = 0.01$ were standardized. The latter is about right and has the advantage that one further significant figure is obtained in x and H at each surface (they vary as $rdec^{\frac{1}{2}}$). This can be made use of in a test for convergence. The value $r = 1$ however can be inefficient if the initial approximation to x is good; however, this feature was not explored. The term $1/c_i(x)$ in the sum was also fixed, although good reports of $1/c_i^2$ have been given. The techniques given below can also be applied in this case and the important results will be given.

Using ∇ to represent the column vector operator $\partial/\partial x_i$ and ∇^2 to denote the operator $\partial^2/\partial x_i \, \partial x_j$, then differentiating $P(x, r)$ gives

$$\nabla P(x, r) = \nabla F(x) - r \sum_{i \in I} \nabla c_i/c_i^2.$$

At the minimum ξ_r of the surface $P(x, r)$, ∇P is zero so that

$$\nabla F(\xi_r) = r \sum_{i \in I} \nabla c_i(\xi_r)/c_i^2(\xi_r) = \sum_{i \in I} \alpha_i \nabla c_i(\xi_r)$$

which follows on defining quantities α_i dependent upon r. But at the solution ξ_0 of the nonlinear program

$$\nabla F(\xi_0) = \sum_{i \in I} \lambda_i \nabla c_i(\xi_0)$$

where λ_i are constant Lagrange multipliers. Thus

$$\lim_{r \to 0} r/c_i^2(\xi_r) = \lim_{r \to 0} \alpha_i = \lambda_i$$

and for small r, $c_i(\xi_r) = r^{\frac{1}{2}}/\lambda_i^{\frac{1}{2}}$ and

$$P(\xi_r, r) = F(\xi_r) + r^{\frac{1}{2}} \sum_{i \in I} \lambda_i^{\frac{1}{2}}.$$

Fiacco and McCormick argue that this implies that $P(\xi_r, r)$ and hence ξ_r should be expressible as a polynomial in $r^{\frac{1}{2}}$, and this is borne out in practice. (If the term $1/c_i^2$ is used to define P then a similar analysis implies that a polynomial in $r^{2/3}$ would be appropriate.)

Fiacco and McCormick also show how a triangular array can be set up to give first order, second order, and so on, estimates of the solution. These are then used to predict a minimum of the next surface which is used to start the minimization process for that surface. It is necessary to have results for the first and second surfaces before any acceleration can be made. In practice we have found that no more than second-order acceleration is necessary and also that the process is stable as regards rounding error.

If the vector ∇P is differentiated again the matrix of second derivatives of the composite function

$$\nabla^2 P(\mathbf{x}, r) = \nabla^2 F - r \sum_{i \in I} (2 \nabla c_i \, \nabla c_i'/c_i^3 + \nabla^2 c_i/c_i^2)$$

is obtained, which can be rearranged at the surface minima as

$$\nabla^2 P(\xi_r, r) = \nabla^2 F - \sum_{i \in I} \alpha_i \, \nabla^2 c_i - 2r^{-\frac{1}{2}} \sum_{i \in I} \alpha_i^{3/2} \, \nabla c_i \, \nabla c_i'.$$

As $r \to 0$ so $\xi_r \to \xi_0$ and $\nabla^2 F$, α_i, ∇c_i and $\nabla^2 c_i$ tend rapidly to their values at the solution of the nonlinear program. Hence $\nabla^2 P$ consists of a constant term $\nabla^2 F - \sum \alpha_i \, \nabla^2 c_i$, and a term $2r^{-\frac{1}{2}} \sum \alpha_i \, \nabla c_i \, \nabla c_i'$ which goes to infinity. Hence if there are k effective constraints, then $\nabla^2 P$ will have k eigenvalues, which vary as $r^{-\frac{1}{2}}$, with eigenvectors in the subspace spanned by vectors ∇c_i $(i \in I)$, and $n-k$ eigenvalues which are constant with eigenvectors orthogonal to the others. Now H is an approximation to the inverse of $\nabla^2 P$, so it should have the same eigenvectors, with k eigenvalues which tend to zero as $r^{\frac{1}{2}}$ and $n-k$ which are constant. Following Fletcher [5] it can be shown that the limit of H as $r \to 0$ is

$$H_0 = \Gamma^+ - \Gamma^+ N (N' \Gamma^+ N)^{-1} N' \Gamma^+$$

where
$$\Gamma = \nabla^2 F - \sum_i \lambda_i \, \nabla^2 c_i,$$

Γ^+ denotes the generalized inverse of Γ (this caters for Γ singular, as can occur in some problems), and N denotes the $n \times k$ matrix with columns ∇c_i $(i \in I)$.

If the conditioning of the problem is now considered, these results bear out Murray [4] in that the condition number of $\nabla^2 P$ (ratio of largest to smallest eigenvalues) will tend to infinity if $k \neq n$. Hence the determination of search directions by solving $(\nabla^2 P) \mathbf{s} = -\mathbf{g}$ (as Fiacco and McCormick do) will become more and more ill-conditioned. However in Davidon's method H is used only in making matrix multiplications, and as the elements of H have a finite limit, so the use of Davidon's method is well conditioned. If $1/c_i{}^2$ is used in defining P then it can be shown that the same k eigenvalues of H will approach zero as $r^{1/3}$.

To test these ideas Rosenbrock's post office parcel problem (1960) was used in three forms (PP1, PP2, PP3) and also the first three test problems (TP1, TP2, TP3) in the nonlinear programming study of Colville [6]. First

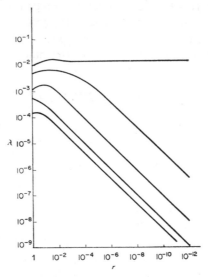

FIG. 1. Eigenvalues of H matrix. (TP1).

of all, it was noticeable how rapidly and predictably the anticipated behaviour in the eigenvalues of H set in, even though H is only an approximation to the true $(\nabla^2 P)^{-1}$. Figures 1 and 2 for TP1 and TP2 show this very clearly, and similar results have been obtained in all cases. In both these figures $\log_{10}\lambda$ versus $\log_{10} r$ has been plotted such that a slope of $-45°$ corresponds to $r^{1/2}$ behaviour.

The first problem in using this information is to decide which eigenvalues for any value of r, say r_1, correspond to which eigenvalues for the next value of r, say $r_2 = rdec \times r_1$. The simplest way to decide is to permute each eigensolution (given by $HX = X\Lambda$ where X is the matrix whose columns are the eigenvectors and Λ is a diagonal matrix of eigenvalues) so that the elements of Λ are in descending order. Having done this, the eigenvalues for the next value of r, say $r_3 = rdec \times r_2$, can be predicted assuming the sort of behaviour

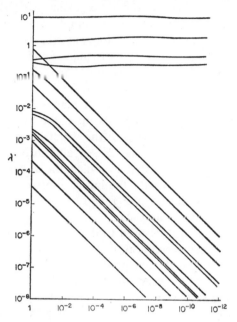

FIG. 2. Eigenvalues of H matrix. (TP2).

shown in diagrams 1 and 2. If the further assumption is made that the eigen-vectors $X(r)$ approach a constant limit (in fact the $n-k$ eigenvectors of H_0 and the k eigenvectors of

$$\sum_{i \in I} \lambda_i^{3/2} \, \nabla c_i \, \nabla c_i'$$

which have non-zero eigenvalues) then the H matrix for r_3 can be predicted by assuming that the eigenvectors are the same as for r_2. That is

$$H(r_3) = X(r_2) \, \Lambda(r_3) \, X'(r_2)$$

where the diagonal elements of $\Lambda(r_3)$ are the predicted eigenvalues for r_3 as above. If this H matrix and the predicted point at which the minimum occurs (as Fiacco and McCormick) are used to start the iteration on the r_3 surface, then an improved rate of convergence should result.

If two corresponding eigenvalues $\lambda(r_1)$ and $\lambda(r_2)$ are given, then how can the eigenvalue $\lambda(r_3)$ be predicted? One way would be to set up a table (as Fiacco and McCormick) and predict the eigenvalue on the basis of some

R. FLETCHER AND A. P. MCCANN

polynomial behaviour. This would not be satisfactory, however, for those eigenvalues which are approaching a constant limit. A more simple and safe way is to assume that the eigenvalues follow a straight line on the $\log(\lambda)/\log(r)$ diagram. That is to say in general $\lambda(r_3)$ is taken as $\lambda^2(r_2)/\lambda(r_1)$. However, if the eigenvalue is increasing, then the assumption is made that asymptotic behaviour to a constant limit is occuring and $\lambda(r_3)$ is taken as $\lambda(r_2)$. Also if the eigenvalue is decreasing more rapidly than would be expected on the $r^{1/2}$ behaviour, then a slope corresponding to $r^{1/2}$ behaviour is assumed, and $\lambda(r_3)$ is taken as $rdec^{1/2} \times \lambda(r_2)$. This ensures that freak results caused by rounding error do not unduly upset the predicted H matrix.

The success of this strategy depends upon the adequacy of the sorting process in pairing up the eigenvalues; the worst cases will be where eigenvalues appear to cross as in Fig. 2. An alternative method for pairing eigenvalues could be implemented in which the eigensolution of $H(r_2)$ is found by the Jacobi method with the initial approximation to $X(r_2)$ taken not as the unit matrix but as $X(r_1)$. This will ensure that corresponding diagonal elements of $\Lambda(r_1)$ and $\Lambda(r_2)$ will have eigenvectors most nearly alike, and this ordering will be used to determine $\Lambda(r_3)$. Because the eigenvectors should not vary rapidly as $r \to 0$, $X(r_1)$ should nearly diagonalize $H(r_2)$ for small r, and the Jacobi process should converge more and more rapidly. This should reduce very substantially the computing time required for the eigensolution so that it becomes a negligible proportion of the whole.

Table I shows the results for TP1, denoting the strategy of doing no acceleration by "A", that of just accelerating by predicting a better point $\mathbf{x}$ (as Fiacco and McCormick) by "B", and that of predicting a better H matrix

TABLE I

Detailed results: for Test Problem 1 [6]. 5 variables, 15 constraints (4 effective).

		Acceleration used					
		None "A"		$\mathbf{x}$ "B"		$\mathbf{x}$ and H "C"	
Surface	r						
1	1	20[a]	58[a]	20	58	20	58
2	10^{-2}	12	40	12	40	12	40
3	10^{-4}	11	25	11	27	9	23
4	10^{-6}	7	22	8	21	7	16
5	10^{-8}	7	22	7	14	4	10
6	10^{-10}	7	22	3	7	3	6
7	10^{-12}	7	20	2	6	2	4
8	10^{-14}	11	37	1	4	1	2
9	10^{-16}	7	23	3	7	3	4

[a] Numbers in each entry are "linear searches" and "function evaluations", as required for that surface.

as well by "*C*". The results for the first two surfaces are common to all strategies because at least two minima are required before information about the third can be predicted. Note the considerable improvement from *A* to *B*

TABLE II

Overall results on various problems

Problems	PP1		PP3		TP1		TP2		TP3	
Variables	3		3		5		13		5	
Effective constraints	1		3		4		11		5	
Performance										
A^a	36^a	141^a	62	214	89	269	244	740	105	332
Overall B	33	105	36	122	66	184	159	468	61	194
C	29	97	32	107	60	163	148	445	52	157
Surfaces 1 and 2	16	62	17	72	32	98	102	315	36	119
A	20	79	45	142	57	171	142	425	69	213
Surfaces B	17	43	19	50	34	86	57	153	25	75
3 to 9 C	13	35	15	35	28	65	46	130	16	38
Saving per cent										
Overall $A \rightarrow B$	26		43		32		38		42	
$B \rightarrow C$	8		12		11		5		19	
Surfaces $A \rightarrow B$	46		65		50		66		65	
3 to 9 $B \rightarrow C$	19		30		24		15		49	

^{a} Strategies "*A*", "*B*", "*C*": see Table I for notation.

and the lesser but still significant improvement from *B* to *C*. After the ninth surface both *F* and **x** are correct to about 8–9 significant figures.

The results for all the test problems are set out in a less detailed way in Table II. Note that the same behaviour in going from *A* to *B* to *C* still persists. It seems that the effect of acceleration over those surfaces where it is used (that is 3 to 9) gives savings of 50–60 per cent in going from *A* to *B*, and a further 25 per cent or so in going from *B* to *C*. When taking all the surfaces into account, the figures are reduced to about 30–40 per cent and 10–15 per cent respectively.

As performance over surfaces 3 to 9 is improved so more and more of the total computing time is taken in minimizing on the first two surfaces. Given that the minimization method is not changed, the only place to look for further savings is in the linear search. On average 3–4 function evaluations are required for each iteration at this stage. Any improvements obtained in the linear search will cause the relative effect of other acceleration techniques as quoted above to become correspondingly greater.

Each iteration in the Davidon method requires the estimation of the parameter η_i in the iteration

$$\mathbf{x}_{i+1} = \mathbf{x}_i + \eta_i \, \mathbf{s}_i$$

so that the function to be minimized is as small as possible. This is the linear search subprocess. It is inefficient to obtain the parameter to high accuracy and is usually satisfactory to evaluate the function and gradient at two points along $\mathbf{s}_i$ and interpolate on the basis of the function obeying some simple relationship. It is only necessary to ensure that the function does not increase at any iteration for stability purposes. In the library procedure for Davidon's method which was used, the interpolation was based on the function being a cubic polynomial in η.

When the function being minimized is one of the surfaces of the *CRST*, there is more information about the behaviour of the surfaces, in particular the fact that $P(\mathbf{x}, r) \to \infty$ as $\mathbf{x}$ approaches the boundary of the feasible region.

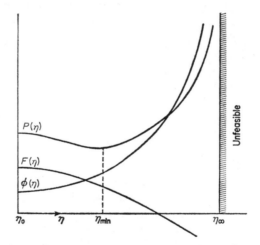

FIG. 3. Variation of penalty functions along a line.

Figure 3 shows this in a typical case. It will be seen that the contribution from $F(x)$ is slowly varying, whilst the infinite behaviour comes from the inverse term $\phi(x) = r \sum 1/c_i$ involving the constraints. If a quadratic function were to be minimized subject to linear constraints, then the behaviour of $P(\mathbf{x}, r)$ with respect to η would be

$$P(\eta) = a + b\eta + c\eta^2 + \frac{d}{\eta_\infty - \eta}. \tag{2}$$

It was decided to attempt to base the interpolation part of the linear search routine on this relationship rather than on a cubic polynomial.

The information available for the interpolation is the values of the functions F and ϕ and also their derivatives with respect to η (F' and ϕ') at two feasible points along s_i, η_0 and η_1. The parameters to be determined are b, c, d and η_∞, the intersection of s_i with the boundary of the feasible region. In fact there is more information available than is necessary to do this, and it was decided to use only the derivatives of F and ϕ, mainly on grounds of convenience. In the first place, ϕ_0' (i.e. $\phi'(\eta_0)$) and ϕ_1' are used to estimate d and η_∞ on the basis of $\phi(\eta) = d/(\eta_\infty - \eta)$, and similarly F_0' and F_1' are used to evaluate b and c from $F(\eta) = a + b\eta + c\eta^2$. Then the equation for the minimum of P, $(\partial P(\eta)/\partial \eta = 0)$, which can be rearranged to give a cubic in η, is solved using the Newton-Raphson iteration, taking the mid-point between the trial points as the starting point. Naturally functions of more parameters could be taken to approximate $\phi(\eta)$ and $F(\eta)$ in view of the added information available. Although the results of the current investigation are very satisfactory, we think that this might be exploited to advantage in future work.

It should be noted that interpolation using Eqn (2) can fail or may be unsatisfactory in some circumstances. The contribution of ϕ to $P(\mathbf{x}, r)$ may not increase or decrease monotonically in which case it is not possible to estimate η_∞ from the trial points. Similarly, the search direction s may be almost parallel to the boundaries of the feasible region yielding large and inaccurate estimates of η_∞. Finally, the Newton-Raphson iteration may fail to converge if the starting point is not sufficiently good. The alternative strategy adopted in the event of such failures may have a decisive effect in determining the success or failure of the method. In our case it was observed that the nonpolynomial interpolation appeared most likely to fail in the later stages of minimization when a reasonable approximation to the minimum for the surface had been obtained and at this stage a simple quadratic interpolation gave very satisfactory convergence. Consequently, quadratic interpolation using the gradients at the two trial points was used when the interpolation described above failed. Such a simple alternative is not entirely satisfactory as appears from the results described below.

In an early version of the program, η_0 was taken as zero, and η_1 as a unit step along s_i. If η_1 proved infeasible, then the step was progressively halved until a feasible point resulted. The values of η_i estimated from the subsequent interpolation (or extrapolation) proved sufficiently good on the smaller test problems to give more rapid convergence than the library version. In some larger problems this was not so and it was judged desirable to have two points bracketing the minimum before estimating η_i by interpolation. Thus, if η_0 and η_1 did not bracket the minimum in the above scheme, then a new point 90 per cent of the distance to η_∞ was examined. This process was repeated until two suitable points bracketing the minimum were obtained. Cases arose where a unit step along s_i did not intersect the boundary of the feasible region and either η_∞ could not be estimated from the trial points or the estimated value was large (> 100 say). In these cases if, at the new point at unit distance along s_i, the value of $P(x, r)$ had decreased and the gradient was still downhill, this point was taken as η_0 and a new trial point was taken at twice the distance along s_i. This process was repeated until the minimum was bracketed, or until a value of η_∞ was obtained which could be used as described above for bracketing the minimum.

Results using this linear search are presented as Strategy D in Table III where the performance is compared with that using Davidon's linear search (Strategies A, B, C). The comparison is presented for the first two response

TABLE III

Improvements to the linear search

Problem	Surface	Strategy "D"		Strategies A, B, C	
PP1	1	8^a	24^a	10	40
	2	6	14	6	22
PP2	1	9	27	8	36
	2	5	11	5	18
PP3	1	11	40	11	52
	2	7	16	5	20
TP1	1	11	37	20	58
	2	16	46	12	40
TP2	1	not solved		78	245
	2	see text		24	70
TP3	1	15	45	23	74
	2	9	21	13	45

[a] See Table I for notation.

surfaces only as differences rapidly became negligible as the acceleration devices "homed in" on the solution. With the five problems where a comparison is possible, the new search gave savings in function evaluations of the order of 30 per cent which is quite worthwhile.

The method failed to locate the minimum with TP2 in a reasonable amount of time. In this problem the starting point was very close to the boundary of the feasible region and the initial gradient was very large. Since the first search direction is that of steepest descent a unit step along this direction gave a point at a great distance from the starting point where the gradient of the penalty term of (x) was of opposite sign to that at the starting point so that the non-polynomial interpolation failed immediately. The use of quadratic interpolation so early in the search gave a very poor estimate of the minimum along the first search direction, leading to several more early failures of the non-polynomial interpolation and subsequently to extremely slow progress towards the minimum.

On the other hand the library program (Strategy A) was based on cubic interpolation which provided far better estimates of the minimum and as a consequence much more rapid convergence towards the minimum. It is clear then that the slow convergence on this problem is caused not by the non-polynomial interpolation but by resorting to quadratic rather than cubic interpolation when the non-polynomial behaviour cannot be employed. The remedy, of resorting to cubic interpolation in these cases, is obvious. However, it would also be worth investigating the effect of adding further non-polynomial terms in Eqn (2) so that the number of cases in which the non-polynomial interpolation fails can be reduced.

References

1. C. W. Carroll (1961). The created response surface technique for optimizing nonlinear restrained systems. *J. Ops. Res. Soc. Am.* **9** (2), 169.
2. R. Fletcher and M. J. D. Powell (1963). A rapidly convergent descent method for minimization. *Comput. J.* **6** (2), 163.
3. A. V. Fiacco and G. P. McCormick (1964). The sequential unconstrained minimization technique for nonlinear programming, a primal-dual method. *Mgmt. Sci.* **10** (2), 360.
4. W. Murray (1969). Ill-conditioning in barrier and penalty functions arising in constrained nonlinear programming. *In* "Proceedings of the Sixth International Symposium on Mathematical Programming", Princeton University, August 1967. Princeton, New Jersey, to be published.
5. R. Fletcher (1968). "Programming under Linear Equality and Inequality Constraints". I.C.I. Ltd, Management Services, Report MSDH/68/19, 1968.
6. A. R. Colville (1968). "A Comparative Study on Nonlinear Programming Codes." IBM New York Scientific Center, Report 320-2949.

Discussion

ANON. I would like to raise three points. Firstly, what would be considered a large nonlinear programming problem in the current state of the art? Secondly, we have some large problems of around fifty variables and constraints, and we find that in using these acceleration techniques we invariably go out of the feasible region. The constraints which we are dealing with are very critical and define a very narrow feasible region. We tend to waste a lot of time with the Davidon-Fletcher-Powell method in trying to find a step size which does not violate a constraint. Finally, concerning the Fiacco and McCormick type techniques, the speakers have not referred to a problem which I think is most important and that is in the determination of the initial feasible point. Although Fiacco and McCormick suggest methods, our experience has been that to get an initial feasible point may take more time than the subsequent solution of the problem.

FLETCHER. How many of your fifty constraints are linear?

ANON. They are all nonlinear.

FLETCHER. Then you certainly have a large problem. I would say about twenty-five variables and nonlinear constraints would constitute a large problem, requiring runs of say 15–20 mins on a large fast computer.

McCANN. With regard to the problem of leaving the feasible region, it was because of these difficulties that we adopted the non-polynomial interpolation. When you are searching for a minimum the region where the function is varying rapidly is in this narrow band near the constraints. Hence we thought that the genearal interpolation routine normally used in library routines would be unlikely to be successful. Thus we introduced a term which varies in the same way as the penalty function. Of course there are difficulties when one is moving parallel to the con-straint as we mentioned, because of the difficulties of locating the boundary of the region, and some criteria as whether to use a higher order formula or just to use simple polynomial interpolation would be useful.

14. A Constrained Minimization Method with Quadratic Convergence

B. A. MURTAGH and R. W. H. SARGENT

Imperial College, London, England

1. Introduction

Various methods have been put forward for solving nonlinear programming problems, but the most generally promising one still seems to be the "gradient-projection" method due to Rosen [1]. This is based on the steepest descent method for minimization, coupled with orthogonal projection of the gradient into a linear manifold approximating any constraints encountered.

For minimization without constraints, steepest descent methods have largely given way to methods aimed at quadratic convergence, such as Fletcher and Powell's [2] well-known modification of Davidon's [3] method. It is therefore natural to seek methods of combining these more efficient minimization methods with Rosen's gradient projection technique in order to deal with constraints, and Goldfarb and Lapidus [4] have already put forward such a method.

The Fletcher–Powell method constructs successive approximations to the inverse of the Hessian matrix of the function to be minimized by seeking the minimum of this function along successive search directions. Their method relies on the fact that at the end of each step the function gradient is orthogonal to the direction of search, and that the Hessian matrix is symmetric so that the search directions are mutually conjugate with respect to it. This is inconvenient for use with gradient projection, since in general the gradient is not orthogonal to the search direction at the point where this enconters a constraint. As a result, a new conjugate direction cannot be set up and the inverse of the Hessian matrix cannot be updated at such pionts; a new sequence of conjugate directions must therefore be started each time the active constraint set is changed. In addition, the Goldfarb–Ltapidus method involves orthogonal projection of the Hessian inverse into the current constraint set, so that all accumulated information orthogonal to this set is lost.

215

The present paper considers a class of methods which makes it possible to update the Hessian inverse for steps of arbitrary length and direction. This makes them particularly useful for use with gradient projection, but they can also be advantageous for minimization without constraints since they do not rely on accurate determination of minima along successive directions.

2. Minimization without Constraints

2.1. *Basis of the Method*

The principle of solution is to assume a local quadratic approximation to the function and construct successive approximations to the inverse of the Hessian matrix, using these to estimate the position of the minimum at each step.

Thus we consider a function, $f(x)$, of n variables, with continuous second derivatives, and expand it in a Taylor series with a remainder term of second order:

$$f_k = f_{k-1} + g_{k-1}^T p_k + \tfrac{1}{2} p_k^T H(x_{k-1} + \theta p_k) p_k \quad (0 \leqslant \theta \leqslant 1) \tag{1}$$

where $f_k = f(x_k)$, $g_{k-1} = g(x_{k-1})$ is the gradient of the function at point x_{k-1}, $p_k = x_k - x_{k-1}$, and H is the symmetric Hessian matrix.

The change in the gradient is then given by

$$q_k = g_k - g_{k-1} = H p_k. \tag{2}$$

Now if $x_k = \hat{x}$, the position of the minimum, we have $g(\hat{x}) = 0$ and provided that H is non-singular, Eqn (2) yields

$$\hat{x} = x_{k-1} - H^{-1} g_{k-1}. \tag{3}$$

We are therefore interested in constructing successive approximations, S_k, to the inverse of the Hessian matrix, and it seems reasonable to require that the approximation at any step should satisfy Eqn (2) for all steps so far made:

$$S_k q_j = p_j \quad (1 \leqslant j \leqslant k) \tag{4}$$

After n steps we can form the $n \times n$ matrices, P_n and Q_n with columns p_j $(1 \leqslant j \leqslant n)$ and q_j $(1 \leqslant j \leqslant n)$ respectively. Eqn (4) then gives

$$S_n Q_n = P_n. \tag{5}$$

Now if the function $f(x)$ is adequately represented by a quadratic function over these n steps, the Hessian matrix H will not vary, and if further H and Q_n are non-singular, we have from Eqns (2) and (5):

$$S_n = P_n \cdot Q_n^{-1} = H^{-1}. \tag{6}$$

We can then obtain the position of the minimum from Eqn (3).

2.2. *Construction of the Matrices* S_k

A number of methods suggest themselves for the construction of S_k to satisfy Eqn (4) and here we consider three related methods.

Method 1. The most obvious method is based on Eqn (5) written in the form

$$S_k = P_k R_k \qquad (7)$$

where $R_k = Q_k^{-1}$.

We start with arbitrary matrices P_0 and Q_0, subject only to the condition that Q_0 is non-singular, and at each step successively replace corresponding columns with the calculated vectors, p_k and q_k:

$$P_k = P_{k-1} + (p_k - P_{k-1} e_j) e_j^T \qquad (8)$$

$$Q_k = Q_{k-1} + (q_k - Q_{k-1} e_j) e_j^T. \qquad (9)$$

The corresponding inverse, R_k, is readily computed by the recursion formula:

$$R_k = R_{k-1} + \frac{(e_j - R_{k-1} q_k) e_j^T R_{k-1}}{e_j^T R_{k-1} q_k}. \qquad (10)$$

We also have:

$$\det |Q_k| = e_j^T R_{k-1} q_k . \det |Q_{k-1}|. \qquad (11)$$

Now if $\det |Q_{k-1}| \neq 0$, R_{k-1} is non-singular and $R_{k-1} q_k$ cannot be null unless q_k itself is null. Otherwise there will be at least one non-zero element of the vector $R_{k-1} q_k$, so that we can choose e_j to ensure that R_k is non-singular. It would seem desirable to choose e_j so that $e_j^T R_{k-1} q_k$ is the element of greatest absolute value, which makes $\det |Q_k|$ as large as possible, but for general functions this may cause the same column to be replaced repeatedly. Since the quadratic is only a local approximation, it is equally important to retain the most recent information and a suitable compromise is to replace the oldest column for which $e_j^T R_{k-1} q_k$ does not fall below a prescribed lower bound, chosen to ensure that a singular Q_k does not arise through rounding errors.

It is convenient to write Eqn (10) in the form:

$$R_k = R_{k-1} + (e_j - R_{k-1} q_k) . y_k^T \qquad (12)$$

with

$$y_k^T = \frac{e_j^T R_{k-1}}{e_j^T R_{k-1} q_k}. \qquad (13)$$

Substitution of Eqns (8), (12) and (13) into Eqn (7) then yields a simple recursion formula for S_k

$$S_k = S_{k-1} + (p_k - S_{k-1} q_k) . y_k^T. \qquad (14)$$

It is not therefore necessary to construct the P_k and Q_k explicitly: we merely start with arbitrary non-singular matrices S_0 and R_0 and up-date them using Eqns (12)–(14). If points are rejected for which $q_k = 0$ and e_j is chosen as explained above, all successive R_k are non-singular, and the S_k satisfy Eqn (4) for all vectors p_j and q_j implicit in the current P_k and Q_k respectively.

Method 2. Since H and H^{-1} are symmetric, it is of interest to investigate the possibility of constructing symmetric S_k to satisfy Eqn (4). Since any symmetric matrix can be expressed as a sum of symmetric simple products of type $z.z^T$, we investigate the recursion formula

$$S_k = S_{k-1} \pm z_k . z_k^T \qquad (15)$$

where S_{k-1} is symmetric and z_k is to be chosen to satisfy Eqn (4). Setting $j = k$ in Eqn (4) and substituting in Eqn (15) we have

$$p_k = S_k q_k = S_{k-1} q_k \pm z_k . z_k^T q_k$$

and
$$q_k^T p_k = q_k^T S_{k-1} q_k \pm (q_k z_k)^2. \qquad (16)$$

From Eqns (15) and (16)

$$S_k = S_{k-1} + \frac{(p_k - S_{k-1} q_k) . (p_k - S_{k-1} q_k)^T}{q^T (p_k - S_{k-1} q_k)} . \qquad (17)$$

We may note that this is also of the form of Eqn (14), where y_k is now given by

$$y_k = \frac{(p_k - S_{k-1} q_k)}{q_k^T (p_k - S_{k-1} q_k)} . \qquad (18)$$

It remains to demonstrate that S_k, constructed according to Eqn (17), satisfies Eqn (4) for all preceding steps as well as the current step, and it turns out that this is only true if $f(x)$ may be represented by a quadratic function over the steps in question. The proof is by induction.

We accordingly suppose that

$$S_{k-1} q_j = p_j \quad (1 \leqslant j < k) \qquad (19)$$

From Eqns) (17 and (18)

$$S_k q_j = S_{k-1} q_j + y_k . (p_k^T q_j - q_k^T S_{k-1} q_j)$$

and using Eqn (19)

$$S_k q_j = p_j + y_k (p_k^T q_j - q_k^T p_j) \quad (1 \leqslant j < k) \tag{20}$$

Now if the quadratic approximation to $f(x)$ is valid, we have from Eqn (2)

$$q_k^T p_j = p_k^T H p_j = p_k^T q_j. \tag{21}$$

The second term in Eqn. (20) therefore vanishes and we have $S_k q_j = p_j$ for $1 \leqslant j < k$, as required.

Now S_k has been constructed so that $S_k q_k = p_k$, and in particular $S_1 q_1 = p_1$ for arbitrary S_0. This completes the proof.

Method 3. Both of the above methods construct S_k by a recursion of the form

$$S_k = S_{k-1} + (p_k - S_{k-1} q_k) \cdot y_k^T. \tag{22}$$

To satisfy Eqn (4) it is sufficient that

$$y_k^T q_j = \delta_{jk} \quad 1 \leqslant j \leqslant k. \tag{23}$$

We note in passing that for this to be possible at each step, q_k must not be linearly dependent on the preceding vectors, q_j.

The Gram–Schmidt orthogonalization procedure now enables us to construct a set of y_k to satisfy Eqn (23), according to the scheme:

$$u_k = q_k - \sum_{i=1}^{k-1} z_i z_i^T q_k, \qquad z_k = \frac{u_k}{(u_k^T u_k)^{\frac{1}{2}}}, \qquad y_k^T = \frac{z_k^T}{z_k^T q_k} \tag{24}$$

with $u_1 = q_1$.

After n steps we have n orthonormal vectors z_k, and the procedure must be supplemented by a rule to decide which of these should be replaced at each step. A rule analogous to that used in Method 1 is to replace the oldest z_i for which $z_i^T q_k$ has an acceptably large magnitude.

Essentially all three methods are methods for solving the set of nonlinear equations: $g(x) = 0$, for which H is the Jacobian matrix. Methods 1 and 3 make no assumptions about the nature of H, other than that it is nonsingular, and they may therefore be used for solving general sets of nonlinear equations. From this point of view, Method 3 is analogous to the version of the "generalized-secant" method proposed by Barnes [5], except that it constructs approximations to H^{-1} rather than H and so avoids the solution of a set of linear equations at each step.

In all three methods the matrix S_k must be stored, and Methods 1 and 3 further require the storage of a second auxiliary matrix (the R_k or the orthogonal vector set, z_i). It will be seen later that the flexibility introduced by the use of R_k in Method 1 can be used to advantage in dealing with constraints.

Method 2 on the other hand has the advantage of requiring less storage space and less computation per step but cannot be used for solving sets of equations with nonsymmetric H. There do not seem to be any special advantages attached to Method 3, so that this method is not further discussed.

Broyden [6] has discussed a more general class of methods based on the recursion

$$S_k = S_{k-1} + p_k y_k^T - S_{k-1} q_k \cdot z_k^T \tag{25}$$

This includes the Fletcher–Powell method as well as the three methods presented here, and Broyden mentions the generalized-secant method and Method 2 above but does not discuss computational performance.

2.3. Choice of Steps, p_k

It is logical to choose the step p_k according to Eqn (3), replacing H^{-1} by its current approximation, S_{k-1}. However, if this is a poor approximation such a step may well lead to an increase in the function, and it is wiser to allow for a modification of the distance along the direction of the predicted minimum according to the formula

$$p_k = -\alpha_{k-1} S_{k-1} g_{k-1} \tag{26}$$

where α_{k-1} is a suitably chosen scalar.

If we ensure that the function $f(x)$ initially decreases as we move from x_{k-1} along the direction p_k, we can always choose an α_{k-1} with $0 < |\alpha_{k-1}| \leqslant 1$ such that $f(x_k) < f(x_{k-1})$. We must therefore have $g_{k-1}^T p_k < 0$, and so long as $g_{k-1}^T S_{k-1} g_{k-1} \neq 0$ we can always choose the sign of α_{k-1} so that this is so.

In Section 5 we shall show that in order to achieve convergence we need to make these conditions slightly stronger, according to the relations

$$|g_k^T S_k g_k| \geqslant \delta \cdot \|g_k\| \cdot \|S_k g_k\| \tag{27}$$

$$f(x_k) - f(x_{k+1}) \geqslant \varepsilon \cdot \alpha_k g_k^T S_k g_k \tag{28}$$

where δ and ε are fixed positive constants.

Evidently at the starting point, x_0, we have $g_0 \neq 0$, so that by choosing S_0 to be positive definite we ensure that $g_0^T S_0 g_0 \neq 0$. We can now ensure that condition (27) is satisfied for all k, by choosing α_k sufficiently small at each step. Equation (28) can also be satisfied by choosing α_k sufficiently small, but it is not possible to say which of the two conditions is the more stringent.

We must also ensure that successive steps eventually span the whole space of x, which requires that y_k in the recursion formula for S_k is not orthogonal to g_{k-1}. For suppose that we have

$$y_k^T g_{k-1} = 0 \tag{29}$$

then from Eqns (23) and (29)

$$y_k^T g_k = 1 \tag{30}$$

and from Eqns (22), (26) and (30)

$$p_{k+1} = -\alpha_k S_k g_k = \frac{\alpha_k(1-\alpha_{k-1})}{\alpha_{k-1}} p_k. \tag{31}$$

From Eqns (22) and (26) we also have

$$\det |S_k| = \det |S_{k-1}| \cdot y_k^T S_{k-1}^{-1} p_k = -\det |S_{k-1}| \cdot y_k^T g_{k-1} \alpha_{k-1}. \tag{32}$$

Thus if S_0 is non-singular and we ensure that $y_k^T g_{k-1} \neq 0$ at each step, all successive S_k are non-singular.

All the methods break down if $q_k = 0$, for q_k occurs in the denominator of y_k in each case. This means that S_{k-1} cannot be updated, and since $g_k = g_{k-1}$ the direction of search is also unchanged. In general of course there will only be a finite set of points for which $q_k = 0$ so that a change of α_{k-1} will usually be sufficient to avoid the problem. However it may arise that $g(x)$ is constant for all points along the line of search and in this case the simplest course of action is to make a step in the steepest descent direction without updating S_{k-1} or the auxiliary matrices. It may be noted that this also constitutes a failing case for the Fletcher–Powell method.

With these precautions the procedures will generate a sequence of points x_k such that $f(x_k) \leqslant f(x_{k-1})$, where the equality occurs only if $g_{k-1} = 0$, indicating that x_{k-1} is a stationary point. A local search about x_{k-1}, for example along each of the coordinate directions, will then either confirm that this is a minimum or provide a new point with a smaller function value, from which the procedure can be continued.

We must now examine how these conditions can be met in practice for the proposed methods.

2.4. *Procedure for Method* 1

We consider first the choice of α_k to satisfy Eqn (27). At step $(k-1)$, $g_{k-1}^T S_{k-1} g_{k-1}$ satisfied Eqn (27), and since $g(x)$ is a continuous function of x, there is a finite neighbourhood about x_{k-1} for which $|g_k^T S_{k-1} g_k|/\|g_k\| \cdot \|S_{k-1} g_k\|$ satisfies Eqn (27) and we can choose α_{k-1} to give an x_k in this neighbourhood Now from Eqn (14) we have

$$g_k^T S_k g_k = g_k^T S_{k-1} g_k + g_k^T (p_k - S_{k-1} q_k) \cdot y_k^T g_k. \tag{33}$$

If the second term on the right-hand side is of such sign and magnitude as to prevent $g_k{}^T S_k g_k$ from satisfying the condition, we can always make $y_k{}^T g_k = 0$ and hence guarantee that condition (27) is satisfied for a sufficiently small α_{k-1}.

From Eqn (23) we see that $y_k{}^T g_k = 0$ makes $y_k{}^T g_{k-1} = -1$, which also conveniently satisfies the necessary condition for spanning the space. Thus, when corrections according to Eqns (12)–(14) with the normal choice of e_j result in either $y_k{}^T g_{k-1} = 0$ or condition (27) not being satisfied we can make further corrections according to the formulae

$$R_k' = R_k + (e_m + R_k g_{k-1}) z_k{}^T$$
$$S_k' = S_{k-1} + (p_k - S_{k-1} q_k)(y_k')^T \tag{34}$$

where $\qquad z^T = -\dfrac{e_m^T R_k}{e_m^T R_k g_{k-1}} \qquad$ and $\qquad (y_k')^T = (e_j + e_m)^T R_k'.$

Note that the result $y_k{}^T g_{k-1} = 0$ leads to a simpler correction

$$S_k' = S_k + (p_k - S_{k-1} q_k) z_k{}^T.$$

The vector e_m is chosen by the same rule as e_j, but with the restriction that $m \neq j$. The combined correction factor y_k' which generates S_k' from S_{k-1} is then equivalent to y_k in Eqn (14) and is orthogonal to g_k as required.

If in fact this condition is not necessary to satisfy Eqn (27), we may be able to obtain $y_k{}^T g_{k-1} \neq 0$ more simply by continuing the search for an e_j which yields this result as well as a suitably large value of $e_j{}^T R_{k-1} q_k$ in the first correction.

It is now possible to give a complete statement of the procedure for step k, given $x_{k-1}, g_{k-1},$ and S_{k-1}.

(1) Choose the sign of α_{k-1} to make $\alpha_{k-1} g_{k-1}^T S_{k-1} g_{k-1}$ positive.

(2) Starting with $|\alpha_{k-1}| = 1$, and computing p_k from Eqn (26), determine a point x_k with $0 \leqslant |\alpha_{k-1}| \leqslant 1$ to satisfy (28).

(3) Compute $g_k = g(x_k)$, and if it is zero proceed to the local search procedure for confirmation of a minimum. Otherwise go to step 4.

(4) If $|g_k{}^T S_{k-1} g_k| / \|g_k\| \cdot \|S_{k-1} g_k\| < \delta$, return to step 2 and seek a smaller value of $|\alpha_{k-1}|$.

(5) If $|\alpha_{k-1}|$ becomes less than ε', return to the value of α_{k-1} first found in step 2 and reset S_k. Otherwise compute $y_k{}^T, R_k,$ and S_k from Eqns (12)–(14), choosing e_j so that $|e_j{}^T R_{k-1} q_k| \geqslant \varepsilon'$ and if possible $|y_k{}^T g_{k-1}| \geqslant \varepsilon'$.

(6) If such a choice of e_j is not possible, or if the resulting values of g_k and S_k do not satisfy (27), correct R_k and S_k using Eqn (34).

In this procedure ε' is a small positive constant chosen to ensure that no difficulties arise through rounding errors. The norms required in Eqn (27) are most conveniently taken as maximum modulus norms ("∞-norm" according to Ostrowski [7]).

A convenient method of carrying out step 2 is to use as many iterations as required of a search procedure for minimizing along a line. A suitable quadratic method which uses only the initial gradient, g_{k-1}, and successive function evaluations is described in the Appendix.

2.5. Procedure for Method 2

For Method 2 we can show that, if S_0 is chosen to be positive definite and we can always choose α_{k-1} so that $y_k^T g_{k-1} < 0$, then all S_k are positive definite. The proof is by induction, so we suppose that S_{k-1} is positive definite and $g_{k-1} \neq 0$.

We note first that $g_{k-1}^T S_{k-1} g_{k-1} > 0$ so that α_{k-1} must be positive to obtain $g_{k-1}^T p_k < 0$. Now consider the matrix $S(t)$ defined by:

$$S(t) = S_{k-1} + t(p_k - S_{k-1} q_k) y_k^T \tag{35}$$

where t is a scalar in the interval $0 \leqslant t \leqslant 1$. From Eqn (35)

$$\det |S(t)| = \det |S_{k-1}| \cdot \{(1-t) - t y_k^T g_{k-1} \alpha_{k-1}\}. \tag{36}$$

Since $\det |S_{k-1}| \neq 0$ and $y_k^T g_{k-1} < 0$ by hypothesis and $\alpha_{k-1} > 0$, it follows that the right-hand side of Eqn (36) cannot vanish, and hence that $S(t)$ is non-singular, over the interval $0 \leqslant t \leqslant 1$. Now according to a theorem given by Carathéodory [8], a matrix $S(t)$ is positive definite in each interval $t_1 \leqslant t \leqslant t_2$ for which $S(t)$ is non-singular, provided that this interval contains a point t_0 for which $S(t_0)$ is positive definite. The point $t = 0$ is such a point in the interval $0 \leqslant t \leqslant 1$, and $S(t)$ is therefore positive definite over this interval. In particular $S_k = S(1)$ is positive definite and this completes the proof.

To examine the conditions for $y_k^T g_{k-1} < 0$ we consider separately the numerator and denominator of y_k. Substituting from Eqn (26) we have

$$g_{k-1}^T (p_k - S_{k-1} q_k) = (1 - \alpha_{k-1}) g_{k-1}^T S_{k-1} g_{k-1} - g_{k-1}^T S_{k-1} g_k \tag{37}$$

$$q_k^T (p_k - S_{k-1} q_k) = (2 - \alpha_{k-1}) g_{k-1}^T S_{k-1} g_k - (1 - \alpha_{k-1}) g_{k-1}^T S_{k-1} g_{k-1}$$

$$- g_k^T S_{k-1} g_k. \tag{38}$$

Now since S_{k-1} is positive definite, $g_{k-1}^T S_{k-1} g_{k-1} > 0$ and $g_{k-1}^T S_{k-1} g_k$ decreases from this value as α_{k-1} increases from zero, and becomes zero at the point where $f(x_k)$ has a stationary value (either a minimum or a point of inflection) along the line.

If the stationary value occurs when α_{k-1} is less than unity, then at this point and for a finite range of α_{k-1} greater than this, Eqn (37) shows that the numerator of y_k is positive whilst Eqn (38) shows that the denominator is negative. Thus y_k is finite and $y_k{}^T g_{k-1} < 0$ as required over this range of α_{k-1}.

If the stationary value of α_{k-1} is greater than unity it is not possible to predict a definite range of α_{k-1} which satisfies both of these conditions. However, if a suitable value of α_{k-1} does not exist we can avoid difficulty by simply writing $S_k = S_{k-1}$ rather than using Eqn (17).

The procedure for step k, given x_{k-1}, S_{k-1} and $g_{k-1} \neq 0$, is

(1) Starting with $\alpha_{k-1} = 1$ and computing p_k from Eqn (26), determine a point x_k with $0 < \alpha_{k-1} \leqslant 1$ to satisfy Eqn (28).

(2) Compute $g_k = g(x_k)$, and if it is zero proceed to the local search procedure for confirmation of a minimum. Otherwise go to step 3.

(3) Compute $q_k{}^T(p_k - S_{k-1} q_k)$, and if it has a magnitude greater than ε', compute y_k and $y_k{}^T g_{k-1}$. Check that $y_k{}^T g_{k-1} < -\varepsilon'$, and that condition (27) is satisfied.

(4) If any of these tests fail, return to the value of α_{k-1} first found in step 1 and reset S_k. Otherwise compute S_k from Eqn (17).

As a practical point we have

$$g_k{}^T S_k g_k = g_k{}^T S_{k-1} g_k + \frac{\{g_k{}^T (p_k - S_{k-1} q_k)\}^2}{\{q_k{}^T (p_k - S_{k-1} q_k)\}} . \tag{39}$$

It is therefore possible to compute $g_k{}^T S_k g_k$ in step 3 without actually updating S_{k-1}.

3. Minimization with Constraints

3.1. *The Projection Operator*

We now consider the problem

$$\text{minimize} \quad f^0(x) \tag{40}$$

$$\text{subject to} \quad f^j(x) \leqslant 0 \quad (1 \leqslant j \leqslant m^*) \tag{41}$$

If the constraints are linear, they can be written in the form

$$(g^j)^T x - b_j \leqslant 0 \quad (1 \leqslant j \leqslant m^*) \tag{41a}$$

(We assume that the equations have been scaled so that the vectors g^j, $1 \leqslant j \leqslant m^*$, are normalized.)

Suppose that the function $f^0(x)$ is quadratic and that the constrained minimum point lies on the intersection of m linearly independent hyperpl. given by a matrix equation of the form

$$G_m{}^T x - b = 0. \tag{42}$$

This intersection forms an $(n-m)$-dimensional subspace orthogonal to the m-dimensional subspace spanned by the normals, $g^j, 1 \leqslant j \leqslant m$, which form the columns of G_m.

The minimum point in the constraint space is characterized by a gradient orthogonal to this space

$$\bar{g}^0 = G_m \lambda \tag{43}$$

where λ is an arbitrary m-vector.

Suppose we stepped to the unconstrained minimum point $\hat{x}$ from any point x_{k-1}

$$\hat{x} - x_{k-1} = \hat{p} = -H^{-1} g^0_{k-1}. \tag{44}$$

Now this point, $\hat{x}$, transgresses the set of m constraints associated with Eqn (42), and we therefore seek $\bar{x}$ to satisfy Eqn (43), and also

$$G_m{}^T (\bar{x} - x_{k-1}) + f_{k-1} = 0 \tag{45}$$

where

$$f_{k-1} = f^j(x_{k-1}) \qquad (1 \leqslant j \leqslant m). \tag{46}$$

Now,

$$(\bar{x} - x_{k-1}) = H^{-1} (\bar{g}^0 - g^0_{k-1}) \tag{47}$$

and using Eqn (43)

$$\bar{x} - x_{k-1} = H^{-1} (G_m \lambda - g^0_{k-1}). \tag{48}$$

λ is determined from Eqn (45)

$$G_m{}^T (\bar{x} - x_{k-1}) = (G_m{}^T H^{-1} G_m) \lambda - G_m H^{-1} g^0_{k-1} = -f_{k-1} \tag{49}$$

$$\lambda = (G_m{}^T H^{-1} G_m)^{-1} (G_m{}^T H^{-1} g^0_{k-1} - f_{k-1}). \tag{50}$$

Note that the inverse exists, since H is positive definite.

Using Eqn (48), the step, $\bar{p}$, is thus given by

$$\bar{p} = \bar{x} - x_{k-1} = H^{-1} \{G_m (G_m{}^T H^{-1} G_m)^{-1} (G_m{}^T H^{-1} g^0_{k-1} - f_{k-1}) - g^0_{k-1}\} \tag{51}$$

and using Eqn (44)

$$\bar{p} = \hat{p} - H^{-1} G_m (G_m{}^T H^{-1} G_m)^{-1} (G_m{}^T \hat{p} + f_{k-1}). \tag{52}$$

Note that, if x_{k-1} is in the subspace, then $f_{k-1} = 0$; in which case Eqn (51) reduces to

$$\bar{p} = Pp = -PH^{-1} g^0_{k-1} \tag{53}$$

where

$$P = \{I - H^{-1} G_m (G_m H^{-1} G_m)^{-1} G_m{}^T\}. \tag{54}$$

P is idempotent and is therefore a projection operator, which projects any vector into the $(n-m)$-dimensional subspace formed by the intersection of the m constraining hyperplanes.

If H^{-1} is replaced by its current approximation, S_{k-1}, and S_{k-1} is such that the inverse $(G_m{}^T S_{k-1} G_m)^{-1}$ exists, then Eqns (44) and (52) become

$$p_k = \hat{p}_k - M_{m, k-1}(G_m{}^T \hat{p}_k + f_{k-1}) \tag{55}$$

where
$$\hat{p}_k = -S_{k-1} g^0_{k-1} \tag{56}$$

and
$$M_{m, k-1} = S_{k-1} G_m (G_m{}^T S_{k-1} G_m)^{-1}. \tag{57}$$

Equations (53) and (54) become

$$p_k = -\alpha_{k-1} P_{m, k-1} S_{k-1} g^0_{k-1}. \tag{58}$$

where
$$P_{m, k-1} = \{I - S_{k-1} G_m (G_m{}^T S_{k-1} G_m)^{-1} G_m{}^T\}. \tag{59}$$

Note that $P_{m, k-1}$ also projects any vector into the intersection of constraining hyperplanes. $P_{m, k-1}$ becomes an orthogonal projection operator analogous to Rosen's if we set $S_{k-1} = I$.

3.2. *Updating required for* S_{k-1}

At each step, S_{k-1} is updated to S_k by Methods 1 or 2 of Section 2. When constraints are present, we must then also update $(G_m{}^T S_{k-1} G_m)^{-1}$, $M_{m, k-1}$, and possibly $P_{m, k-1}$. The two methods show advantages and disadvantages at this stage, each requiring different treatment.

Method 1. With this method we can use the degrees of freedom in the choice of y_k to make unnecessary any recursive updating of these matrices.

Each time we add a new constraint to the basis, we not only effectively put q_k in column j of Q_{k-1} by updating R_{k-1}, but we also put g^i (corresponding to the new constraint i) in some other column of Q_k after the next step using Eqn (34). This involves an extra recursion of R_k, but note that this is required only when we add a new constraint to the basis, and not at every step. On dropping a constraint (say i') from the basis, the column of Q_{k-1} containing $g^{i'}$ simply becomes eligible for replacement by q_k, along with the others.

This procedure ensures that
$$y_k{}^T G_m = 0 \tag{60}$$
hence, from Eqn (14)

$$S_k G_m = S_{k-1} G_m \tag{61}$$

so that
$$(G_m{}^T S_k G_m)^{-1} = (G_m{}^T S_{k-1} G_m)^{-1}$$

$$P_{m, k} = P_{m, k-1} \tag{62}$$

$$M_{m, k} = M_{m, k-1}.$$

Note that although the constraint normals occupy m columns of ⸺ this does not prevent quadratic convergence in a subspace, since at most $(n - m)$ steps are needed to predict the minimum point. Further, from Eqns (14) and (61), we have $S_k q_j = S_{k-1} q_j$ for any previous vector q_j implicit in R_{k-1} and orthogonal to the subspace.

Method 2. Using this method of updating S_{k-1}, y_k is uniquely determined and we cannot impose the conditions (60). We must therefore use recursion formulae. However, since the correction applied to S_{k-1} is in the form of a simple product, we can update the inverse moment matrix without difficulty.

$$(G_m{}^T S_k G_m)^{-1} = \{G^T{}_m S_{k-1} G_m + G_m{}^T (p_k - S_{k-1} q_k) y_k{}^T G_m\}^{-1}$$

$$= (G_m{}^T S_{k-1} G_m)^{-1} - \frac{v_k{}^T y_k{}^T G_m (G_m{}^T S_{k-1} G_m)^{-1}}{1 + y_k{}^T G_m v_k} \tag{63}$$

where
$$v_k = (G_m{}^T S_{k-1} G_m)^{-1} G_m{}^T (p_k - S_{k-1} q_k). \tag{64}$$

Note that this recursion must be applied at every step.

There is no simple recursion for updating $P_{m, k-1}$. However, the search strategy in which $P_{m, k-1}$ is used requires, at each step, explicit formulation of the vector

$$\lambda = (G_m{}^T S_k G_m)^{-1} G_m{}^T S_k g_k{}^0. \tag{65}$$

It is little extra computation to form p_{k+1} from

$$p_{k+1} = S_k (G_m \lambda - g_k{}^0) \tag{66}$$

than it is to form it from

$$p_{k+1} = -P_{m, k} S_k g_k{}^0. \tag{67}$$

Thus there is little justification for even storing the matrix $P_{m, k}$, let alone updating it at each step.

Again for $M_{m, k-1}$, there is no simple recursion applicable. Having updated S_{k-1} and the inverse moment matrix, it is probably better to compute p_k in Eqn (55) by successive premultiplication of the vector $(G_m{}^T \hat{p}_k + f_{k-1})$ by the relevant matrices.

3.3. *Adding and Dropping Constraints*

For updating the inverse moment matrix, $(G_m{}^T S_k G_m)^{-1}$, we use recursion formulae analogous to those used by Rosen.

3.3.1. *Adding Constraint to Basis*

$$(G_{m+1}^T S_k G_{m+1}) = A = \begin{bmatrix} A_{11} & A_{12} \\ A_{21} & A_{22} \end{bmatrix}. \tag{68}$$

where $A_{11} = G_m{}^T S_k G_m$

$\quad A_{12} = G_m{}^T S_k g^{m+1}$

$\quad A_{21} = (g^{m+1})^T S_k G_m$

$\quad A_{22} = (g^{m+1})^T S_k g^{m+1}. \tag{69}$

Write:
$$A_0 = A_{22} - A_{21} A_{11}^{-1} A_{12} \tag{70}$$

which is a scalar. Hence

$$(G_{m+1}^T S_k G_{m+1})^{-1} = B = \begin{bmatrix} B_{11} & B_{12} \\ B_{21} & B_{22} \end{bmatrix}. \tag{71}$$

where $B_{11} = A_{11}^{-1} + A_{11}^{-1} A_{12} A_0^{-1} A_{21} A_{11}^{-1}$

$\quad B_{12} = -A_{11}^{-1} A_{12} A_0^{-1}$

$\quad B_{21} = -A_0^{-1} A_{21} A_{11}^{-1}$

$\quad B_{22} = A_0^{-1}. \tag{72}$

The only inverse matrix required is

$$A_{11}^{-1} = (G_m{}^T S_k G_m)^{-1}$$

which is known.

Assuming that g^{m+1} is linearly independent of the columns of G_m, we must still show that the new inverse always exists, that is

$$\det | G_{m+1}^T S_k G_{m+1}| \neq 0. \tag{73}$$

Note that the result follows immediately for Method 2 of updating S_{k-1}, as S_k remains positive definite. (Note also that $A_{12} = A_{21}^T$ for Method 2.) This is not so for Method 1, but since we have a certain degree of arbitrariness in the choice of S_k we can take explicit precautions to satisfy Eqn (73). By a Cauchy expansion of the determinant:

$$\det |G_{m+1}^T S_k G_{m+1}| = A_0 . \det |G_m{}^T S_k G_m|$$

where
$$A_0 = (g^{m+1})^T P_{m,k} S_k g^{m+1}. \tag{74}$$

Now if it should happen that $A_0 = 0$, we then make a small step, without adding the hyperplane, and update S_k again. We then have:

$$A_0 = (g^{m+1})^T P_{m,k+1} S_k g^{m+1} + (g^{m+1})^T P_{m,k+1}(p_{k+1} - S_k q_{k+1}) y_{k+1}^T g^{m+1}$$
$$= c_1 + c_2 y_{k+1}^T g^{m+1}, \tag{75}$$

where $c_1 = 0$, and c_2 is invariant on further choice of y_{k+1}, since $P_{m, k+1} = P_{m, k}$ for Method 1.

If the determinant is again zero for this new step, we can impose a further condition on y_{k+1}:

$$y_{k+1}^T g^{m+1} = 1. \tag{76}$$

However, in the further event that $c_2 = 0$ over a range of α_k, we have a degenerate case. This is only a remote possibility and the simplest procedure seems to be to reset S_k to a new positive definite matrix, for example the identity matrix.

3.3.2. *Dropping Constraint from Basis*

$$(G_m^T S_k G_m)^{-1} = A^{-1} = B = \begin{bmatrix} B_{11} & B_{12} \\ B_{21} & B_{22} \end{bmatrix}. \tag{77}$$

Solving for
$$A_{11}^{-1} = (G_{m-1}^T S_k G_{m-1})^{-1}$$
$$= B_{11} - B_{12} B_{22}^{-1} B_{21}. \tag{78}$$

It can be shown, by using a permutation matrix, that this formula also applies for the jth constraint to be dropped, with the mth and jth rows and columns interchanged.

If Method 1 for updating S_{k-1} was being used, it is possible to update the matrix $M_{m, k}$ conveniently on adding and dropping constraints:

(1) Adding constraint to basis:

$$M_{m+1, k} = [M_{m, k} - h_{m+1}(g^{m+1})^T M_{m, k}, h_{m+1}], \tag{79}$$

where
$$h_{m+1} = \frac{S_k g^{m+1} - M_{m, k} G_m^T S_k g^{m+1}}{(g^{m+1})^T (S_k g^{m+1} - M_{m, k} G_m^T S_k g^{m+1})}. \tag{80}$$

(2) Dropping constraint from basis:

$$M_{m-1, k} = M'_{m, k} + \frac{M_{m, k} e_j (g^j)^T M'_{m, k}}{1 + (g^j)^T M_{m, k} e_j}, \tag{81}$$

where $M'_{m, k}$ is the matrix $M_{m, k}$ with the jth column deleted. Note that the jth column can be any column, corresponding to the constraint to be dropped.

3.4. *Search Strategies*

3.4.1. *Constrained Step (Linear Constraints)*

This is analogous to Rosen's strategy for linear constraints. If x_k is an interior point, we search in the unconstrained direction.

$$p_{k+1} = -\alpha_k S_k g_k^0. \tag{82}$$

If x_k is on a set of active constraints, we project the unconstrained direction into the intersection of hyperplanes which form the constraint boundaries

$$p_{k+1} = -\alpha_k P_{m,\,k} S_k g_k^0. \tag{83}$$

A criterion, similar to that used by Rosen, for dropping a hyperplane from the constraint basis can also be developed for the operator $P_{m,\,k}$.

When

$$\|P_{m,\,k} S_k g_k^0\| \leqslant \tfrac{1}{2}\lambda_j/|b_{jj}| \tag{84}$$

then

$$\|P_{m-1,\,k} S_k g_k^0\| \geqslant \|P_{m,\,k} S_k g_k^0\|. \tag{85}$$

λ_j is the jth and algebraically largest element of the vector λ in Eqn (65), and b_{jj} is the corresponding jth diagonal element of $(G_m^T S_k G_m)^{-1}$. The inequality (85) indicates the need to drop the jth constraint from the basis.

From Eqn (43) it will be noted that λ is, in fact, the vector of Lagrange multipliers corresponding to the m active constraints. The Kuhn–Tucker conditions for a constrained overall minimum at x_k can be expressed as:

$$P_{m,\,k} S_k g_k^0 = 0 \tag{86}$$

$$\lambda_j \leqslant 0 \quad (1 \leqslant j \leqslant m) \tag{87}$$

A basic step of the algorithm, starting at point x_k, is as follows.

(1a) If Method 1 is used, and a new constraint, i, was incorporated in the basis for the previous step p_k, put g^i in the j^*th column of Q_k by using the recursion formula (10) on g^i and R_k.

$$j^* = \{j^* | e_{j*}^T R_k g^i \geqslant e_j^T R_k g^i, \quad j \notin J\}$$

where J is the set of columns occupied by either q_k, or $g^1 \dots g^m$. Then update S_{k-1} using Eqn (22).

(1b) If Method 2 is used, update S_{k-1} to S_k using Eqn (17), then update $(G_m^T S_{k-1} G_m)^{-1}$ to $(G_m^T S_k G_m)^{-1}$ using Eqn (63).

(2) If x_k is on the boundary of a new constraint add g^{m+1} to the constraint basis, update $(G_m^T S_k G_m)^{-1}$ to $(G_{m+1}^T S_k G_{m+1})^{-1}$ using Eqn (71), and put $m = m+1$.

(3) Form

$$\lambda = (G_m^T S_k G_m)^{-1} G_m^T S_k g_k^0. \tag{88}$$

(4) Form

$$p_{k+1} = S_k (G_m \lambda - g_k^0). \tag{89}$$

(5) Form

$$\beta = \tfrac{1}{2}\lambda_j/|b_{jj}| \quad (j \equiv \gamma_j > \lambda_i \quad 1 \leqslant i \leqslant m). \tag{90}$$

(6) Stop if $\|p_{k+1}\| \leqslant \varepsilon$ and $\beta \leqslant \varepsilon$, where ε is some small number; the Kuhn–Tucker conditions for an overall minimum are satisfied.

(7) If $\|p_{k+1}\| \leqslant \beta$, drop the constraint corresponding to λ_j from the basis, update $(G_m^T S_k G_m)^{-1}$ to $(G_{m-1}^T S_k G_{m-1})^{-1}$ using Eqn (78), and put $m = m - 1$. Return to step 3.

(8) Step to x_{k+1}: Calculate

$$\alpha_k{}^* = \min_i \{b_i - (g^i)^T x_k)/p_{k+1}^T g^i\} \quad (m+1 \leqslant i \leqslant m^*) \tag{91}$$

= position of nearest constraint.

Put $\alpha_k = \min (1.0, \alpha_k{}^*)$.

Step $x' = x_k + \alpha_k p_{k+1}$

if $f(x') < f(x_k)$, put $x_{k+1} = x'$; otherwise use quadratic interpolation to adjust α_k accordingly. Set $k = k + 1$ and return to step 1.

Degeneracy and linear dependence of hyperplanes can be handled in a similar fashion to Rosen.

3.4.2. Unconstrained Step (Linear and Nonlinear Constraints)

An alternative strategy, which may be more suitable for nonlinear constraints, is to step first to the unconstrained minimum point, and then test for constraint violations. The advantage of the constrained step strategy with linear constraints is that there is, at most, only one constraint added to, and one dropped from, the basis at each step. This avoids repeated recursion on the inverse moment matrix; however this advantage is not so obvious for nonlinear constraints, since the vectors g^i forming the columns of G_m change with position.

A basic step of the algorithm, starting at point x_k, is as follows.

(1) Update S_{k-1} to S_k, using Methods 1 or 2 of Section 2.

(2) Form

$$\hat{p}_{k+1} = -S_k g_k{}^0. \tag{92}$$

(3) Search for α_k such that $f(x)$ is an approximate minimum along the direction $\hat{p}_{k+1}$, set $\hat{p}_{k+1} = -\alpha_k S_k g_k{}^0$, and step

$$\hat{x}_{k+1} = x_k + \hat{p}_{k+1}. \tag{93}$$

(4) Test for constraint violations at $\hat{x}_{k+1}$, and if there are none go to step 9.

Calculate $f^j(\hat{x}_{k+1}) \quad (1 \leqslant j \leqslant m^*)$

(5) Form the set J of violated constraints†

$$f^j(\hat{x}_{k+1}) > 0 \quad (1 \leqslant j \leqslant m) \tag{94}$$

For any nonlinear constraint in J, make a linear approximation to it about the base point x_k. Hence define the subspace formed by the intersection of the linear approximations to the constraint set J

---

† When more than n constraints are violated, Lagrange multipliers are used to decide which constraints to drop from the violated set. When endeavouring to add a constraint to a violated set which is already full, the constraint j with the algebraically largest λ_j (with respect to the new constraint gradient $g_k{}^{m+1}$) is replaced.

$$G_{m,\,k}^T (x - x_k) + f_k = 0 \tag{95}$$

where $\qquad\qquad f_k = f^j(x_k) \qquad (1 \leqslant j \leqslant m) \tag{96}$

and $G_{m,\,k}$ is the $n \times m$ matrix whose columns are the gradients of the constraint functions

$$g^j(x_k) \quad (1 \leqslant j \leqslant m) \tag{97}$$

(6) Project $\hat{p}_{k+1}$ onto this intersection

$$p_{k+1} = \hat{p}_{k+1} - M_{m,\,k}(G_{m,\,k}^T \hat{p}_{k+1} + f_k) \tag{98}$$

where $\qquad\qquad M_{m,\,k} = S_k G_{m,\,k}(G_{m,\,k}^T S_k G_{m,\,k})^{-1}. \tag{99}$

(7) If there are bounds on the variables (or some other linear constraints outside the nonlinear constraints such that the feasible region is enclosed in a region bounded by linear constraints), step

$$x_{k+1} = x_k + p_{k+1} \tag{100}$$

otherwise, step

$$x_{k+1} = x_k + \alpha_k p_{k+1} \tag{100a}$$

where α_k is now adjusted to ensure $f(x_{k+1}) < f(x_0)$.

(8) Test x_{k+1} for constraint violations. If any new constraints are violated, add these to the set $J\dagger$ and return to step 5.

(9) Test for convergence. If $\|p_{k+1}\| \geqslant \varepsilon$, set $k = k+1$ and return to step 1.

(10) Form $\qquad\qquad \lambda = (G_{m,\,k}^T S_k G_{m,\,k})^{-1} G_{m,\,k}^T S_k g_k^0. \tag{101}$

Stop if $\lambda < \varepsilon$, the Kuhn–Tucker conditions for an overall minimum are satisfied. if any λ_j is positive, drop the constraint corresponding to the largest λ_j and reform λ_j repeat this process until all $\lambda_j < \varepsilon$, then return to step 6.

The inverse moment matrix, $(G_{m,\,k}^T S_k G_{m,\,k})^{-1}$ and $M_{m,\,k}$ are updated accordingly by recursion when any constraint, j, is added to or dropped from the active set J. Also, for nonlinear constraints, since $g^j(x_{k+1})$ differs from $g^j(x_k)$, we must use the same recursion (Eqns 71 and 78) when we replace the appropriate column in $G_{m,\,k}$.

4. Finite Convergence for Quadratic Functions

4.1. *Minimization without Constraints*

THEOREM 1. *Suppose we have a set of steps, p_i, $1 \leqslant i \leqslant k$, not necessarily consecutive, which satisfy the equation*:

$$S_k q_i = p_i \quad (1 \leqslant i \leqslant k) \tag{102}$$

with S_k non-singular.

---

† See footnote on p. 231.

Then, for a quadratic function, a step according to:

$$p_{k+1} = -S_k g_k \tag{103}$$

is either linearly independent of the steps, p_i, or attains a stationary point of the function.

Proof. Suppose that p_{k+1} is linearly dependent on the p_i:

$$p_{k+1} = \sum_{i=1}^{k} \beta_i p_i, \tag{104}$$

where at least one β_i is non-zero.

Since the function is quadratic:

$$q_i = H p_i. \tag{105}$$

From equations (104) and (105):

$$q_{k+1} = H p_{k+1} = H \sum_{i=1}^{k} \beta_1 p_i = \sum_{i=1}^{k} \beta_i q_i. \tag{106}$$

From equations (102), (104) and (106)

$$p_{k+1} = \sum_{i=1}^{k} \beta_i p_i = \sum_{i=1}^{k} \beta_i S_k q_i = S_k q_{k+1}. \tag{107}$$

From equations (103) and (107), since S_k is non-singular:

$$-g_k = q_{k+1} = g_{k+1} - g_k$$

whence $\qquad\qquad g_{k+1} = 0.$

This is the condition for attaining a stationary point, which is thus a necessary condition for linear dependence of p_{k+1} on the set p_i, $1 \leqslant i \leqslant k$.

COROLLARY 1.1. *The stationary point is attained in at most $(n+1)$ steps.*

This result follows immediately from the fact that the maximum possible number of linearly independent steps is n.

Although step (103) is always the first estimate for p_{k+1}, the proposed methods subsequently modify this by the factor $\alpha_k \neq 1$ if the function increases. However, this does not change the direction of the step, which therefore remains linearly independent of the set of p_i. The theorem therefore shows that we still build up a set of linearly independent p_i step by step.

No information is rejected for Method 2, and none is rejected in the first n steps for Method 3, so the Corollary is immediately applicable to these two methods. For Method 1 however we have to show that no vector q_i is subsequently eliminated in updating R_k and S_k.

Now if we suppose that g_k is linearly dependent on the set q_i, $1 \leqslant i \leqslant k$:

$$g_k = -\sum_{i=1}^{k} \beta_i q_i. \tag{108}$$

Then from equations (102), (103) and (108):

$$p_{k+1} = -S_k g_k = \sum_{i=1}^{k} \beta_i S_k q_i = \sum_{i=1}^{k} \beta_i p_i. \tag{109}$$

Thus if we do not step to a stationary point with p_{k+1}, the gradient g_k cannot be linearly dependent on the set q_i, $1 \leqslant i \leqslant k$. Since Q_k is non-singular, there must therefore be a non-zero element $e_m^T R_k g_k$ among the rows which do not correspond to the set q_i, and our algorithm will not eliminate one of the q_i if we need to incorporate g_k in Q_k.

To obtain the same result for q_{k+1} we need to assume in addition that H is non-singular. Then if q_{k+1} is linearly dependent on the set q_i

$$q_{k+1} = \sum_{i=1}^{k} \beta_i q_i \tag{110}$$

we have, using Eqn (105):

$$p_{k+1} = H^{-1} q_{k+1} = \sum_{i=1}^{k} \beta_i H^{-1} q_i = \sum_{i=1}^{k} \beta_i p_i. \tag{111}$$

Then again the attainment of a stationary point is a necessary condition for linear dependence of q_{k+1} on the set q_i, $1 \leqslant i \leqslant k$. Therefore there is a non-zero element $e_j^T R_k q_{k+1}$ among the rows which does not correspond to the set q_i, and our algorithm will not eliminate one of these in updating R_k and S_k. It then follows from Corollary 1 that we attain the minimum in at most $(n+1)$ steps.

We may note at this point that Broyden [6] was only able to prove finite convergence of our Method 2 for non-singular H, and wrongly asserted that this was a necessary condition for the generalized-secant methods.

4.2. Minimization with Linear Constraints

THEOREM 2. *Suppose we have a set of steps p_i ($1 \leqslant i \leqslant k$) not necessarily consecutive, which satisfy Eqn (102) with S_k non-singular. Then for a quadratic function, $f^0(x)$, a step according to*

$$p_{k+1} = S_k G_m (G_m^T S_k G_m)^{-1} (G_m^T S_k g_k^0 - f_k) - S_k g_k^0 \tag{112}$$

is either linearly independent of the steps, p_i, or else attains a stationary point in the space formed by the intersection of the constraint set

$$G_m^T (x - x_k) + f_k = 0.$$

Proof. As in Theorem 1 we suppose that p_{k+1} is linearly dependent on the p_i, so that Eqns (104)–(107) hold. From Eqn (112) and (107), since S_k is non-singular

$$G_m (G_m{}^T S_k G_m)^{-1} (G_m{}^T S_k g_k{}^0 - f_k) - g_k{}^0 = q_{k+1} = g^0{}_{k+1} - g_k{}^0. \qquad (113)$$

Thus $\qquad g^0{}_{k+1} = G_m (G_m{}^T S_k G_m)^{-1} (G_m{}^T S_k g_k{}^0 - f_k) = G_m \lambda \qquad (114)$

where $\qquad\qquad \lambda = (G_m{}^T S_k G_m)^{-1} (G_m S_k g_k{}^0 - f_k).$

Eqn (114) shows that $g^0{}_{k+1}$ is orthogonal to the intersection of the constraint set $G_m{}^T (x - x_k) + f_k = 0$, which is the condition for a stationary point in this intersection.

Hence the attainment of a stationary point in the intersection of the constraint set is a necessary condition for linear dependence of p_{k+1} on the set $p_i,\ 1 \leqslant i \leqslant k$.

COROLLARY 2. *The stationary point in the intersection of m linear constraints is attained in at most $(n - m + 1)$ steps in this intersection.*

COROLLARY 3. *If the intersection of a set of linear constraints has been spanned by previous steps then a step according to Eqn (112) attains immediately a stationary point in this intersection.*

Again the modification of Eqn (112) by a factor $\alpha_k \neq 1$ if the step is not successful does not change the direction of the step and we build up a set of linearly independent p_i within the intersection of the constraints step by step.

The properties of the three methods carry over from the unconstrained case, so that Corollaries 2 and 3 apply to them, subject in the case of Method 1 to non-singularity of the Hessian matrix.

5. Overall Convergence

Although a non-singular Hessian matrix was implicit in the basis of the methods presented, we have already seen that this assumption is unnecessary for minimization of quadratic functions, and it is also possible to prove ultimate convergence to a stationary point for a more general class of functions. The proofs are applicable to both methods developed here and are analogous to a proof given by Ostrowski [7] for a similar class of descent methods. However, the conditions used here have the advantage that they may be computationally verified.

5.1. *Minimization without Constraints*

THEOREM 3. *Consider an open bounded, n-dimensional set Ω in R^n with boundary S. The function $f(x)$ is defined and continuous on $\Omega + S$, and has a fixed value, C, on S, a value less than C for all $x \in \Omega$, and a value greater than C for all*

$x \notin \Omega + S$. Further, $f(x)$ has continuous second derivatives everywhere on $\Omega + S$, and there is a Λ such that

$$\|H(x)\| \leqslant \Lambda \quad x \in \Omega + S. \tag{115}$$

Starting at any point $x_0 \in \Omega + S$ with $g(x_0) \neq 0$, we generate a sequence $x_1 x_2 \ldots x_k x_{k+1} \ldots$ according to the rule

$$p_{k+1} = x_{k+1} - x_k = -\alpha_k S_k g_k \tag{116}$$

where the S_k are chosen to be non-singular and to satisfy the conditions

$$\rho \leqslant \|S_k\| \leqslant \sigma \tag{117}$$

and

$$|g_k^T S_k g_k| \geqslant \delta \cdot \|g_k\| \cdot \|S_k g_k\| \tag{118}$$

where δ, ρ, σ are fixed, positive numbers.

Then we can always choose a finite, non-zero α_k such that

$$f(x_k) - f(x_{k+1}) \geqslant \varepsilon \, \alpha_k g_k^T S_k g_k > 0 \tag{119}$$

with ε again a fixed positive number.

With all α_k chosen to satisfy (119), the sequence $x_1 x_2 \ldots x_k \ldots$ lies in Ω and tends to

$$\Omega^* = \{x | x \in \Omega; \quad g(x) = 0\}$$

in the sense that the distance of x_k from Ω^* tends to zero

$$|x_k, \Omega^*| \to 0 \quad k \to \infty \tag{120}$$

Proof. Expanding $f(x)$ about x_k in a Taylor series with a remainder of second order

$$f(x_{k+1}) = f(x_k) + g_k^T p_{k+1} + \tfrac{1}{2} p^T_{k+1} \cdot H(x_k + \theta p_{k+1}) \cdot p_{k+1} \quad (0 \leqslant \theta \leqslant 1) \tag{121}$$

Then, using Eqn (116) and choosing the sign of α_k to make $\alpha_k g_k^T S_k g_k$ positive, we obtain

$$f(x_k) - f(x_{k+1}) = \alpha_k g_k^T S_k g_k \left\{ 1 - \tfrac{1}{2} \frac{|a_k| \cdot g_k^T S_k^T H(x_k + \theta p_{k+1}) \cdot S_k g_k}{|g_k^T S_k g_k|} \right\} \tag{122}$$

Now the right-hand side of Eqn (122) has a positive limit as $\alpha_k \to 0$, so that for sufficiently small α_k we have $f(x_{k+1}) < f(x_k) \leqslant C$. Thus x_{k+1} remains in Ω, and so does the argument of $H(x_k + \theta p_{k+1})$.

For sufficiently small α_k, we can therefore use Eqns (115) and (117) in Eqn (122) to obtain

$$f(x_k) - f(x_{k+1}) \geqslant \alpha_k g_k^T S_k g_k \left\{ 1 - \frac{\sigma \Lambda}{2\delta} \cdot |\alpha_k| \right\} \tag{123}$$

So long as $|\alpha_k|$ remains less than $2\delta/\sigma\Lambda$ we have $f(x_{k+1}) < C$ and x_{k+1}

remains in Ω. We can therefore always choose a finite, non-zero α_k such that $\alpha_k g_k{}^T S_k g_k > 0$ and

$$|\alpha_k| < \frac{2\delta(1-\varepsilon)}{\sigma\Lambda} \tag{124}$$

to generate a point $x_{k+1} \in \Omega$ which satisfies condition (119).

With all α_k so chosen, the sequence $f(x_k)$ is monotonically decreasing and convergent and we have

$$f(x_k) - f(x_{k+1}) \to 0 \qquad k \to \infty \tag{125}$$

Now from Eqns (118) and (119)

$$f(x_k) - f(x_{k+1}) \geqslant \delta\varepsilon \cdot |\alpha_k| \cdot \|g_k\| \cdot \|S_k g_k\| \tag{126}$$

and since all α_k are non-zero it follows from Eqn (125) that either $\|g_k\| \to 0$ or $\|S_k g_k\| \to 0$ as $k \to \infty$. But all the S_k are non-singular and bounded below (cf. 117), so that either result implies that

$$g_k \to 0 \qquad k \to \infty \tag{127}$$

Further, from Eqns (116) and (117)

$$\|p_{k+1}\| \leqslant |\alpha_k| \cdot \|S_k\| \cdot \|g_k\| \leqslant \sigma \cdot |\alpha_k| \cdot \|g_k\| \tag{128}$$

and it follows that

$$p_{k+1} \to 0 \qquad k \to \infty \tag{129}$$

Now suppose that the sequence x_k does not tend to Ω^*. Then since the sequence x_k is in $\Omega + S$, a closed, bounded region of R^n, we can find a subsequence, x_i for which

$$|x_i, \Omega^*| \to d > 0.$$

But this subsequence in turn contains a subsequence converging to a point, ζ say, contained in Ω, so that

$$|\zeta, \Omega^*| = d > 0. \tag{130}$$

But $g_k \to 0$, so that we have $g(\zeta) = 0$ and ζ belongs to Ω^*. This contradicts Eqn (130), so that the sequence x_k must tend to Ω^*, and the proof is complete.

We note further that if the basic procedure converges to a stationary point which is not a (weak) local minimum, then the local search procedure will restart the basic procedure at a point with a lower function value, which thus excludes convergence to the same point again. We shall therefore ultimately converge to a (weak) local minimum.

It is to be expected that conditions (117) and (118) will usually be satisfied when S_k is generated from the standard rules, using an α_k which satisfies

Eqn (119). If this is not the case, however, we can always choose a new S_k which *does* satisfy these conditions. For example, Eqn (117) can always be satisfied by appropriate scaling of the elements of S_k, whilst for Eqn (118) S_k can be reset to a suitable positive definite matrix.

It is not in fact necessary to compute $\|S_k\|$ directly, for it is sufficient to replace the conditions (117) by the more easily computed conditions

$$\rho\|g_k\| \geqslant \|S_k g_k\| \leqslant \sigma g_k\|. \tag{131}$$

5.2. *Minimization with Constraints*

If the feasible region is a closed, bounded region Ω in R^n, whose boundary S is made up of a set of linear manifolds, we do not need to make the special assumptions concerning the value of the function $f(x)$ as in Theorem 3, for the projection algorithm ensures that the sequence of points x_k remains in $\Omega + S$.

However, the sequence is generated either by steps p_{k+1} according to Eqn (116), or by steps according to

$$p_{k+1} = -\alpha_k\{P_{m,k} S_k g_k^0 + M_{m,k} f_k\} \tag{132}$$

for different sets of constraints.

The subsequence of points generated by use of Eqn (116) is either finite, in which case it contains no limit point and will terminate if the procedure happens to find a stationary point, or it is infinite, in which case Theorem 3 may be applied to the subsequence, showing that it converges to Ω^* in the sense of the theorem.

We may similarly consider the subsequences of points generated by Eqn (132) for each set of constraints. Again if such a subsequence is finite it will terminate if it happens to find a stationary point in the intersection of the constraints. If it is infinite, we must examine the question of convergence.

Now if we use an anti-zigzagging precaution of the type used by Zoutendijk [9], we shall have $f_k = 0$ after a few steps according to Eqn (132) for a given set of constraints, and can then use the following theorem.

THEOREM 4. *Consider an open, bounded, n-dimensional region Ω in R^n with boundary S. The function $f(x)$ is defined and has continuous second derivatives on $\Omega + S$, and there is a Λ such that*

$$\|H(x)\| \leqslant \Lambda \qquad x \in \Omega + S \tag{133}$$

The linear manifold, S', described by the equation

$$G_m^T x - b = 0 \tag{134}$$

forms a portion of the boundary S. Starting at any point $x_0 \in S'$ with $g(x_0)$

not orthogonal to S' we generate a sequence $x_1 x_2 \ldots x_k \ldots$ according to the rule

$$p_{k+1} = -\alpha_k P_{m, k} S_k g_k{}^0 \tag{135}$$

where

$$P_{m, k} = I - S_k G_m (G_m{}^T S_k G_m)^{-1} G_m{}^T \tag{136}$$

and the S_k are chosen to be non-singular and to satisfy the conditions

$$\rho \leqslant \|S_k\| \qquad \|P_{m, k} S_k\| \leqslant \sigma \tag{137}$$

$$|y_k{}^{0T} P_{m, k} S_k g_k{}^0| \geqslant \delta . \|g_k\| . \|P_{m, k} S_k g_k{}^0\| \tag{138}$$

where δ, ρ, σ are fixed, positive numbers.

Then we can always choose a finite non-zero α_k such that

$$f(x_k) - f(x_{k+1}) \geqslant \varepsilon \alpha_k g_k{}^{0T} P_{m, k} S_k g_k{}^0 \geqslant 0 \tag{139}$$

with ε again a fixed positive number.

With all the α_k chosen to satisfy Eqn (139), the sequence $x_1 x_2 \ldots x_k \ldots$ lies in S' and tends to a point in $S^* = \{x | x \in S'; P_{m, k} S_k g_k{}^0 = 0\}$ in the sense that the distance of x_k from S^* tends to zero.

Proof. Since $x_0 \in S'$ the rule given by Eqns (135) and (136) ensures that the whole sequence $x_1 x_2 \ldots x_k \ldots$ lies in S'. We can accordingly expand $f(x_{k+1})$ in a Taylor series with remainder of second order, and by a method entirely analogous to that used in Theorem 3 show that Eqn (125) is true and a finite, non-zero α_k can be chosen to satisfy Eqn (139).

Then from Eqns (138) and (139)

$$f(x_k) - f(x_{k+1}) \geqslant \delta \varepsilon . |\alpha_k| . \|g_k{}^0\| . \|P_{m, k} S_k g_k{}^0\|. \tag{140}$$

From Eqn (125) it then follows that either $\|g_k{}^0\| \to 0$ or $\|P_{m, k} S_k g_k{}^0\| \to 0$, and both of these imply that

$$P_{m, k} S_k g_k{}^0 \to 0 \qquad k \to \infty \tag{141}$$

Further, from Eqn (135)

$$\|p_{k+1}\| = |\alpha_k| . \|P_{m, k} S_k g_k{}^0\| \tag{142}$$

and hence

$$p_{k+1} \to 0 \qquad k \to \infty \tag{143}$$

As in Theorem 3 we can prove from these results that the sequence x_k must tend to S^*.

It was earlier shown (*cf.* Eqn 66) that $P_{m, k} S_k g_k{}^0 = 0$ (with $\|S_k\| \neq 0$) implies that $g_k{}^0$ is orthogonal to the manifold S', so that we do indeed converge to a constrained stationary point in S'.

Conditions (137) and (138) can again be satisfied if necessary by resetting S_k. Again it is sufficient to replace conditions (137). by the more easily computed conditions

$$\rho||G_m\lambda - g_k{}^0|| \leqslant ||S_k(G_m\lambda - g_m{}^0)|| \leqslant \sigma||G_m\lambda - g_k{}^0|| \qquad (144)$$

with λ given by Eqn (65).

Unfortunately, we cannot force convergence in this way for nonlinear constraints, since it is possible to reach points outside the feasible region. Such points may give a lower value of $f(x)$ than the solution, so that a condition like (139) would prevent the return of the sequence to the feasible region, and cannot therefore be imposed.

Guided by the above considerations, however, it seems reasonable either to enclose the actual feasible region in a region bounded by linear constraints and add these to the set m^*, or to enclose it in a contour as described in Theorem 3. Then if x_k transgresses any constraints we replace condition (139) by the condition that $f(x_{k+1}) \leqslant C^*$, the best feasible value obtained so far.

6. Numerical Examples

Example 1 (*Unconstrained*). Rosenbrock's Valley.

Minimize $f^0(x) = 100(x_1{}^2 - x_2)^2 + (1 - x_1)^2.$

Initial point $(-1\cdot2, 1\cdot0).$

The results, in comparison with Steepest Descent and Fletcher–Powell, are shown in Table I below. Both methods are considerably better than Steepest Descent, although only Method 2 shows up as comparable to Fletcher–Powell for this example. Fletcher and Powell's [2] published results for their method and Steepest Descent do not give the number of function evaluations but only the number of steps. In order to obtain a direct comparison, the search along successive directions was also continued to an approximate minimum in our methods although of course the expected gain is precisely in avoiding the need for this.

TABLE I

Step	Steepest Descent	Fletcher–Powell	Method 1	Method 2
0	24·200	24·200	24·200	24·200
3	3·704	3·687	3·760	3·352
6	3·339	1·605	2·260	1·441
9	3·077	0·745	1·384	0·683
12	2·869	0·196	0·488	0·248
15	2·689	0·012	0·253	0·050
18	2·529	1×10^{-8}	0·010	5×10^{-8}
21	2·383	—	2×10^{-4}	—
24	2·247	—	5×10^{-11}	—

Example 2. Quadratic with Linear Constraints.

Minimize

$$f^0(x) = \sum_{j=1}^{n} a_j x_j + \tfrac{1}{2} \sum_{i=1}^{n} \sum_{j=1}^{n} c_{ij} x_i x_j$$

subject to

$$-x_i \leqslant 0 \quad (i = 1, \ldots n)$$

$$\sum_{j=1}^{n} g_{ij} x_j - b_i \leqslant 0 \quad (i = n+1 \ldots m)$$

$n = 4$, $m = 7$. Initial Point $(\tfrac{1}{2}, \tfrac{1}{2}, \tfrac{1}{2}, \tfrac{1}{2})$.

$i/^j$	1	2	3	4
a_j	−1	−3	1	−1

c_{ij}	1	2	3	4
1	2	0	−1	0
2	0	1	0	0
3	−1	0	2	1
4	0	0	1	1

$i/^j$	1	2	3	4	b_i
g_{ij}					
5	1	2	1	1	5
6	3	1	2	−1	4
7	0	−1	−4	0	−1·5

The results are shown in Table II.

TABLE II

	Method 1			Method 2		
Step	Number of Function Calculations	Number of Constraints in Basis	$f^0(x)$	Number of Function Calculations	Number of Constraints in Basis	$f^0(x)$
0	1	0	−1·250	1	0	−1·250
1	2	0	−2·882	2	0	−2·882
2	3	1	−4·081	3	1	−4·044
3	4	2	−4·124	4	2	−4·120
4	5	1	−4·409	5	1	−4·397
5	6	2	−4·682	6	2	−4·682

In this example, the constrained overall minimum was reached in $n+1$ steps. This will not occur in general, however, even for quadratic functions, since we may not be in the correct subspace at the nth step.

Example 3. Cubic Example of Goldfarb and Lapidus.

Minimize

$$f^0(x) = \sum_{j=1}^{n} a_j x_j + \sum_{j=1}^{n} \sum_{i=1}^{n} c_{ij} x_i x_j + \sum_{j=1}^{n} d_j x_j^3$$

subject to

$$-x_i \leqslant 0 \quad (i = 1 \dots n)$$

$$\sum_{j=1}^{n} g_{ij} x_j - b_i \leqslant 0 \quad (i = n+1 \dots m)$$

$n = 5$, $m = 15$. Initial point $(0, 0, 0, 0, 1)$.

i/j	1	2	3	4	5	
a_j	−15	−27	−36	−18	−12	
c_{ij}						
1	30	−20	−10	32	−10	
2	−20	39	−6	−31	32	
3	−10	−6	10	−6	−10	
4	32	−31	−6	39	−20	
5	−10	32	−10	−20	30	
d_j	4	8	10	6	2	

i/j	1	2	3	4	5	b_i
g_{ij}						
6	16	−2	0	−1	0	40
7	0	2	−0	−0·4	−2	2
8	3·5	0	−2	0	0	0·25
9	0	2	0	4	1	4
10	0	9	2	−1	2·8	4
11	−2	0	4	0	0	1
12	1	1	1	1	1	40
13	1	2	3	2	1	60
14	−1	−2	−3	−4	−5	−5
15	−1	−1	−1	−1	−1	−1

The results, in comparison with Goldfarb–Lapidus, are shown in Table II1.

TABLE III

| | Goldfarb–Lapidus | | | Method 1 | | | Method 2 | | |
Step	Number of function calculations	Number of constraints in basis	$f^0(x)$	Number of function calculations	Number of constraints in basis	$f^0(x)$	Number of function calculations	Number of constraints in basis	$f^0(x)$
0	1	0	+ 20·0000	1	0	+ 20·0000	1	0	+ 20·0000
1	2	1	− 23·8967	2	1	− 23·8967	2	1	− 23·8967
2	3	2	− 25·1972	3	2	− 25·2247	3	2	− 25·1909
3	5	3	− 25·2605	4	2	− 28·7506	4	2	− 30·6838
4	6	2	− 28·5235	5	3	− 28·9891	5	3	− 32·0117
5	7	3	− 29·6326	6	3	− 31·0703	6	4	− 32·348664
6	8	4	− 32·0615	7	3	− 32·1685	7	4	− 32·348679
7	10	3	− 32·1134	8	3	− 32·3469			
8	11	3	− 32·3353	9	4	− 32·348678			
9	13	4	− 32·348679	10	4	− 32·348679			

Both Methods 1 and 2 compare favourably with Goldfarb–Lapidus, especially in the number of function calculations required. In particular, note that there is a marked drop in function value at step 3 for both methods. A hyperplane was dropped from the constraint basis (and one added) for this step. Although this is not a particularly good example to illustrate the point, since it happens only once, it does indicate the advantage of not having to discard information orthogonal to any given subspace.

Method 1, while having the computational advantage of not having to use the recursion formulae repeatedly, does have the disadvantage that for general functions we are suppressing information by storing the constraint normals g^i in the columns of Q_k.

Example 4. Unconstrained Step Strategy for Nonlinear Constraints
A simple example to illustrate the idea of this strategy is to minimize a quadratic form subject to a spherical constraint.

Minimize $\qquad f^0(x) = (x_1 - x_2)^2 + ((x_1 + x_2 - 10)/3)^2 + (x_3 - 5)^2$

subject to $\qquad f'(x) = x_1{}^2 + x_2{}^2 + x_3{}^2 - 48 \leqslant 0$
$$-4\cdot5 \leqslant x_1, x_2 \leqslant 4\cdot5$$
$$-5\cdot0 \leqslant x_3 \leqslant 5\cdot0.$$

Initial point $(-5, 5, 0)$.

The progress of the solution, together with the coefficients of the linearized constraint equation, $\sum_i g_i' x_i - b' \leqslant 0$, is shown in Table IV.

TABLE IV

Step	i	Point x_i	Coefficients g_i'	Step p_i	Number of Active Constraints	
1	1	0·9392	−1·0	5·9392	0	
	2	0·2486	1·0	−4·7514	b'	= 9·80
	3	2·6727	0·0	2·6727	$f(x)$	= 14·5217
2	1	1·7733	1·8785	0·8341	1	
	2	1·7652	0·4972	1·5165	b'	= 56·0871
	3	5·0000	5·3453	2·3273	$f(x)$	= 4·6391
3	1	4·0720	3·5466	2·2987	2	
	2	4·1975	3·5303	2·4324	b'	= 79·2603
	3	5·0000	10·0000	0·0	$f(x)$	= 0·3485
4	1	3·6762	0·8144	−0·3958	1	
	2	3·6718	0·8395	−0·5258	b'	= 10·7201
	3	4·6437	1·0000	−0·3563	$f(x)$	= 0·9085
5	1	3·6502	0·7352	−0·0260	1	
	2	3·6503	0·7344	−0·0215	b'	= 9·6560
	3	4·6208	0·9287	−0·0229	$f(x)$	= 0·9533

Step 6 $\quad \| p \| = 5 \times 10^{-4}$

APPENDIX

In many problems, calculation of the gradient is an expensive step. When searching along the direction, $p_{k+1} = -\alpha_k S_k g_k$, from x_k, it is convenient to use only function values and make a quadratic interpolation.

Write:
$$f(x_{k+1}) = f(x_k - \alpha S_k g_k) = \phi(\alpha) \tag{1}$$

and assume $\phi(\alpha)$ can be represented by

$$\phi(\alpha) = a + b\alpha + c\alpha^2. \tag{2}$$

For $\alpha = 0$:
$$\phi(0) = f(x_k) = f_k = \phi_0 = a \tag{3}$$

and
$$\partial\phi/\partial\alpha = -(g_k)^T S_k g_k = b. \tag{4}$$

For $\alpha = \alpha_r$:
$$\phi(\alpha_r) = \phi_r = a + b\alpha_r + c\alpha_r^2. \tag{5}$$

For a minimum along the line:

$$\partial\phi/\partial\alpha = 0 = b + 2c\alpha. \tag{6}$$

Hence choose:
$$\alpha_{r+1} = \frac{b}{2c}. \tag{7}$$

But from Eqn (5):
$$c = \frac{\phi_r - a - b\alpha_r}{\alpha r^2} \tag{8}$$

and substituting for a and b from Eqns (3) and (4)

$$\alpha_{r+1} = \frac{1}{2} \frac{\alpha_r^2 (g_k)^T S_k g_k}{(\phi_r - \phi_0) + \alpha_r \cdot (g_k)^T S_k g_k}. \tag{9}$$

Thus we may use Eqn (9) iteratively to generate successive points. along the line, and if $g_k^T S_k g_k > 0$ and $\phi_r > \phi_0$, we have $0 < \alpha_{r+1} < \alpha_r$ for all r

Acknowledgements

The authors are indebted to Martin J. Leigh for many helpful and illuminating discussions, in particular concerning the proofs of convergence. B. A. Murtagh wishes to acknowledge financial support from the Commonwealth Scholarship Commission of the Association of Commonwealth Universities.

References

1. J. B. Rosen (1960). The gradient projection method for nonlinear programming. Part 1. Linear constraints. *J. Soc. Ind. Appl.. Math.* **8**, 181.
2. R. Fletcher and M. J. D. Powell (1963). A rapidly convergent descent method for minimization. *Comput. J.* **6**, 163.

3. W. C. Davidon (1959). "Variable Metric Method for Minimisation" A.E.C. Research and Development Report, ANL-5990 (Rev.)
4. D. Goldfarb and L. Lapidus (1967). "A Conjugate Gradient Method for Nonlinear Programming". Paper presented at A.I.Ch.E. 61st National Meeting, Houston, February, 1967.
5. J. G. P. Barnes (1965). An algorithm for solving nonlinear equations based on the secant method. *Comput. J.* **8**, 66.
6. C. G. Broyden (1967). Quasi-Newton methods and their application to function minimization. *Math. Comp.* **21**, 368.
7. A. M. Ostrowski (1966). "The Solution of Equations and Systems of Equations". Pure and Applied Mathematics Series, Vol. 9, pp 195-199. Academic Press, New York and London.
8. C. Carathéodory (1967). "Calculus of Variations and Partial Differential Equations of the First Order". Vol. 2, p.190. Holden Day.
9. G. Zoutendijk (1960). "Methods of Feasible Directions", p. 72, Elsevier, Amsterdam,

Discussion

G. P. McCAULEY (University of Birmingham). In Professor Sargent's Method 1 on Rosenbrock's Valley, are the number of entries to the function evaluation routine as many as were made in the Fletcher–Powell method?

SARGENT. Each step involves a search for a minimum along a direction involving a number of function evaluations. The method of cubic interpolation as used by Fletcher and Powell was also used here so the comparison should follow the results in Table I.

FLETCHER. The reason for only comparing the methods at iterations 0, 3, 6, and so on in this Table is because a comparison was originally also made with Powell's method (1962) a cycle of which required three iterations in two variables.

15. An Algorithm for Constrained Minimization

W. MURRAY

National Physical Laboratory, Teddington, England

1. Introduction

In the last few years considerable attention has been paid to the solution of constrained optimization problems *via* Penalty and Barrier Functions, particularly by Fiacco and McCormick [1–3]. The more successful of these two approaches so far has been with Barrier Functions and some reasons for this are given in [4]. Unfortunately Barrier Functions deal only with inequality constraints and it becomes necessary to employ Penalty Functions if we wish to deal with equality constraints.

Penalty and barrier functions suffer from a number of difficulties. For instance, the problem becomes progressively more ill-conditioned as the solution is approached, and when constraints are linear no special advantage is taken of this linearity. An additional difficulty with penalty functions is that convergence depends on the initial penalty parameter r_0 being sufficiently small. Although one can write elaborate procedures to determine the "best" r_0 it may be that the information is not available for it to be a good choice. Even given a good initial estimate and a suitable r_0 as can be seen from Fig. 1 we may still fail to converge to the desired local minimum.

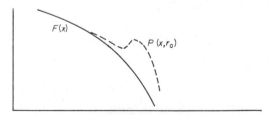

FIG. 1. Possible behaviour of penalty function $P(x, r_0)$ for an objective function $F(x)$.

In addition to all this we are solving a discrete set of problems and all the information accumulated in solving one cannot easily be employed to solve the next. By exploiting the properties of penalty functions and their relationship with the Lagrangian function we derive an algorithm that continuously alters r. In this way we overcome many of the disadvantages mentioned while still retaining many of the advantages.

2. Properties of the Penalty Function

The problem of concern is

$$\min \{F(x)\} \qquad x = (x_1, ..., x_n)$$
$$g_i(x) \geqslant 0 \qquad i = 1, 2, ..., m \qquad (1)$$
$$F(x), g_1(x) \varepsilon C^2$$

For simplicity of presentation we neglect equality constraints.

The penalty function we shall use is

$$P(x, r_k) = F(x) + r_k^{-1} g^T g \qquad (2)$$

g being the vector of violated constraints.

Pietrzykowski in [5] proved the following theorem.

THEOREM 1. *Let $\overset{*}{x}$ be a solution of the problem* (1) *and $\overset{*}{x}_k$ a locally strong minimum of* (2). *Then a necessary and sufficient condition that there exists a sequence $\overset{*}{x}_0, \overset{*}{x}_1, \overset{*}{x}_2, ...,$ for which*

$$\lim \overset{*}{x}_k = \overset{*}{x}$$

is that $r_0, r_1, ...,$ be a sequence of positive numbers converging to zero.

We note that multiplying $g_i(x)$ in Eqn (2) by a positive scalar has no effect on the theorem.

The dual of (1) can be written

$$\left. \begin{array}{c} \max\limits_{x,u} \left\{ \phi(x, u) = F(x) - \sum\limits_{i=1}^{m} u_i g_i(x) \right\} \\[2mm] \nabla_x \phi(x, u) = 0, \qquad u_i \geqslant 0, \qquad i = 1, ..., m \end{array} \right\} \qquad (3)$$

Let
$$\overset{*}{u}_{k, i} = -2r_k^{-1} g_i(\overset{*}{x}_k), \qquad g_i(\overset{*}{x}_k) < 0$$
$$= 0, \qquad g_i(\overset{*}{x}_k) \geqslant 0 \qquad (4)$$

hence
$$\overset{*}{u}_{k, i} \geqslant 0, \qquad i = 1, ..., m$$

Differentiating $\phi(x, u)$ with respect to x we obtain

$$\nabla_x \phi(x, u) = \nabla_x F(x) - \sum_{i=1}^{m} u_i \nabla_x g_i(x)$$

Substituting from (4) gives

$$\nabla_x \phi(\overset{*}{x}_k, \overset{*}{u}_k) = \nabla_x F(\overset{*}{x}_k) + \sum_{i \in I} 2r_k^{-1} g_i(\overset{*}{x}_k) \nabla_x g_i(\overset{*}{x}_k)$$

where I is the set of indices of the violated constraints.

Since $\overset{*}{x}_k$ is a minimum of $p(x, r_k)$ then

$$\nabla_x P(\overset{*}{x}_k, r_k) = 0.$$

$(\overset{*}{x}_k, \overset{*}{u}_k)$ is therefore dual feasible

$$\phi(\overset{*}{x}_k, \overset{*}{u}_k) = 2P(\overset{*}{x}_k, r_k) - F(\overset{*}{x}_k) \tag{5}$$

hence
$$\lim_{k \to \infty} \phi(\overset{*}{x}_k, \overset{*}{u}_k) = F(\overset{*}{x}).$$

Differentiating Eqn (2) we obtain

$$\nabla_x P(x, r_k) = \nabla_x F(x) + 2r_k^{-1} A^T g \tag{6}$$

where A is the Jacobian of g.

$$\nabla_x P(x+\varepsilon, r_k) = \nabla_x F(x) + B\varepsilon + 2r_k^{-1} \left(\sum_{i \in I} (g_i(x) + \varepsilon^T \nabla_x g_i(x)) \right)$$
$$\times (\nabla_x g_i(x) + G_i \varepsilon) + \ldots$$

where B is the Hessian matrix of $F(x)$ and G_i is the Hessian matrix of $g_i(x)$.

Let
$$\|\varepsilon\| = \|g(x)\| = O(r_{k-1}) \tag{7}$$

then
$$\nabla_x P(x+\varepsilon, r_k) = \nabla_x F(x) + 2r_k^{-1} (A^T g + A^T A \varepsilon) + O(r_{k-1}).$$

Let $x+\varepsilon$ be such that

$$\| \nabla_x F(x) + 2r_k^{-1} (A^T g + A^T A \varepsilon) \| = O(r_{k-1}). \tag{8}$$

Let x be such that
$$\| \nabla_x P(x, r_{k-1}) \| = O(r_{k-1}). \tag{9}$$

Hence
$$A\varepsilon = - \left(1 - \frac{r_k}{r_{k-1}} \right) g + O(r_{k-1} r_k). \tag{10}$$

Now Eqn (8) is true for $x + \varepsilon = \overset{*}{x}_k$, and Eqn (9) is true for $x = \overset{*}{x}_{k-1}$. Murray [4] has shown that under certain assumptions Eqn (7) is also true for $x = \overset{*}{x}_{k-1}$, and indeed this follows from Eqn (4) if we assume $\overset{*}{u}$ finite.

We can use Eqn (10) to obtain estimates of $F(\overset{*}{x}_k), P(\overset{*}{x}_k, r_k)$ and $g(\overset{*}{x}_k)$

$$F(\overset{*}{x}_k) \doteq F(\overset{*}{x}_{k-1}) + \varepsilon^T \nabla_x F(\overset{*}{x}_{k-1}) = F(\overset{*}{x}_{k-1}) - \frac{2}{r_{k-1}} g^T A\varepsilon$$

$$= F(\overset{*}{x}_{k-1}) + \frac{2}{r_{k-1}} \left(1 - \frac{r_k}{r_{k-1}} \right) g^T g. \tag{11}$$

Similarly

$$P(\overset{*}{x}_k, r_k) \doteq P(\overset{*}{x}_{k-1}) + \frac{1}{r_{k-1}} \left(1 - \frac{r_k}{r_{k-1}} \right) g^T g. \tag{12}$$

Combining Eqns (11) and (12)

$$F(\overset{*}{x}_k) - F(\overset{*}{x}_{k-1}) \doteq 2(P(\overset{*}{x}_k, r_k) - P(\overset{*}{x}_{k-1}, r_{k-1}))$$

and in particular for $r_k = 0$

$$F(\overset{*}{x}) \doteq 2P(\overset{*}{x}_{k-1}, r_{k-1}) - F(\overset{*}{x}_{k-1}). \tag{13}$$

This is exactly the expression for $\phi(\overset{*}{x}_{k-1}, \overset{*}{u}_{k-1})$ from Eqn (5).

In practice these estimates have proved very close to the actual values obtained. It would be useful therefore to employ Eqn (10) in a more continuous and direct fashion.

THEOREM 2. *Let S_k be the set of points that satisfy Eqns (7) and (9) for a sequence r_k, $k = 0, 1, 2, \ldots$, which are positive and converge to zero.*

$$\hat{S}_k = \left\{ x + \varepsilon \mid x \in S_{k-1}, A\varepsilon \geqslant -\left(\frac{1 - r_k}{r_{k-1}} \right) g, \|\varepsilon\| \leqslant r_{k-1} \right\}$$

$$\overset{*}{S}_k = \left\{ y \mid y \in S_k, \|\nabla_y P(y, r_k)\| \leqslant \max_y \frac{r_k}{r_{k-1}} \|\nabla_y P(y, r_k)\| \right\}$$

Then a sufficient condition that $S_k = \overset{*}{S}_k$ and $\lim_{k \to \infty} \overset{*}{S}_k = \overset{*}{x}$ is $O(r_k) \geqslant O(r_{k-1}^2)$.

Proof. For $y \in \hat{S}_k$

$$g(y) = O(g(x) + A\varepsilon + r_{k-1}^2)$$

$$= O\left(\frac{r_{k-1}}{r_k} g(x) + r_{k-1}^2 \right)$$

$$g(y) = O(r_k). \tag{14}$$

From Eqn (8) for $y \in \hat{S}_k$ $\nabla_y P(y, r_k) = O(r_{k-1})$

for $y \in \overset{*}{S}_k$ $\nabla_y P(y, r_k) = O(r_k). \tag{15}$

Equations (14) and (15) are precisely the conditions for $y \in S_k$.

For $y \in \overset{*}{S}_k$ $\lim_{k \to \infty} \nabla_y P(y, r_k) = \lim_{k \to \infty} g(y) = 0$

therefore $\lim_{k \to \infty} \overset{*}{S}_k = \overset{*}{x}.$

3. Quadratic Approximation

The normal procedure to determine the minimum of some function $\theta(x)$ is to make a quadratic approximation to $\theta(x)$ and then search along the direction which passes through the minimum of that quadratic. We form a quadratic approximation by making an initial estimation of the inverse Hessian matrix of $\theta(x)$ and we then form a new approximation from the old in the light of information obtained within an iteration. Precisely which recurrence relation one should use has been the subject of research by a number of authors [6]. For our part we have tried about six such recurrence relations with a number of modifications to ensure the resulting matrix is positive definite and symmetrical. At this time we prefer not to comment on preferences but will be contented to assume that it is possible to approximate $\theta(x+p)$ by

$$\theta(x+p) \doteq \theta(x)+p^T\nabla_x\theta(x)+\tfrac{1}{2}p^TBp$$

where B is in some sense an approximation to the Hessian matrix of $\theta(x)$.

Since we already have some conditions on a profitable direction of search we shall choose p such that

$$\min_{p} \{\tfrac{1}{2}p^TBp+f^Tp\} \tag{16}$$

subject to

$$Ap \geqslant -\left(1-\frac{r_k}{r_{k-1}}\right)g \tag{17}$$

where $f = \nabla_x\theta(x)$. Our requirement for the choice of $\theta(x)$ is that for p chosen subject to Eqn (17), $P(x+p, r_k)$ is reduced if $\theta(x+p)$ is reduced. We have considered the following functions; our principle experience has been with (b).

(a) $P(x, r_k)$

(b) $F(x)$

(c) $\bar{P} = F(x) - \sum_{i \in I} g_i(x)$

It can be shown that if $\Theta(x)$ is chosen to be either (a), (b) or (c) the solution to (16) is unaltered if $f = \nabla_x F(x)$. The choice of (c) is possibly a little obscure.

Consider the first neglected terms in our approximation of $\nabla_x P(x, r_k)$ in Section 2.

$$\left(B+2r_k^{-1}\sum_{i \in I}g_i(x+\varepsilon)\,G_i\right)\varepsilon$$

where B and G_i are as defined there.

As will be seen in Section 4 our g_i are modified by a factor $\hat{u}_i$ at each iteration so that

$$\lim_{k \to \infty} -2r_k^{-1} g_i = 1.$$

At the minimum therefore the Hessian matrix of $\bar{P}$ approximates to the neglected terms. The advantage of choosing (b) or (c) is that they are independent of r.

Whatever the choice of $\theta(x)$ the linear search is made subject to $P(x, r_k)$. If $p^T \nabla_x P(x, r_k) > 0$ it can be shown there exists an r $(0 < r < r_k)$ for which $p^T \nabla_x P(x, r) < 0$. Alternatively a new p can be computed for a smaller r. As will be seen in Section 4 this entails very little extra work.

4. Quadratic Programming (QP)

In order to determine a direction of search we are required to solve the following QP

$$\min_p \{\tfrac{1}{2} p^T B p + f^T p\}$$

$$Ap \quad \geqslant -b \tag{18}$$

where $b = (1 - r_k/r_{k-1}) g$.

The dual can be written [7]

$$\min_{z, u} \{\tfrac{1}{2} z^T B z + b^T u\}$$

$$A^T u - Bz = f, \qquad u \geqslant 0.$$

If $\hat{p}$ is the solution to Eqn (18) then

$$\hat{p} = B^{-1}(A^T \hat{u} - f)$$

where $\hat{u}$ is the solution of

$$\min_u \{\tfrac{1}{2} u^T Q u - c^T u\} \tag{19}$$

$$u \geqslant 0$$

$$Q = AB^{-1}A^T$$

and

$$c = (AB^{-1}f - b).$$

Let

$$Q\overset{*}{u} = c. \tag{20}$$

If any $\overset{*}{u}_i < 0$ then this corresponds to a prediction that at least one of the constraints will leave the violated set. Although this happens it does so infrequently; furthermore, it is even rarer for more than one constraint to leave the violated set in the same iteration.

At each iteration we modify the constraints in the violated set as follows

$$g_i = \hat{u}_i g_i.$$

Hence as we approach the solution

$$\hat{u} \to e$$

where

$$e = \sum e_i$$

and e_i is the ith column of the identity matrix. If for some reason it was required that the ratio $g_i(0)/g_j(0) = 1, i, j \in I$ then the required multiplier is $\hat{u}_i{}^{\ddagger}$.

The following is a brief outline of a modification of the method of Theil and van de Panne [8] applied to Eqn (19).

Determine the lower triangular matrix L

where

$$LL^T = Q.$$

Define

$$\Delta \overset{*}{u} = \overset{*}{u} - e.$$

Solve

$$LL^T \Delta \overset{*}{u} = c - Qe.$$

If $\overset{*}{u}_i \geqslant 0$ then $\hat{u} = \overset{*}{u}.$

If this is not so our procedure is based on the following results.

(a) If only one constraint (ith) is not binding then

$$\hat{u} = \overset{*}{u} - \overset{*}{u}_i q_i/q_{i,\,i}. \tag{21}$$

(b) If only two constraints (ith, jth) are not binding then

$$\hat{u} = \overset{*}{u} - (1/\beta)\,(q_{j,\,j}\,\overset{*}{u}_i - q_{i,\,j}\,\overset{*}{u}_j)\,q_i - (1/\beta)\,(q_{i,\,i}\,\overset{*}{u}_j - q_{i,\,j}\,\overset{*}{u}_i)\,q_j \tag{22}$$

where $LL^T q_i = e_i$ and $\beta = q_{i,\,i}\,q_{j,\,j} - q^2{}_{i,\,j}.$

If three or more constraints are not binding then we content ourselves with approximating $\hat{u}$ by $\bar{u}$. Various schemes have been tried for choosing $\bar{u}$ but unfortunately it has proved difficult to construct meaningful tests in order to compare them; for instance, it has not been possible to determine whether it is better to insist $\bar{u}_i \geqslant 0$. It is hoped to construct cases where two constraints leave the violated set and then form $\bar{u}$ by implementing only Eqn (21). From this we may be able to draw inferences for the more complex case.

4.1. *Univariate Search*

A univariate search procedure has the advantages, that, (i) we can employ our estimate e to $\hat{u}$ and, (ii) we can vary the effort in determining $\bar{u}$ depending on the effort required to compute the various functions involved. The result of a single search is given by

$$\bar{u}_j = \max\,(0, \bar{u}_j - (Q_j\bar{u} + c_j)/Q_{j,\,j}) \tag{23}$$

Q_j being the jth row of Q, and $\bar{u}$ the current estimate.

Our experience with this procedure within our algorithm is limited. We have, however, used it successfully on independent QP of up to fifty variables.

4.2. *Linear Constraints*

If the constraints are linear then, except when there is a change in the violated set, $A_{i+1}^{-1} = A_i = A$. In this event we can derive a recurrence relation for Q_i^{-1}. If there is an alteration in the violated set it is a simple matter to determine the effect on Q_{i+1}^{-1} by applying similar rules to those given in [9]. Say, for instance, that the recurrence relation for H_i is that given by Fletcher and Powell in [10], that is

$$H_{i+1} = H_i + \lambda_i \frac{p_i p_i^T}{p_i^T y_i} - \frac{H_i y_i y_i H_i^T}{y_i^T H_i y_i} . \tag{24}$$

Pre-multiplying by A and post-multiplying by A^T gives

$$Q_{i+1} = Q_i + \frac{\lambda_i b_i b_i^T}{p_i y_i} - \frac{d_i d_i^T}{y_i^T H_i y_i} . \tag{25}$$

where $d_i = AH_i y_i$.

Applying Householders modification rule first to

$$Q_i' = Q_i + \frac{\lambda_i b_i b_i^T}{p_i^T y_i}$$

and then to

$$Q_{i+1} = Q_i' - \frac{d_i d_i^T}{y_i^T H_i y_i}$$

gives the result

$$Q_{i+1}^{-1} = Q_i^{-1} + (\alpha^2 \beta^2 \gamma - \beta) w_i w_i^T + \gamma v_i v_i^T - \alpha \beta \gamma (w_i v_i^T + v_i w_i^T) \tag{26}$$

where

$$v_i = Q_i^{-1} d_i \qquad w_i = Q_i^{-1} b_i.$$

$$\alpha = v_i^T d_i \qquad \beta = \lambda_i / (p_i^T y_i + \lambda_i b_i^T w_i)$$

$$\gamma = 1/(y_i^T H_i y_i - d_i^T v_i + \beta \alpha^2).$$

5. Penalty Parameters

The penalty parameters r_i occur directly only in the linear search procedure. The choice of r_0 can therefore be left until an estimate to u (u_0) and a direction of search p_0 has been determined. Clearly we would like to choose $r_0 = \|\overset{*}{x} - (x_0 + p)\|$ but since $\overset{*}{x}$ is unknown we have to approximate this quantity. We can of course easily form a lower bound on r_0 if $\|g(x_0 + p)\|$ or $\|g(x_0)\|$ are large. If $\|g\|$ is small then $p^T \nabla_x F(x_0)$ is a measure of the gradient projection of $F(x)$ on the boundary and will be in general only be small when x_0 is close to $\overset{*}{x}$. In addition one would not expect r to be small if $\|p\|$ was large. Having modified g we choose r_0 such that

$$p_0{}^T \nabla_x P(x_0 + p_0, r_0) = 0 \tag{27}$$

f this is consistent with the above statements, and

$$p_0{}^T \nabla_x P(x_0, r_0) < 0. \tag{28}$$

If the latter statement is untrue then r_0 is reduced until

$$p_0{}^T \nabla_x P(x_0, r_0) = -\|\nabla_x P(x_0, \hat{\ })\| \tag{29}$$

where

$$p_0{}^T \nabla_x P(x_0, \hat{\ }) = 0.$$

As we mentioned in Section 1, it is not difficult to construct examples for which any method for the selection of r_0 will fail. Usually these are just the problems that are in any event difficult to solve without the handicap of a poorly chosen r_0. Extra caution is taken during the initial iteration to ensure the additional information gained does not conflict with our choice of r_0, if need be $r_k > r_{k-1}$.

Several strategies have been tried for reducing r_k, for instance, we have reduced r_k only after we have found $\overset{*}{x}_k$. The most successful procedure was to reduce r_k slowly at first and then to relate the reduction according to the progress made. If necessary r_k is reduced to ensure $p_k{}^T \nabla_x P(x_k, r_k) < 0$.

6. Comments and Observations

We have run variants of this algorithm for over a year. Our concern has not been with obtaining comparisons with other methods, but with trying to analyse the various processes involved. Our principle experience has been with $\theta(x) = F(x)$ and the comments are restricted to this case. In addition to the quadratic version, we have also developed a program based only on a linear approximation to $F(x)$. This would be necessary if $F(x) \notin C^2$ and might be desirable if a simpler program were required. Although the rate of convergence with this procedure was not always spectacular, to attempt to find the minimum of $P(x, r_k)$ by a first-order technique would be disastrous. It is perhaps instructive to think of the procedures as being similar to Kelley's cutting-plane method [11], except that our cuts are made to coincide with the "valley" of $P(x, r_k)$.

Many authors have pointed out that it is rather fatuous to make a quadratic approximation to the objective function without making a similar approximation to the constraints. It is not difficult to demonstrate that the solution may in fact depend on the curvature of the constraints. With $\theta(x) = F(x)$ we are not making any apparent approximation to the curvature of g, but we have not found any additional difficulties with problems whose solution depends on the curvature. An explanation as to why this may be so can perhaps be seen from comparing the path taken by the procedure to that taken by a projection method.

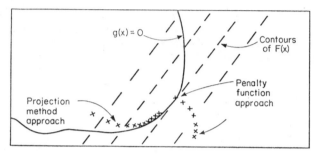

FIG. 2.

From Eqn (6) we have

$$\nabla_x P(\overset{*}{x}(r), r) = \nabla_x F(\overset{*}{x}(r)) + 2r^{-1}A^T g = 0. \tag{30}$$

Differentiating with respect to r and rearranging the terms gives

$$\left(rB + 2A^T A + 2\sum_{i \in I} g_i\,(\overset{*}{x}(r))G_i \right) \frac{d\overset{*}{x}(r)}{dr} = -\nabla_x F(\overset{*}{x}(r)) \tag{31}$$

B and G_i being defined in Section 2.

Letting $r \to 0$ in Eqn (31)

$$2A^T A \frac{d\overset{*}{x}}{dr} = -\nabla_x F(\overset{*}{x}) = -A^T u$$

or
$$2Ay = -u \tag{32}$$

where
$$\frac{d\overset{*}{x}}{dr} = y.$$

Since $u_i \neq 0$, y does not lie on *any* of the hyperplanes $\nabla_x g_i\,(0)\,x = 0$, $i \in I$. For small r

$$\overset{*}{x}(r) = \overset{*}{x} + ry + O(r^2). \tag{33}$$

Substituting for $\overset{*}{x}(r)$ in Eqn (30)

$$\nabla_x P(\overset{*}{x}(r), r) = \nabla_x F(\overset{*}{x}) + rBy +$$

$$2r^{-1} \sum_{i \in I} (ry^T \nabla_x g_i(\overset{*}{x}) + \frac{r^2}{2} y^T G_i y) \times (\nabla_x g_i(\overset{*}{x}) + rG_i y) + O(r^2) = 0. \quad (34)$$

Substituting from Eqn (32) we get on cancellation

$$r\left(B - \sum_{i \in I} u_i G_i\right) y + rA^T d + O(r^2) = 0 \quad (35)$$

where
$$d_i = y^T G_i y$$

We must have therefore

$$\left(B - \sum_{i \in I} u_i G_i\right) y + A^T d = 0$$

or
$$y = -\Phi^{-1} A^T d \quad (36)$$

where
$$\Phi = B - \sum_{i \in I} u_i G_i.$$

From Eqn (32)
$$-2A\Phi A^T d = -u$$

hence
$$d = \tfrac{1}{2}(A\Phi^{-1}A^T)^{-1} u \quad (37)$$

and
$$y = -\tfrac{1}{2}\Phi^{-1} A^T (A\Phi^{-1}A^T) u. \quad (38)$$

If $N(A)$ denotes the null space of A, then for $v \in N(A)$

$$v^T \Phi y = 0. \quad (39)$$

For some vector z

$$w = (\overset{*}{x} + rz) \in S(r)$$

if and only if
$$2Az = -u + O(r).$$

Therefore for small r, if $w \in S(r)$ then w does not lie on any of the hyperplanes, $\nabla_x g_i(0)x = 0$, $i \in I$. Moreover there exist a $w \in S(r)$ such that $(\overset{*}{x} - w) \perp N(A)$.

Acknowledgements

The work described above has been carried out at the National Physical Laboratory

258 W. MURRAY

References

1. A. V. Fiacco and G. P. McCormick (1964). The sequential unconstrained minimization technique for nonlinear programming: a primal-dual method. *Mgmt. Sci.* **10** (2), 360.
2. A. V. Fiacco amd G. P. McCormick (1964). Computational algorithm for the sequential unconstrained minimization technique for nonlinear programming. *Mgmt. Sci.* **10**, 601.
3. A. V. Fiacco and G. P. McCormick (1965). "Extensions of the Sequential Unconstrained Minimization Technique (SUMT) for Nonlinear Programming". Paper given at the 1965 American Meeting of the Inst. of Management Science, Feb., 1965, San Francisco.
4. W. Murray (1967). Ill-conditioning in barrier and penalty functions arising in constrained nonlinear programming. *In* Proceedings of the Sixth International Symposium on Mathematical Programming, Princeton University, Aug. 1967.
5. T. Pietrzykowski (1962). "On a Method of Approximative Final Conditional Maximums". Inst. Maszyn Matematcyoznych PAN, Algorythmy VI, 1962.
6. C. B. Broyden (1967). Quasi-Newton methods and their application to function minimization. *Math. Comput.* **21** (99), 368.
7. W. S. Dorn (1960). Duality in quadratic programming. *Q. appl. Math.* **18**, 155.
8. H. Theil and C. van de Panne (1960). Quadratic programming as an extension of classical quadratic maximization. *Mgmt. Sci.* **7** (1), 1.
9. J. B. Rosen (1960). The gradient projection method for nonlinear programming. Part 1. Linear Constraints. *J. Soc. ind. appl. Math.* **8** (1), 181.
10. R. Fletcher and M. J. D. Powell (1963). A rapidly convergent descent method for minimization. *Comput. J.* **6**, 163.
11. J. E. Kelley, Jr. (1960). The cutting-plane method for solving convex programs. *J. Soc. ind. appl. Math.* **8** (4), 703.

16. Nonlinear Least Squares Fitting using a Modified Simplex Minimization Method

W. SPENDLEY

I.C.I. Ltd, Billingham, England

1. Introduction

The Simplex Search Technique was first put forward by Spendley and co-workers [1] in the context of Evolutionary Operation, that is the empirical optimization of plant performance in the presence of error. Because of this genesis, no detailed consideration was given at the time to its more general use as a technique for numerical maximization or minimization, although some tentative observations were listed. In particular, there was no discussion of the question of scaling or rescaling to accommodate to the local geometry of the surface being explored; although the authors were aware of the problem, they had not at that time any positive comments to offer. An elegant solution to the rescaling problem was advanced by Nelder and Mead in 1965 [2] which was in keeping with the basic simplicity of the original concept. With their modifications the technique becomes an effective and, more particularly, a robust procedure for function maximization or minimization*. The latter property is especially desirable in any procedure intended for general use—for example, as a computer subroutine to be utilized in an arbitrary context.

In any problem of numerical maximization or minimization, location of the stationary point is facilitated by the fitting of an approximating quadratic, and certain suggestions were put forward by Spendley and co-workers for the fitting of such quadratics. In the least squares fitting with which Nelder and Mead were primarily concerned, the device is of special value since the form of the approximating quadratic gives directly the variances and covariances of the final parameter estimates, insofar as these

---

* See Box [3]. Of the four Direct Search Methods which he compared, only two, the Rosenbrock Method [4] and the Nelder and Mead Simplex Procedure did not give rise to failures in particular instances. Efficiencies of the two procedures were about the same, with possibly some advantage to the Simplex Method for smaller numbers of dimensions.

can be assumed normally distributed. For this reason Nelder and Mead terminated their search with the fitting of a quadratic approximation to values of the sums of squares at $\frac{1}{2}(n+1)(n+2)$ points in the parameter space, following a suggestion made in Section 6.4 of [1]. The number $\frac{1}{2}(n+1)(n+2)$ is, in general, the minimum number of function values necessary to define the $\frac{1}{2}(n+1)(n+2)$ constants of the fitted quadratic. In Section 6.5 of [1], however, it is commented that this number of points is not necessary in the special case of least squares fitting; in this instance function values at only $(n+1)$ points—for example the vertices of an n-dimensional simplex—are needed. The comment seems to have been overlooked by Nelder and Mead, but is nevertheless of some importance since it implies that, in the special context of least squares fitting, simplex methods having quadratic convergence may be developed. It is with such development that the present paper is concerned.

2. Theory

2.1. *General*

The theory of Generalized Least Squares is outlined in a number of recent papers, for example Powell [5]. It is required to find that set of parameter values $\theta_1, \theta_2, ..., \theta_n$ ($\boldsymbol{\theta}$, say) which will minimize

$$S = \sum_{k=1}^{m} [f_k(\boldsymbol{\theta})]^2 \qquad (m \geqslant n) \tag{1}$$

where $f_k(\boldsymbol{\theta})$ will typically be of the form

$$f_k(\boldsymbol{\theta}) = \{y_k - \phi(\boldsymbol{\theta}, \mathbf{x}_k)\} \tag{2}$$

that is, the deviation of an observed value y_k from a predicted value $\phi(\boldsymbol{\theta}, \mathbf{x}_k)$ involving both the parameter values $\boldsymbol{\theta}$ and appropriate values $\mathbf{x}_k$ of the independent variables.

Let $\hat{S}$, corresponding to the point $\hat{\boldsymbol{\theta}}$, be the minimum attainable value of the sum of squares, and let a correction vector $\boldsymbol{\delta}$ be defined by

$$\boldsymbol{\delta} = \hat{\boldsymbol{\theta}} - \boldsymbol{\theta}. \tag{3}$$

Now
$$\frac{\partial S}{\partial \theta_i} = 2 \sum_{k=1}^{m} f_k(\boldsymbol{\theta}) \frac{\partial f_k}{\partial \theta_i}. \tag{4}$$

Over a region in which S may be assumed quadratic in the θ_i, $f_k(\boldsymbol{\theta})$ must similarly be linear in the θ_i. Hence

$$\frac{\partial f_k}{\partial \theta_i} = \text{const} = C_i^{(k)} \tag{5}$$

say, and,

$$f_k(\hat{\boldsymbol{\theta}}) = f_k(\boldsymbol{\theta}) + \sum_{i=1}^{n} \delta_i C_i^{(k)}. \tag{6}$$

Substituting in Eqn (4) we have

$$\left(\frac{\partial S}{\partial \theta_j}\right)_{\boldsymbol{\theta}=\hat{\boldsymbol{\theta}}} = 2\sum_{k=1}^{m}\left[f_k(\boldsymbol{\theta}) + \sum_{i=1}^{n}\delta_i C_i^{(k)}\right].C_j^{(k)} \tag{7}$$

$$= 0 \qquad \text{(for all } j\text{)} \tag{8}$$

since $S(\boldsymbol{\theta} = \hat{\boldsymbol{\theta}}) = \hat{S}$ is a minimum.

Hence, rearranging, we have

$$\sum_{i=1}^{n} \delta_i \sum_{k=1}^{m} C_i^{(k)}.C_j^{(k)} = -\sum_{k=1}^{m} f_k(\boldsymbol{\theta}).C_j^{(k)} \qquad \text{(for all } j\text{)} \tag{9}$$

or in matrix notation

$$\boldsymbol{\Gamma}\boldsymbol{\delta} = \mathbf{F} \tag{10}$$

$$\boldsymbol{\delta} = \boldsymbol{\Gamma}^{-1}\mathbf{F} \tag{11}$$

where the ijth element of the matrix $\boldsymbol{\Gamma}$ is

$$\sum_{k=1}^{m} C_i^{(k)}.C_j^{(k)}$$

and the jth element of the column vector $\mathbf{F}$ is

$$-\sum_{k=1}^{m} f_k(\boldsymbol{\theta}).C_j^{(k)}.$$

Thus, given $\mathbf{F}$ and $\boldsymbol{\Gamma}$, the correction vector $\boldsymbol{\delta}$ may be determined. It is to be noted that only $(n+1)$ sets of quantities are involved in Eqn (11), viz the $C_i^{(k)}$ and the $f_k(\boldsymbol{\theta})$. It is for this reason (and only for sums of squares surfaces) that the quadratic approximation may be developed from function values at only $(n+1)$ points in the parameter space, rather than the $\frac{1}{2}(n+1)(n+2)$ which would in general be necessary.

We note that, (i) the variation of S with $\boldsymbol{\delta}$ is given by

$$S = \hat{S} + \boldsymbol{\delta}' \boldsymbol{\Gamma} \boldsymbol{\delta}$$
$$= \hat{S} + \boldsymbol{\delta}' \mathbf{F} \tag{12}$$

so that, given $\boldsymbol{\delta}$ and S, it is possible to predict $\hat{S}$, the value of the sum of squares at the predicted solution point, and (ii) if σ^2 denotes the residual variance of the $f_k(\theta)$ values, the variance–covariance matrix of the parameter estimates δ (hence $\hat{\theta}$) is given by

$$V = \sigma^2 \, \Gamma^{-1}. \tag{13}$$

The latter result is most readily apprehended from the similarity between Eqns (9) and those defining a conventional multiple regression.

2.2. *Application to Simplex Search*

Function values at only $(n+1)$ points are needed to define a quadratic approximation and, moreover, there is no requirement in the general theory for any orthogonality of the coordinate directions describing the parameter space. Now any simplex along the path of a simplex search provides both a design of $(n+1)$ points at which function values must necessarily—for the purposes of the search—be determined, and its own set of coordinate axes. Superimposition of a quadratic approximation is thus both a natural and a not very demanding step.

Let θ_0 denote the coordinates of the vertex of the current simplex giving the lowest sum of squares S_0, and let θ_i, S_i $(i = 1, \dots, n)$ denote the coordinates and sums of squares for the remaining vertices. Then we may define a new coordinate system, ξ_i say, having origin at θ_0 and with unit coordinates represented by the lines joining θ_0 to the other n vertices θ_i. The transformation is clearly

$$\theta = A\xi + \theta_0 \tag{14}$$

where the separate columns of the matrix A are the vectors of differences $(\theta_i - \theta_0)$.

The overall minimum of the sum of squares surface is at $\hat{\theta}$, which in the new coordinate system becomes $\hat{\xi}$, where

$$\hat{\xi} = A^{-1}(\hat{\theta} - \theta_0) \tag{15}$$

and the quadratic approximation (12), *viz*

$$S = \hat{S} + (\hat{\theta} - \theta)' \, \Gamma(\hat{\theta} - \theta)$$

transforms to

$$S = \hat{S} + (\hat{\xi} - \xi)' \, A' \, \Gamma \, A(\hat{\xi} - \xi). \tag{16}$$

For brevity, write $A' \, \Gamma \, A = \Omega$. Then the elements of the matrix Ω are given by

$$\Omega_{ij} = \sum_{k=1}^{m} M_i^{(k)} . M_j^{(k)} \tag{17}$$

where $M_i{}^{(k)}$ denotes

$$\frac{\partial f_k(\theta)}{\partial \xi_i}.$$

But, by reason of the transformation used,

$$M_i{}^{(k)} = f_k(\theta_i) - f_k(\theta_0) \qquad (i = 1, ..., n) \tag{18}$$

and is derivable directly from the function values computed for the points θ_0, θ_i. Thus the matrix Ω may readily be assembled.

In terms of the ξ-coordinates the correction vector is given by $\Omega^{-1}F_0$, where the subscript attached to F implies that the elements of F are to be evaluated at θ_0. Thus, transforming back to the θ-coordinates once again, we have

$$\delta = A\Omega^{-1}F_0 \tag{19}$$

and

$$V = \sigma^2(A\Omega^{-1}A'). \tag{20}$$

3. Application

While the theory shows that a quadratic approximation may readily be derived at any stage, it does not specify an operating procedure. The quadratic fitting is an adjunct to the basic simplex procedure, rather than an essential part of it, and as such the use to be made of it is largely at the user's discretion. Our own use of it to date has been somewhat tentative, and may be summarized as follows.

(1) After N sums of squares evaluations, perform a quadratic fit, and compute the sum of squares at the indicated solution point.

(2) If the sum of squares so computed is less than the minimum sum of squares for the current simplex, that is if the quadratic fit is "successful", incorporate the new point in the simplex and restart the simplex procedure. Perform a new quadratic fit after a further $1 \cdot 5N$ sums of squares evaluations.

(3) If the quadratic fit is unsuccessful, that is it fails to yield a sum of squares at the indicated solution point less than the current minimum, ignore the new point and carry out a further quadratic fit after $3N$ further sums of squares evaluations.

(4) Whenever a quadratic fit is performed and a sum of squares at the indicated solution point is calculated, compare this with the predicted value (Eqn 12). If for two successive quadratic fits the indicated solution points agree within a preset accuracy, and the agreement of the predicted and directly calculated sums of squares is also satisfactory, the search is terminated.

Otherwise, it is continued till convergence is achieved either on a similar test or on the tests appropriate to the basic simplex procedure.

Typically, we have taken $N = 3n$, where n is the number of dimensions of search, although there are some grounds for believing a rather smaller multiple of n might be better (see Section 4.2). Both N and the multipliers 1·5 and 3 are of course arbitrary to a considerable extent. The rationale underlying the initial choice was that the system of simplexes should be given time to "settle down" and adapt to the local geometry before any quadratic fit was attempted, that somewhat longer should be allowed for "settling down" and progress to a more favourable region after the inclusion of a new minimum by the quadratic fit and the possible gross distortion of the system of simplexes, and that after an unsuccessful quadratic fit time should be allowed for substantial progress before a further fit was attempted. Whilst it is believed that these general principles are sound, there is clearly room for further experimentation to establish the most effective implementation of them.

4. Numerical Results

4.1. Verification

As a first check of the procedure, and of the computer program incorporating it, the linear multiple regression example given in Davies [6, p. 221] was attempted. As expected, convergence was reached on the second quadratic fit, giving results in full agreement with those quoted by Davies.

4.2. Frequency of Quadratic Fitting

The example used in this connection arises from some work on least squares fitting of reaction rate data. One equation—the revised Temkin equation [7]—for synthesis of ammonia over promoted iron catalysts gives the reaction rate as

$$r = \frac{k(N_2)^{1-\alpha}\left[1 - \dfrac{(NH_3)^2}{K_p\,(N_2)\,(H_2)^3}\right]}{\left[\dfrac{L}{(H_2)} + \dfrac{(NH_3)^2}{K_p\,(N_2)\,(H_2)^3}\right]^{1-\alpha}\left[1 + \dfrac{L}{(H_2)}\right]^{\alpha}} \tag{21}$$

where (N_2), and so on, denote the partial pressures of the reactants and k is of

the form

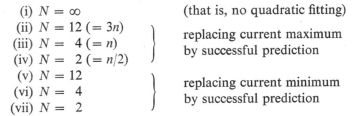

$$k = k_s \exp \left\{ - \frac{E}{R} \left(\frac{1}{T} - \frac{1}{T_s} \right) \right\}. \tag{22}$$

The parameters to be estimated are k_s, E, L and α.

Fitting was carried out on 110 experimentally determined reaction rates. Seven strategies on the lines of those described above were examined, viz.

(i) $N = \infty$ (that is, no quadratic fitting)

(ii) $N = 12 \, (= 3n)$

(iii) $N = 4 \, (= n)$ } replacing current maximum by successful prediction

(iv) $N = 2 \, (= n/2)$

(v) $N = 12$

(vi) $N = 4$ } replacing current minimum by successful prediction

(vii) $N = 2$

Three starting points were used, each with a different set of initial step-lengths. All searches terminated at the same solution point and, wherever a quadratic fit was introduced, by satisfaction of the criteria appropriate to this.

The results obtained (numbers of sums-of-squares evaluations) are shown in Table I.

TABLE I

Comparison of Alternative Strategies

Strategy	Starting point 1	2	3	Average	Notes
(i)	222	231	231	228	No QF
(ii)	158	85	217	153·3	Replacing
(iii)	185	204	173	187·3	Maximum
(iv)	79	78	189	115·3	
(v)	175	85	194	151·3	Replacing
(vi)	78	101	155	111·3	Minimum
(vii)	66	78	168	104	

With increase in the frequency of quadratic fitting, there is—with the exception of the somewhat surprising results for strategy (iii)—a progressive diminution of the mean number of sums of squares evaluations needed for convergence. This however is at the expense of additional computation within the procedure and, while the effect of this is dependent both on the dimensionality of the search (since the time for a matrix inversion varies as n^3) and on the difficulty of function evaluation, the diminishing returns evidenced by comparisons of strategies (vi) and (vii) suggest that little is gained by going to any greater frequency than that implied by $N = n$. It is to be noted that strategies involving replacement of the current minimum do somewhat better than those involving replacement of the maximum—presumably because of the greater distortion of simplexes inherent in the latter. The standard program in use by us is in fact based on replacement of the minimum, although the original reason for this was a fear of premature collapse of the system.

4.3. *Relative Performance of the Method*

This is assessed by reference to one of the currently conventional yardsticks— the trigonometrical functions of Fletcher and Powell [8]. This facilitates comparison with the results derived by Box [3] and in particular with those for the unmodified simplex procedure. The criterion adopted for convergence was in this instance the same as that used by Box, viz agreement to better than 0·0001 between each α_j and the corresponding value reached in the search.

Results for three distinct surfaces in each of 5, 10 and 20 dimensions are shown in Table II.

TABLE II

Performance on Fletcher and Powell's Trigonometric Functions

Case	Number of dimensions		
	5	10	20
1	131	166	1321
2	63	256	421
3	85	606	>1700

The third search in twenty dimensions was terminated because the computation over-ran the time allotted, not because of any failure to converge. However, the rate of progress on termination was slow.

For comparison, Box's Table III is reproduced below. Here N relates to the Nelder and Mead simplex procedure; explanations of the other captions may be found in Box's paper.

TABLE III
Comparisons in 5, 10 and 20 dimensions from Box [3])

Method	Number of Dimensions		
	5	10	20
DSC	303	2,269	5,183
	281	938	5,924
	307	1,378	8,254
R	465	1,210	10,208
	465	1,258	4,681
	388	1,298	8,411
	384		
N	229	752	6,970
	195	962	12,100
	298	970	10,426
P	104	329	1,519
	103	369	2,206
F	354	1,639	4,200
	288	2,860	7,854
	216	1,276	12,348
D	114	396	1,764
	138	319	1,428
			2,541
PSS	21	35	46
	22	34	65
B	42		
	41	NA	NA
	37		

The acceleration resulting from the quadratic fitting is clear and in one instance (20 dimensions, case 2) quite remarkable. Overall, the performance of the modified simplex method now appears comparable with that of other general methods having quadratic convergence, for example D (Davidon) and P (Powell's 1964 method), and some further improvment may still be possible by increasing the frequency of fitting, since all the results quoted in Table II were obtained with $N = 3n$. The performance in no way approaches that of methods PSS or B, but against this must be set the risk of occasional failure with these methods, as illustrated by Box.

5. A Note on Constraints
In work on kinetic modelling, constraints on the search parameters are relatively frequent; for example, adsorption equilibrium constants cannot be negative. We have dealt with such constraints by the simple expedient of setting equal to the boundary value any parameter which seeks to violate a

constraint, and continuously monitoring to detect when all simplex points arrive at the constraint. At this stage, the least favourable point is eliminated and the search is continued in one fewer dimensions. This prevents singularity of the matrices to be inverted, and is preferable to transformations which merely restrict approach to one side, or remove the constraint to infinity. The supposition is of course that the solution point lies in the plane of the constraint; given this, however, the device is clearly applicable to more general linear constraints on the parameter values than those with which we have been concerned.

6. Conclusions

It has been shown that, within the general area of nonlinear least squares fitting, the incorporation of quadratic acceleration procedures into a Nelder and Mead simplex search is both conceptually and practically straightforward. With their aid, the performance of the method appears broadly comparable with that of other general methods having quadratic convergence. Moreover, since the quadratic fitting is an adjunct to, rather than an essential part of, the basic procedure, the robustness of the latter is largely unaffected.

Some work remains to be done to establish a "best" general purpose strategy, but this is a matter of detail rather than principle.

References

1. W. Spendley, G. R. Hext and F. R. Himsworth (1962). Sequential application of simplex designs in evolutionary operation and optimization. *Technometrics*, **4**, 441.
2. J. A. Nelder and R. Mead (1965). A simplex method for function minimization. *Comput. J.* **7**, 308.
3. M. J. Box (1966). A comparison of several current optimization methods, and the use of transformations in constrained problems. *Comput. J.* **9**, 67.
4. H. H. Rosenbrock (1960). An automatic method for finding the greatest or least value of a function. *Comput J.* **3**, 175.
5. M. J. D. Powell (1965). A method for minimizing a sum of squares of nonlinear functions without calculating derivatives. *Comput J.* **7**, 303.
6. O. L. Davies (ed) (1957). "Statistical Methods in Research and Production" (3rd edition, revised). Oliver and Boyd, London and Edinburgh.
7. M. I. Temkin, N. M. Morozov and E. N. Shapatina (1963). *Kinet. Katal.* **4**, 565.
8. R. Fletcher and M. J. D. Powell (1963). A rapidly convergent descent method for minimization. *Comput. J.* **6**, 163.

Discussion

McCann. In the strategy when the minimum point in the Simplex is replaced, what happens to the shape of the Simplex as you proceed to the ultimate solution of the problem? As you are not moving the Simplex along bodily, are you leaving some of the original points in at the far end?

Spendley. The Simplex requires some time to adjust itself to the precise metric of the surface. We get a prediction of the lower point and we replace one of the Simplex points by this, and allow it to proceed by reflection and contraction in the usual way. Part of the reason why we are more successful in replacing the minimum point rather than the maximum is that in this way we impose less distortion on the shape of the Simplex. The maximum is the next point to be eliminated anyway.

Powell. Have you tried the very simple idea of applying the ordinary Simplex algorithm but deciding which vertex to discard on the basis of having the sum of squares reduced at the new vertex? You have the information available which would enable you to do this.

Spendley. No, we have not tried this.

Beale. On the question of discarding points, I thought you would discard the geometrically nearest point to avoid distortion of the Simplex. Did you consider this?

Spendley. To determine the geometrically nearest point would involve additional calculation. The minimum point would be the one which was nearest in terms of the numerical response and we would hope that this would be the nearest geometrically.

Anon. Would it not be better to use transformations to deal with the upper and lower bound constraints rather than the approach which you mention?

Spendley. The approach that I take it you mean is that if $l \leqslant x \leqslant u$ then we take a new variable y defined by $x = l + (u - l) \sin^2 y$ and search on y instead of x. Now this is a periodic function in y and it is quite possible to have a step of π in the y direction in which case the whole method breaks down. We have had this happen.

Davies. Our experience has been similar with Davidon's method and we find that on extrapolation when searching for a minimum, the same sort of thing can happen. It requires a great deal of care to ensure that this does not upset the algorithm.

Beale. I agree with these remarks and think that it would only be a good idea to use transformations in a tremendously complicated function which cannot be made any worse. It would seem that transformations are a particularly unfortunate thing to combine with the Simplex method.

Spendley. I think this may well be true. We have not used transformations as I said, and are happier to let the Simplex run up against the constraint and just discard one point.

Powell. I think I have constrained my patience long enough in hearing people say this is not a good idea. The idea of using these transformations is a very fine one. At Harwell we have a number of good programs for unconstrained optimization and virtually none for the constrained problem. Yet our physicists have got their results on these problems by using transformations with the unconstrained routines all in the space of a week or so from the initiation of the problem. They have proved extremely useful and are a very valuable contribution.

BARD. We have compared the number of function evaluations of this method with SUMT (McCormick's Sequential Unconstrained Maximization Technique) and the results are almost exactly the same. Although it cannot be decided on the basis of this, one has to apply ingenuity each time a transformation is used and this is an important point.

FLETCHER. I think we may be missing the point here in that transformations (in particular the $y = \log(x)$ one for $x \geqslant 0$) would be fatal when the solution actually lies on that constraint. There are a lot of good general methods for solving the problem but transformations are not one of them.

BEALE. In my experience there are many problems with simple upper and lower bounds on many of the variables and I do believe that not enough effort has gone into deciding how to incorporate these very simple constraints into unconstrained optimization problems. I am thinking on the lines, for example, of putting a constraint back to its bound when it exceeds it, as Mr Spendley described.

R. P. IBBETT (Imperial College). Encouraged by what Mr Powell has to say about transformations, I ran a sequence of six tests on a model using Rosenbrock's technique with his method of applying constraints, and also using Rosenbrock's method with transformations for the constraints. I found that the latter technique yielded a better final criterion.

17. Least Distance Programming

A. W. TUCKER*

Princeton University, Princeton, New Jersey, U.S.A.

Let $M = C^{\tau}C$; this occurs if, and only if, the (real) matrix M is square, symmetric, and positive semidefinite. Then

$$w = C^{\tau}\lambda+\gamma, \quad \lambda = Cz, \quad z^{\tau}w = 0, \quad z\geqslant0, \quad w\geqslant0$$

are necessary and sufficient conditions [1] that λ and z be optimal solutions of the quadratic programs

$$\text{minimize } \phi(\lambda) = \tfrac{1}{2}\lambda^{\tau}\lambda \text{ for } w = C^{\tau}\lambda+\gamma\geqslant0,$$

$$\text{minimize } f(z) = \tfrac{1}{2}z^{\tau}Mz+z^{\tau}\gamma \text{ for } z\geqslant0$$

and that (z, w) be an *equilibrium point* [2] of the system

$$w = Mz+\gamma.$$

The ϕ-program concerns the least distance from the origin ($\lambda = 0$) to a point λ of the closed convex polyhedral set $\{\lambda|C^{\tau}\lambda+\gamma\geqslant0\}$; this program has a unique optimal solution if the polyhedral set is not empty. Note that

$$\phi(\lambda)+f(z) = \tfrac{1}{2}(\lambda-Cz)^{\tau}(\lambda-Cz)+z^{\tau}w$$

where $w=C^{\tau}\lambda+\gamma$. Hence $\phi+f\geqslant0$ if $z\geqslant0$, $w\geqslant0$, and $\phi+f = 0$ if $\lambda = Cz$, $z^{\tau}w = 0$.

The *foot of the perpendicular* from the origin ($\lambda = 0$) to the linear manifold $\{\lambda|A^{\tau}\lambda+\alpha = 0\}$ (where A is a submatrix of C consisting of linearly independent columns of C and α is the corresponding subvector of γ) corresponds to the *complementary basic solution* [3] of $w = Mz+\gamma$ (where $M=C^{\tau}C$) that has $w_j = 0$ for columns of A and $z_j = 0$ otherwise. This foot of the perpendicular is the point of the polyhedral set $\{\lambda|C^{\tau}\lambda+\gamma\geqslant0\}$ at least distance

* This work was supported by the Office of Naval Research (Contract NONR 1858-21).

271

from the origin if, and only if, the complementary basic solution has no negative components.

Using the non-singular square submatrix $A^\tau A$ of the square matrix $M = C^\tau C$ as *block pivot* [4], there arises a pair of quadratic programs, *dual* in the sense of Cottle [5], equivalent to minimizing f for $z \geqslant 0$ and to maximizing $-\phi = -\frac{1}{2}z^\tau M z$ for $w = Mz + \gamma \geqslant 0$. If A corresponds to a non-negative complementary basic solution, then these Cottle programs have obvious optimal solutions.

The matters here summarized are treated more fully in [6] and [7]. The author acknowledges gratefully his many discussions of these matters with Dr Philip Wolfe and Dr T. D. Parsons.

Note added in proof: The minimization of any convex quadratic function under linear constraints can be made a "hybrid" of least distance programming and linear programming [8].

References

1. H. W. Kuhn and A. W. Tucker (1951). Nonlinear programming. *In* Procedings 2nd Berkeley Symposium on Mathematical Statistics and Probability. (J. Newman ed.), pp. 481-492. University California Press, Berkeley.
2. C. E. Lemke (1965). Bimatrix equilibrium points and mathematical programming. *Mgmt. Sci.* **11**, 681-689.
3. G. B. Dantzig and R. W. Cottle (1967). Positive (semi-) definite programming. *In* "Nonlinear Programming". (J. Abadie, ed.) pp. 55-73. John Wiley, New York.
4. T. D. Parsons (1966). "A Combinatorial Approach to Convex Quadratic Programming". Ph.D. Thesis, Princeton University, 1966.
5. R. W. Cottle (1963). Symmetric dual quadratic programs. *Q. appl. Math.* **21**, 237-243.
6. A. W. Tucker (1968). A least-distance approach to quadratic programming. *In* "Mathematics of the Decision Sciences", Part 1 (G. B. Dantzig and A. F. Veinott, Jr., eds.), Lectures in Applied Mathematics, Vol. 11, pp. 163-176. American Mathematical Society, Providence, R. I.
7. A. W. Tucker (1969). Least distance programming. *In* Proceedings of the International Symposium on Mathematical Programming, Princeton 1967 (H. W. Kuhn, ed.) to be published by John Wiley, New York.
8. T. D. Parsons and A. W. Tucker (1969). Duality geometry of quadratic programs (Abstract). *Notices of the American Mathematical Society*, **16**, 191.

18. Sufficient Conditions for the Convergence of a Variable Metric Algorithm

D. GOLDFARB

Courant Institute of Mathematical Sciences, New York University,
New York, U.S.A.

1. Introduction

In the past few years, there has been a renewed interest in unconstrained optimization accompanied by the development of several new optimization methods. Among these procedures, the Variable Metric Methods, first introduced by Davidon in 1959 [1], appear to be the most interesting from a mathematical viewpoint. Moreover, Davidon's original method, simplified and reformulated by Fletcher and Powell [2], is one of the most efficient methods in general use for minimizing an unconstrained function of several variables when its first partial derivatives are readily obtained.

Investigators have high regard for the Davidon-Fletcher-Powell method, not only because of past computational experience, but also because of the following two facts: it is stable, that is the value of the objective function is decreased at each step, and it is able to minimize a strictly convex quadratic function in a finite number of steps. Stability alone, however, is not sufficient to guarantee convergence and, as far as the author knows, convergence of this method has not been proved except in the quadratic case.

In an effort to prove convergence for a class of functions larger than strictly convex quadratics, the author was led to investigate other versions of the variable metric method. The discussion in this paper is primarily concerned with a rank-one version of the variable metric method (so named because the variable matrix H is changed by adding to it a matrix of rank-one at each iteration). This method was devised and analysed for the quadratic case by Wolfe [3]. (Rank-one methods have also been investigated by Davidon [4] and Broyden [5]). Although the method as presented may behave in an undesirable fashion when the conditions set forth in this paper are not fulfilled (for example, it need not be stable), it is considered here because it can be simply presented and easily analysed.

273

2. Notation

Variable metric methods are used for finding an unconstrained local minimum of a function $f(x) = f(x_1, ..., x_n)$ of n variables $x_1, ..., x_n$. For simplicity, vector and matrix notation will be used. Thus, $f(x)$ is a scalar function of the n-dimensional column vector x whose components are $x_1, ..., x_n$. The gradient of $f(x)$ at x is denoted by the column vector

$$g(x) = \left(\frac{\partial f(x)}{\partial x_1}, ..., \frac{\partial f(x)}{\partial x_n} \right)'$$

where the superscript (') denotes transposition, while the Hessian of $f(x)$ at x is denoted by

$$G(x) = \left[\frac{\partial^2 f}{\partial x_i \partial x_j} \right].$$

To avoid confusion, subscripts will, henceforth, indicate the iterates. The notations $A \geqslant 0$ and $B > 0$ will be used to indicate that the symmetric matrices A and B are positive semi-definite and positive definite, respectively. By the expressions $A \geqslant B$, $A > B$, $A \leqslant 0$, and $A < 0$ it is meant that $A - B \geqslant 0$, $A - B > 0$, $-A \geqslant 0$ and $-A > 0$.

3. Rank-One Methods

Algorithm.

(1) Initially x_0 and H_0 are chosen.

(2) Given x_k and H_k, calculate $g_k = \nabla f(x_k)$ and $H_k g_k$.

(3) If $\|g_k\| = 0$ terminate. Otherwise, calculate $s_k = - t_k H_k g_k$ and $x_{k+1} = x_k + s_k$ where t_k is chosen as that value of t that minimizes $f(x_k - t H_k g_k)$.

(a) If $H_k y_k = s_k$, where $y_k = g_{k+1} - g_k$ set

$$H_{k+1} = H_k.$$

(b) If $H_k y_k \neq s_k$ and $(H_k y_k - s_k)' y_k \neq 0$ set

$$H_{k+1} = H_k - \frac{(H_k y_k - s_k)(H_k y_k - s_k)'}{(H_k y_k - s_k)' y_k}. \tag{1}$$

(c) If $H_k y_k \neq s_k$ but $(H_k y_k - s_k)' y_k = 0$ set

$$H_{k+1} = H_k.$$

Set $k = k+1$ and return to step (2).

LEMMA 1. *If* $f(x) = f^0 + a'x + \frac{1}{2}x'Gx$, *where* G *is non-singular, and* H_0 *is chosen so that* $H_0 \geqslant G^{-1}$, *then* $H_k \geqslant G^{-1}$ *for all* k. *Similarly, if* $H_0 \leqslant G^{-1}$, *then* $H_k \leqslant G^{-1}$ *for all* k.

Proof. If $H_{k+1} = H_k$, result is trivial. If H_{k+1} is computed from H_k by Eqn (1), after substituting $G^{-1}y_k$ for s_k and pre- and post-multiplying by x, we get:

$$(x, (H_{k+1}-G^{-1})x) = (x, (H_k-G^{-1})x) - \frac{[(x, (H_k-G^{-1})y_k)]^2}{(y_k, (H_k-G^{-1})y_k)} \geqslant 0$$

by Schwarz' inequality and induction. Notice that we can write the last term since it is assumed that $(H_k y_k - s_k)' y_k \neq 0$. The second statement is then obvious if one multiplies each term by -1.

LEMMA 2. *If* $f(x) = f^0 + a'x + \frac{1}{2}x'Gx$, *where* G *is non-singular, and* $H_0 \geqslant G^{-1}$, *then* $H_k \leqslant H_0$ *for all* k. *Similarly if* $H_0 \leqslant G^{-1}$, *then* $H_k \geqslant H_0$ *for all* k.

Proof. By Lemma 1 if $H_0 \geqslant G^{-1}$ then $H_k - G^{-1} \geqslant 0$ for all k. Therefore for Eqn (1)

$$(x, (H_{k+1}-H_k)x) = - \frac{(x, (H_k-G^{-1})y)^2}{(y, (H_k-G^{-1})y_k)} \leqslant 0.$$

When $H_{k+1} = H_k$ the result is trivial. The second statement is again obvious.

LEMMA 3. *If* $f(x)$ *is strictly convex and has bounded second partial derivatives, that is*

(i) $mI \leqslant G(x) \leqslant MI$, *where* $0 < m \leqslant M < \infty$,

(ii) $G^{-1}(x_0) \leqslant H_0 \leqslant KI$, *where* $1/m \leqslant K < \infty$, *and*

(iii) $G(x)$ *is non-decreasing along any path of non-increasing function values (i.e.* $G(x+\delta x) \geqslant G(x)$ *if* $f(x+\delta x) \leqslant f(x)$, $\delta x > 0$*),*
      then $(1/M)I \leqslant H_k \leqslant KI$ *for all* k.

Proof. Define

$$\tilde{G}_k = \int_0^1 G(x_k + ts_k)\, dt. \tag{2}$$

Now

$$\tilde{G}_k s_k = \int_0^1 G(x_k + ts_k)\, dt\, s_k = \int_0^1 G(x_k + ts_k) . s_k\, dt$$

$$= \int_0^1 \frac{d}{dt}\, g(x_k + ts_k)\, dt = g_{k+1} - g_k.$$

Since $G(x) \geqslant mI > 0$ we have

$$s_k = \tilde{G}_k^{-1} y_k.$$

Following the proofs of Lemmas 1 and 2 we can show that

$$(\tilde{G}_k^{-1} \leqslant H_k \leqslant KI) \Rightarrow (\tilde{G}_k^{-1} \leqslant H_{k+1} \leqslant KI).$$

Using the easily proved fact that $(A \geqslant B \geqslant 0) \Rightarrow (B^{-1} \geqslant A^{-1})$ and recalling assumption (iii) and Eqn (2), we have $\tilde{G}_{k+1}^{-1} \leqslant \tilde{G}_k^{-1}$ and hence

$$\tilde{G}_{k+1}^{-1} \leqslant H_{k+1} \leqslant KI.$$

But $(\tilde{G}_k^{-1} \leqslant H_k \leqslant KI)$ is true for $k = 0$ since $\tilde{G}_0^{-1} \leqslant G^{-1}(x_0)$ and the lemma follows since $\tilde{G}_k^{-1} \geqslant (1/M)I$ for all k. Similarly we can prove

LEMMA 4. *If*

(i) $mI \leqslant G(x) \leqslant MI, 0 < m \leqslant M < \infty,$

(ii) $lI \leqslant H_0 \leqslant G^{-1}(x_0), 1/M \geqslant l > 0,$ *and*

(iii) $G(x)$ *is non-increasing along any path of non-increasing function values, then* $lI \leqslant H_k \leqslant (1/m)I$ *for all* k.

Notice that if either of the sets of conditions of Lemma 3 or Lemma 4 are met, then case (c) cannot occur.

Definition. An "optimal variable metric" method is one in which new points are given by

$$x_{k+1} = x_k - t_k H_k g_k \tag{3}$$

where t_k is chosen as that value of t that minimizes $f(x_k - tH_k g_k)$ and where some suitable iteration scheme is given for improving the approximation of H_k to $\tilde{G}_k^{-1}$.

THEOREM 1. *An "optimal variable metric" method will converge to a global minimum of a strictly convex function $f(x)$ whose second partial derivatives are bounded, that is*

$$0 < mI \leqslant G(x) \leqslant MI \tag{4}$$

if $\alpha I \leqslant H_k \leqslant \beta I$ *for all* k, *where* $0 < \alpha \leqslant \beta < \infty$.

Proof. Represent $f(x)$ by the three-term Taylor series expansion about x_k, where $x = x_k + \tau z$,

$$f(x) = f(x_k) + \tau z' g_k + \tfrac{1}{2}\tau^2 z' G(\xi)z \tag{5}$$

and where ξ belongs to the line segment joining x and x_k.
Define z by

$$z = \frac{-H_k g_k}{\|H_k g_k\|}.$$

Substituting this into Eqn (5) and using condition (4) and $\|z\| = 1$, we have

$$f(x) - f(x_k) \leqslant + \tau z' g_k + \tfrac{1}{2} \tau^2 M$$

$$= \tau z' g_k \left(1 + \frac{\tau M}{2 z' g_k} \right). \tag{6}$$

But, from the assumption on H_k

$$z' y_k = - \frac{g_k' H_k g_k}{\|H_k g_k\|} < 0.$$

Therefore, if τ is restricted to $0 < \tau < (2 g_k' H_k g_k / (M \|H_k g_k\|))$ then the right-hand side of Eqn (6) must be strictly negative. Let

$$x'_{k+1} = x_k + \tau' z$$

where

$$\tau' = \frac{g_k' H_k g_k}{M \|H_k g_k\|}.$$

Hence

$$f(x'_{k+1}) - f(x_k) \leqslant - \frac{(g_k' H_k g_k)^2}{2 M \|H_k g_k\|^2} \leqslant - \frac{\alpha^2 \|g_k\|^2}{2 \beta^2 M}$$

where the last inequality follows from

$$(g_k, H_k g_k) \geqslant \alpha \|g_k\|^2 \text{ and } \|H_k g_k\|^2 \leqslant \beta^2 \|g_k\|^2.$$

Since t_k is chosen so that the decrease in $f(x)$ is maximal

$$f(x_{k+1}) - f(x_k) \leqslant - \frac{\alpha^2 \|g_k\|^2}{2 \beta^2 M}.$$

If the algorithm terminates in a finite number of steps the proof is trivial. If the sequence of points $\{x_k\}$ generated by the algorithm is infinite then the corresponding sequence of function values $\{f_k\}$ must have a limit point $\bar{f}$ since $f(x)$ is bounded in $R = \{x | f(x) \leqslant f(x_0)\}$. This implies that in the limit the difference between neighboring f_k tends to zero, that is

$$0 = \lim_{k \to \infty} (f_{k+1} - f_k) \leqslant \lim_{k \to \infty} \left[\frac{-\alpha^2 \|g_k\|^2}{2 \beta^2 M} \right]$$

from which it follows that

$$\lim_{k \to 0} g(x_k) = 0.$$

Using the strict convexity of $f(x)$, it is not very difficult to show that the sequence $\{x_k\}$ converges to a point $\bar{x}$ which is the unique global minimum of $f(x)$ and which satisfies $g(\bar{x}) = 0$.

Lemmas 3 and 4 together with Theorem 1 then imply the following theorem.

THEOREM 2. *Conditions* (i) − (iii) *of Lemma* 3 *are sufficient for the convergence of the algorithm. Conditions* (i)–(iii) *of Lemma* 4 *are sufficient for the convergence of the algorithm.*

The above results are not meant to be exhaustive; rather they are meant to indicate the special role that positive definiteness of the operator $H - G^{-1}$ plays in the analysis of the convergence of variable metric methods. This property will also presumably be crucial in the study of the rate of convergence near the minimum.

4. Davidon-Fletcher-Powell Method

The Davidon-Fletcher-Powell variable metric method is basically the same as the algorithm outlined at the beginning of the previous section except that the three formulae of step (3) for up-dating the matrix H_k are replaced by

$$H_{k+1} = H_k + \frac{s_k \, s_k'}{s_k' \, y_k} - \frac{H_k \, y_k \, y_k' \, H_k}{y_k' \, H_k \, y_k}. \tag{7}$$

Since this method is an "optimal variable metric" method, we know from Theorem 1 that $0 < \alpha I \leqslant H_k \leqslant \beta I$ for all k is a sufficient condition for convergence under the hypotheses of Eqn (4). Since uniform bounds for H_k are needed, it is natural to look for analogs of Lemmas 1 and 2. With this in mind we first prove the following Lemma.

LEMMA 5. *If* $A \geqslant B \geqslant 0$, *then* $S(x, y) = S_A(x, y) - S_B(x, y) \geqslant 0$ *for all* x *and all* y, *other than* y *in the null space of* B, *where*

$$S_Q(x, y) = (x, Qx) - \frac{(x, Qy)^2}{(y, Qy)}.$$

Proof.

$$S(x, y) = (x, Ax) - (x, Bx) - \frac{(x, Ay)^2}{(y, Ay)} + \frac{(x, By)^2}{(y, By)}. \tag{8}$$

According to the hypothesis of the lemma, either

$$(y, (A-B)y) = 0 \quad \text{or} \quad (y, (A-B)y) > 0.$$

In the first case it is trivial to show that $(x, Ay) = (x, By)$, for example using Schwarz' Inequality, $|(x, (A-B)y)|^2 \leqslant (x, (A-B)x)(y, (A-B)y) = 0$. Hence the last two terms in Eqn (8) for $S(x, y)$ are equal and $S(x, y)$ reduces to $S(x, y) = (x, Ax) - (x, Bx) \geqslant 0$. On the other hand, if $(y, (A-B)y) > 0$, then the term $[(x, (A-B)y)^2/(y, (A-B)y)]$ can be added to and subtracted from the right-hand side of Eqn (8) to give

$$S(x, y) = (x, (A-B)x) - \frac{(x, Ay)^2}{(y, Ay)} + \frac{(x, By)^2}{(y, By)} + \frac{(x, (A-B)y)^2}{(y, (A-B)y)}$$

$$- \frac{(x, (A-B)y)^2}{(y, (A-B)y)}.$$

By Schwarz' Inequality with respect to the metric $A-B$ this becomes

$$S(x, y) \geqslant \frac{(x, By)^2}{(y, By)} - \frac{(x, Ay)^2}{(y, Ay)} + \frac{(x, Ay)^2 - 2(x, Ay)(x, By) + (x, By)^2}{(y, (A-B)y)}$$

$$= \frac{[(x, By)(y, Ay) - (x, Ay)(y, By)]^2}{(y, Ay)(y, By)(y, (A-B)y)} \geqslant 0.$$

Notice that $S(x, y) = 0$ if and only if x and y are both in the nullspace of $(A-B)$. If B is strictly positive definite then the double-Schwarz Inequality proved above holds for all x and y.

The analog of Lemma 1 is then:

LEMMA 6. *If $f(x) = f^0 + a'x + \frac{1}{2}x'Gx$, where $G > 0$ and H_0 is chosen so that $H_0 \geqslant G^{-1}$, then $H_k \geqslant G^{-1}$ for all H_k generated by the Fletcher-Powell-Davidon algorithm. Similarly, if $0 < H_0 \leqslant G^{-1}$, then $0 < H_k \leqslant G^{-1}$ for all k.*

Proof. Since $f(x)$ is quadratic, $s_k = G^{-1}y_k$. (G is non-singular by assumption.) Thus from Eqn (7) we have that

$$(x, (H_{k+1} - G^{-1})x) = (x, (H_k - G^{-1})x)$$

$$+ \frac{[(x, G^{-1}y_k)]^2}{(y_k, G^{-1}y_k)} - \frac{[(x, H_k y_k)]^2}{(y_k, H_k y_k)}.$$

But from Lemma 5 if it is assumed that $H_k \geqslant G^{-1}$ then $(x, (H_{k+1} - G^{-1}) x) \geqslant 0$ for all x, that is $H_{k+1} \geqslant G^{-1}$, and the first result of the theorem follows by induction.

It has been shown elsewhere [2] that if $H_0 > 0$ then $H_k > 0$ for all k. The proof that $0 < H_k \leqslant G^{-1}$ for all k if $0 < H_0 \leqslant G^{-1}$ is also obvious if both the right- and left-hand sides of Eqn (7) are subtracted from G^{-1} and the resulting expression pre- and post-multiplied by x.

Attempts at proving the analog of Lemma 2, however, have been unsuccessful.

Acknowledgement

The author is indebted to Professor William C. Davidon and Dr Philip Wolfe for several stimulating discussions. The author also acknowledges support for this paper from the AEC Computing and Applied Mathematics Center, Courant Institute of Mathematical Sciences, New York University, under Contract AT(30-1)-1480 with the U.S. Atomic Energy Commission.

References

1. W. C. Davidon (1959). A.E.C. Research Development Report, ANL-5990.
2. R. Fletcher and M. J. D. Powell (1963). *Br. Comput. J.* 6, 163-168.
3. P. Wolfe, (1967). IBM Working Paper.
4. W. C. Davidon (1968). *Br. Comput. J.* (10, 406–410.
5. C. G. Broyden (1967). *Math. Comp.* 21, 368–381.

Discussion

BARD. You have proved that the points coverge to the minimum; can you prove that the matrices converge to their value at the minimum?

DAVIDON. I have not proved this, nor has Goldfarb, but I think that they do. Of course it is an artificial assumption that as the function approaches the minimum monotonically, then so does the variance. However, I think that the estimates of the variance in rank 1 methods are far better than those in rank 2 methods; the reason being that in the latter a part of the matrix is projected out and then almost an identical part can be added in again in ill-conditioned cases. This causes the matrices to be very much affected by numerical accuracy. In the rank 1 methods only a small change is made to the matrix so that it does not suffer from this difficulty. This is related to the convergence but is not a proof of convergence.

BROYDEN. I don't think that you can ever prove convergence of the variance: for example in an extreme case, what happens if you start at the solution with an arbitrary matrix?

DAVIDON. I agree: one would have to introduce some condition, for instance that a number of steps at least equal to the dimension had been taken.

SARGENT. Are the conditions on the function in Lemma 3 necessary to the theorem or does the theorem depend on Lemma 3.

DAVIDON. The conditions of Lemma 3 are necessary for the proof of Theorem 2: if you have the premises of Lemma 3 you have the conclusion of Theorem 1. If not Thoerem 1 does not have a proof, although we all hope that with conditions weaker than Lemma 3 we shall still get convergence.

SARGENT. How does one get "global minimum" rather than "local minimum" in Theorem 1.

DAVIDON. If the variance is bounded then the function is strictly convex, so the minimum must be a global one.

(Dr. Goldfarb was unable to present his paper in person at the conference. Dr. Davidon did this on his behalf and also gave answers to questions in the discussion. ed.)

19. A Method for Nonlinear Constraints in Minimization Problems

M. J. D. Powell

Mathematics Branch, Atomic Energy Research Establishment, Harwell, England

1. Introduction

We wish to calculate the values of the variables $(x_1, x_2, ..., x_n)$ which cause a given function $F(x_1, x_2, ..., x_n)$ to be least, subject to given constraints on the variables

$$\psi_t (x_1, x_2, ..., x_n) = 0 \quad (t = 1, 2, ..., m). \tag{1}$$

We assume that the problem has a solution, and that the given functions have continuous second derivatives (but note that we do not require the second derivatives to be computed). If an equality constraint is so simple that it can be used to eliminate a variable, the elimination is often worthwhile, so that we have in mind problems in which the functions $\psi_t (\mathbf{x})$ are complicated, and we expect the constraints to provide much interdependence between the variables.

Because successful techniques exist for unconstrained minimization (see for example, Davidon [1], Fletcher and Powell [2] and Powell [3]), the method offered in this paper converts the constrained problem to a sequence of unconstrained minimization problems, having the property that the successive solutions of the unconstrained problems converge to the answer that we require. An example of such an approach is given in a recent paper by Fiacco and McCormick [4], who minimize a function like

$$F(\mathbf{x}) + r_k^{-1} \sum_{t=1}^{m} [\psi_t (\mathbf{x})]^2 \tag{2}$$

for successive values of the positive parameter r_k, chosen so that

$$\lim_{k \to \infty} r_k = 0. \tag{3}$$

283

It should be apparent that, under mild conditions on the functions, the successive minima converge to the solution of our problem. A different example is given by Carroll's [5] Created Response Surface Technique for minimization subject to inequality constraints, $g_t(\mathbf{x}) \geqslant 0$, in which the unconstrained functions include terms like $r_k/g_t(\mathbf{x})$, and it is expected that $g_t(\mathbf{x}) = 0$ at the solution, for some values of t.

We quoted these two examples because they indicate the need for a new method. The point is that they both have a tendency to involve very large numbers, namely r_k^{-1} and $[g_t(\mathbf{x})]^{-1}$, which causes the functions that are minimized to be very sensitive to changes in the variables in a way that makes them difficult to manage. For instance a contour map of an objective function in the case $n = 2$ might indicate deep narrow valleys, and to follow a curved deep narrow valley by a sequence of straight lines without increasing the value of the objective function (which is the usual approach of the methods for unconstrained minimization) would require a large number of steps.

In the new method each unconstrained minimization problem depends on $2m$ parameters, namely $(\sigma_1, \sigma_2, ..., \sigma_m)$ and $(\theta_1, \theta_2, ..., \theta_m)$, for we calculate the vector of variables $\mathbf{x}$, to minimize

$$\phi(\mathbf{x}, \boldsymbol{\sigma}, \boldsymbol{\theta}) = F(\mathbf{x}) + \sum_{t=1}^{m} \sigma_t [\psi_t(\mathbf{x}) + \theta_t]^2. \tag{4}$$

We find that the required solution can usually be obtained for moderate values of the parameters, and the method is attractive because it is so easy to change the parameters to generate a suitable sequence of unconstrained problems. The main difference between the functions (2) and (4) is just the introduction of the parameters $(\theta_1, \theta_2, ..., \theta_m)$, and it is because these parameters are present that it is satisfactory to use moderate values of $(\sigma_1, \sigma_2, ..., \sigma_m)$.

2. The Method

The method is based on the following simple theorem.

THEOREM 1. *If the values of the variables* $\mathbf{x}$ *which minimize* $\phi(\mathbf{x}, \boldsymbol{\sigma}, \boldsymbol{\theta})$ *are* $\xi(\boldsymbol{\sigma}, \boldsymbol{\theta})$, *then* $\xi(\boldsymbol{\sigma}, \boldsymbol{\theta})$ *is a solution to the constrained problem*:

Minimize $F(\mathbf{x})$ *subject to the conditions*

$$\psi_t(\mathbf{x}) = \psi_t(\xi(\boldsymbol{\sigma}, \boldsymbol{\theta})) \qquad (t = 1, 2, ..., m). \tag{5}$$

Proof. If the theorem does not hold, and the variables $\xi^*(\boldsymbol{\sigma}, \boldsymbol{\theta})$ minimize $F(\mathbf{x})$ subject to the conditions (5), then $\phi(\xi^*) < \phi(\xi)$, which is a contradiction.

The theorem is important because it states that we just have to obtain values of the parameters such that

$$\psi_t\left(\xi(\sigma, \theta)\right) = 0 \qquad (t = 1, 2, ..., m) \tag{6}$$

so we hope that an iterative adjustment of the parameters can be made to work. For example if $m = 1$ and σ_1 is fixed, then adjusting θ_1 will provide a line of points $\xi(\sigma_1, \theta_1)$ in the space of the variables and, if this line intersects the surface $\psi_1(x) = 0$, we just have to calculate the corresponding value of θ_1.

We recall that the difficulties of an algorithm derived from expression (2) occur because r_k^{-1} becomes large, so we try to satisfy the system of Eqns (6) by adjusting θ for fixed values of the parameters σ. Since the equations are nonlinear, the adjustment could require much computation. However, a particularly pleasing feature of the new method is that the extremely simple correction

$$\theta_t \leftarrow \theta_t + \psi_t(\xi) \qquad (t = 1, 2, ..., m) \tag{7}$$

happens to work well, provided that the numbers σ_t are sufficiently large.

The justification of the correction (7) is given in the next section. The dominant reason is that the correction to θ calculated by the generalized Newton iteration, which has ultimate quadratic convergence (see for example [6]), is expression (7) except for terms which are $O(\sigma_t^{-1})$. Further by choosing the numbers σ_t to be sufficiently large, we find that the iteration (7) can be made to have linear convergence at as fast a rate as we please.

Therefore the new method is to adjust the parameters by applying the correction (7), unless it happens that $\max_t |\psi_t(\xi)|$ either fails to converge or converges to zero at too slow a rate, when σ is increased in order to improve the convergence. Figure 1 shows a precise algorithm, which is designed to decrease the maximum residual of the system of Eqns (6) by a factor of four on each iteration.

A few comments on the flow diagram (Fig. 1) are needed. k is the iteration number, and c_k is usually set to the least value of $\max_t |\psi_t(\xi)|$ which has been calculated. However, to start the iterative process we set $c_0 = A$, where A is some large positive number exceeding the magnitude of $|\psi_t(x)|$ $(t = 1, 2, ..., m)$. We ask that the functions $F(x)$ and $[\psi_t(x)]^2$ be scaled to have similar magnitudes in order that the initial choice

$$\left. \begin{array}{l} \sigma_t = 1 \\[2mm] \theta_t = 0 \end{array} \right\} \qquad (t = 1, 2, ..., m) \tag{8}$$

is sensible. However we have omitted the range $t = 1, 2, ..., m$ from the first box in order to shorten the presentation. Wherever the subscript "t" is used in the figure the range $t = 1, 2, ..., m$ is implied. If the switch is "down" it indicates that we have just chosen a new value for σ, but if it is "up" we applied the correction (7) on the previous iteration. We continue to apply the correction (7) provided that it gives the required convergence, namely

$$c_k \leqslant \tfrac{1}{4} c_{k-1} \tag{9}$$

but if the condition (9) fails we increase σ. The decision to use a factor ten to increase σ_t is an arbitrary one.

Note that when σ increases we adjust θ so that $\sigma_t \theta_t$ is unchanged. We do this because usually the required values of $\sigma_t \theta_t$ $(t = 1, 2, ..., m)$ are

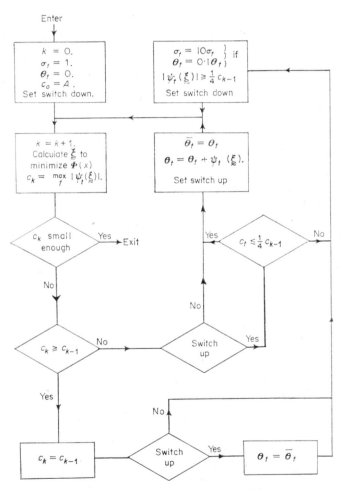

FIG. 1. Flow diagram for the algorithm

independent of σ. This result follows because, letting ξ^*, σ^* and θ^* be the final values of ξ, σ and θ, we have, from the condition

$$\left[\frac{\partial}{\partial x_i}\phi(\mathbf{x})\right]_{\mathbf{x}=\xi} = 0 \qquad (i = 1, 2, ..., m) \tag{10}$$

that

$$\left[\frac{\partial}{\partial x_i}F(\mathbf{x}) + 2\sum_{t=1}^{m}\sigma_t^*\theta_t^*\frac{\partial}{\partial x_i}\psi_t(\mathbf{x})\right]_{\mathbf{x}=\xi^*} = 0. \tag{11}$$

Therefore the final gradient vector of $F(\mathbf{x})$ is a linear combination of the final gradient vectors of the functions $\psi_t(\mathbf{x})$, and the appropriate linear factors are $-2\sigma_t^*\theta_t^*$ $(t = 1, 2, ..., m)$. We state this result as a theorem.

THEOREM 2. *If our problem has a unique solution, and at this solution the gradient vectors of the functions $\psi_t(\mathbf{x})$ are linearly independent, then, for $t = 1, 2, ..., m$, the final value of $\sigma_t\theta_t$ is independent of the parameters.*

To demonstrate the method we calculated the values of the variables $(x_1, x_2, x_3, x_4, x_5)$ that minimize the product $x_1\,x_2\,x_3\,x_4\,x_5$, subject to the conditions

$$\left.\begin{array}{l}\psi_1(\mathbf{x}) = x_1^2 + x_2^2 + x_3^2 + x_4^2 + x_5^2 - 10 = 0\\[4pt]\psi_2(\mathbf{x}) = x_2\,x_3 - 5x_4\,x_5 = 0\\[4pt]\psi_3(\mathbf{x}) = x_1^3 + x_2^3 + 1 = 0\end{array}\right\} \tag{12}$$

However there will be difficulty if the algorithm is applied directly, because for any choice of σ and θ the function $\phi(\mathbf{x})$ is not bounded below. Therefore we let

$$F(x_1, x_2, x_3, x_4, x_5) = \exp(x_1\,x_2\,x_3\,x_4\,x_5). \tag{13}$$

To reduce the residuals of Eqn (6) to less than 10^{-4} required only three iterations, and the resultant values of the vectors σ, θ and ψ are given in

TABLE I

An example in which σ remains fixed

k	σ	θ	ψ
1	1·0000	0·0000	0·0196
	1·0000	0·0000	−0·0185
	1·0000	0·0000	0·0025
2	1·0000	0·0196	0·0004
	1·0000	−0·0185	−0·0005
	1·0000	0·0025	0·0001
3	1·0000	0·0201	0·0000
	1·0000	−0·0190	0·0000
	1·0000	0·0026	0·0000

Table I. The final values of the variables were calculated to be

$$\xi^* = (-1\cdot7172, 1\cdot5957, 1\cdot8272, -0\cdot7636, -0\cdot7637) \qquad (14)$$

and the fact that $\xi_4^* \neq \xi_5^*$ is due to a request for only four decimals accuracy from the procedure used for the unconstrained minimizations [3].

To demonstrate the way in which the parameters σ are adjusted, we solved the same problems after deliberately setting each component of σ to too small a value initially. The progress of the resultant ten iterations is given in Table II.

TABLE II

An example in which σ is adjusted

k	σ	θ	ψ
1	0·0010	0·0000	1·9079
	0·0010	0·0000	−1·3191
	0·0010	0·0000	0·1649
2	0·0010	1·9079	1·4622
	0·0010	−1·3191	−1.1067
	0·0010	0·1649	0·1457
3	0·0100	0·1908	0·6776
	0·0100	−0·1319	−0·5661
	0·0010	0·1649	0·6613
4	0·0100	0·8684	0·3720
	0·0100	−0·6980	−0·3560
	0·0010	0·8262	0·4880
5	0·1000	0·0868	0·0913
	0·1000	−0·0698	−0·0937
	0·0100	0·0826	0·1371
6	0·1000	0·1781	0·0178
	0·1000	−0·1635	−0·0204
	0·0100	0·2197	0·0321
7	0·1000	0·1959	0·0038
	0·1000	−0·1839	−0·0045
	0·0100	0·2518	0·0073
8	0·1000	0·1997	0·0009
	0·1000	−0·1885	−0·0010
	0·0100	0·2590	0·0016
9	0·1000	0·2006	0·0002
	0·1000	−0·1895	−0·0002
	0·0100	0·2607	0·0004
10	0·1000	0·2008	0·0000
	0·1000	−0·1897	0·0000
	0·0100	0·2610	0·0001

The tables show that the method is promising, but we postpone the discussion of its development until Section 4.

3. Theory

In this section we prove some theorems on the convergence of the method. Because some readers may prefer to omit the various proofs, the results are stated first, and the paper is intelligible without the details following the comments on Theorem 5.

First we state that the flow diagram (Fig. 1) describes a process that will finish.

THEOREM 3. *If there exists a solution (ξ^* say) to the constrained minimization problem, and if $F(\mathbf{x})$ is bounded below, then the sequence $\{c_0, c_1, c_2, ...\}$ converges to zero.*

However this theorem does not state that as c_k tends to zero, the sequence of unconstrained minima tends to the solution ξ^*. To prove this result we require more assumptions, and for brevity in the statement of Theorem 4 we call one of them "condition X". We say that condition X holds if, given any neighbourhood, R say, of ξ^*, there exists a number $\varepsilon > 0$ such that we can solve the equations

$$\psi_t(\mathbf{x}) = \alpha_t \ (t = 1, 2, ..., m) \tag{15}$$

for some $\mathbf{x} \in R$, where the numbers α_t are arbitrary, except that they must satisfy $|\alpha_t| \leqslant \varepsilon \ (t = 1, 2, ..., m)$. Note that the condition rules out constraints like $x^2 = 0$.

THEOREM 4. *If condition X holds, and the constrained problem has a unique solution ξ^*, and the successive unconstrained minima are all within a compact region of the space of the variables, S say, and also the functions $F(\mathbf{x})$ and $\psi_t(\mathbf{x})$ are continuous both for $\mathbf{x} \in S$ and at $\mathbf{x} = \xi^*$, then the points ξ converge to the solution ξ^*.*

The next theorem is on the rate of convergence of the method, and it justifies the correction (7). To introduce it some remarks on the generalized Newton iteration are needed.

Remember that we are trying to satisfy the Eqns (6) by adjusting the parameters $\theta_t \ (t = 1, 2, ..., m)$. If θ_t is changed to $\theta_t + \eta_t \ (t = 1, 2, ..., m)$ then, subject to some conditions on second derivatives, the change induced in $\psi_t(\xi)$ is equal to

$$\psi_t(\xi(\sigma, \theta + \eta)) - \psi_t(\xi(\sigma, \theta)) = \sum_{q=1}^{m} J_{tq} \eta_q + O\|\eta\|^2 \tag{16}$$

where J_{tq} is the Jacobian

$$J_{tq} = \frac{d}{d\theta_q} \; \psi_t \left(\xi(\sigma, \theta) \right)$$

$$= \sum_{i=1}^{n} \left[\frac{\partial \psi_t(\mathbf{x})}{\partial x_i} \right]_{\mathbf{x} = \xi} \frac{\partial \xi_i(\sigma, \theta)}{\partial \theta_q}. \tag{17}$$

Therefore second-order convergence is usually obtained if the corrections $\eta_t \; (t = 1, 2, ..., m)$ are calculated to solve the linear equations

$$\sum_{q=1}^{m} J_{tq} \eta_q = -\psi_t \left(\xi(\sigma, \theta) \right) \tag{18}$$

and this is the generalized Newton iteration. Since we employ the correction (7), the two iterations are identical if and only if it happens that $-J$ is the unit matrix. Therefore we are encouraged by the fact that Theorem 5 states that the difference between $-J_{tq}^*$ and δ_{tq} (the Kronecker-delta) becomes small as σ is increased, where J^* is the Jacobian at $\xi = \xi^*$. The proof of the theorem shows that the definition (17) depends on the Hessian

$$G_{ij} = \left[\frac{\partial^2 \phi(\mathbf{x}, \sigma, \theta)}{\partial x_i \, \delta x_j} \right]_{\mathbf{x} = \xi} \tag{19}$$

being positive definite, which condition is usually obtained because $\mathbf{x} = \xi$ minimizes $\phi(\mathbf{x})$.

THEOREM 5. *If the constrained minimization problem has a solution ξ^*, and the first and second derivatives of $F(\mathbf{x})$ and $\psi_t(\mathbf{x})$ are continuous at ξ^*, and the gradient vectors of $\psi_t(\mathbf{x}) \; (t = 1, 2, ..., m)$ are linearly independent at ξ^*, and also the Hessian at ξ^* is positive definite for $\sigma = \bar{\sigma}$, then there exists a number γ, independent of σ, such that for $\sigma_t \geqslant \bar{\sigma}_t \; (t = 1, 2, ..., m)$ the difference between $-J^*$ and the unit matrix is bounded by*

$$|-J_{tq}^* - \delta_{tq}| < \gamma/(\sigma_t - \bar{\sigma}_t). \tag{20}$$

$\bar{\sigma}$ occurs in the statement of the theorem because it is often the case that the Hessian is positive definite only when the constraints are present. From Eqn (16) we obtain that our correction (7) causes the value of $\psi_t(\xi)$ to be replaced by

$$\psi_t(\xi) + \sum_{q=1}^{m} J_{tq}^* \psi_q(\xi) + O \left[\max_t |\psi_t(\xi)| \right]^2 \tag{21}$$

and Theorem 5 shows that the value of expression (21) is bounded by

$$\gamma \sum_{q=1}^{m} |\psi_q(\xi)|/(\sigma_t - \bar{\sigma}_t) + O\,[\max_t |\psi_t(\xi)|]^2 \tag{22}$$

which justifies our statement that arbitrarily fast linear convergence can be obtained by choosing σ to be sufficiently large. Moreover note that in the flow diagram we increase only σ_t if among the numbers $\psi_q(\xi)$ $(q = 1, 2, ..., m)$ only $\psi_t(\xi)$ converges to zero at too slow a rate, and this strategy is also justified by the result (22).

Proof of Theorem 3. The flow diagram shows that the sequence $\{c_0, c_1, c_2, ..., \}$ is monotonic decreasing and is bounded below by zero, so it converges to some limit. For convenience we consider the elements c_q, where

$$q = 3l = 3m(j+1) \tag{23}$$

the numbers l and j being integers. If the test $c_k \leqslant \frac{1}{4} c_{k-1}$ of the flow diagram has been satisfied l times $(1 \leqslant k \leqslant q)$, then

$$c_q \leqslant (\tfrac{1}{4})^l c_0 \tag{24}$$

and otherwise σ has been increased at least l times. We prove the theorem by showing that the operation of increasing σ also limits the value of c_q.

We use the inequality, derived from the definition of ξ,

$$\phi(\xi, \sigma, \theta) \leqslant \phi(\xi^*, \sigma, \theta) \tag{25}$$

for it implies that

$$\sum_{t=1}^{m} \sigma_t\,[\theta_t + \psi_t(\xi)]^2 \leqslant F(\xi^*) - F(\xi) + \sum_{t=1}^{m} \sigma_t \theta_t^2 \leqslant L + \sum_{t=1}^{m} \sigma_t \theta_t^2 \tag{26}$$

where
$$L = F(\xi^*) - \inf_{x} F(x). \tag{27}$$

Therefore, by inspecting the two ways in which the flow diagram adjusts σ_t and θ_t, we see that at the beginning of the kth iteration

$$\sum_{t=1}^{m} \sigma_t \theta_t^2 \leqslant (k-1)\,L \tag{28}$$

whence, for $t = 1, 2, ..., m$, we deduce

$$\theta_t \leqslant \sqrt{[(k-1)\,L/\sigma_t]}. \tag{29}$$

Now the inequalities (26) and (28) also yield that, on the kth iteration,

$$|\theta_t + \psi_t(\xi)| \leqslant \sqrt{(kL/\sigma_t)} \qquad (30)$$

so we obtain

$$|\psi_t(\xi)| < 2\sqrt{(kL/\sigma_t)}. \qquad (31)$$

If σ has been increased at least l times $(1 \leqslant k \leqslant q)$, then there exists a component of σ, σ_τ say, such that

$$\sigma_\tau \geqslant 10^{j+1} \qquad (32)$$

(see Eqn 23). When σ_τ was increased from 10^j to 10^{j+1} we had

$$|\psi_\tau(\xi)| \geqslant \tfrac{1}{4} c_{k-1} \qquad (33)$$

which implies that

$$c_{k-1} \leqslant 4|\psi_\tau(\xi)|$$
$$< 8\sqrt{(kL/10^j)}. \qquad (34)$$

Thus we obtain, for all values of q satisfying Eqn (23), that either the inequality (24) holds or

$$c_q < 8\sqrt{(qL/10^j)}. \qquad (35)$$

Theorem 3 is now proved because both the right-hand sides of the inequalities (24) and (35) tend to zero as q becomes large.

Proof of Theorem 4. From the flow diagram (Fig. 1) we see that the test for convergence will be satisfied only if

$$c_k < c_{k-1} \qquad (36)$$

and we let ξ_j be the jth unconstrained minimum for which the test (36) holds. From Theorem 3 ($F(\mathbf{x})$ is bounded below on S)

$$\lim_{j \to \infty} \max_t |\psi_t(\xi_j)| = 0 \qquad (37)$$

so that, because of the continuity of the functions $\psi_t(\mathbf{x})$, any limit point of the sequence $\{\xi_1, \xi_2, \xi_3, \ldots\}$ satisfies the equality constraints.

Because of the compactness of S at least one limit point must exist. If at a limit point ξ the value of $F(\xi)$ exceeds $F(\xi^*)$, then the required solution has not been obtained, but if $F(\xi) = F(\xi^*)$ then, by the assumed uniqueness of ξ^*, we have reached the required answer. Therefore we prove that for any limit point ξ the inequality

$$F(\xi) \leqslant F(\xi^*) \qquad (38)$$

holds, which is sufficient because $F(\xi) < F(\xi^*)$ contradicts the definition of ξ^*.

Let δ be an arbitrary positive number, let R be a neighbourhood of ξ^* such that

$$|F(\mathbf{x}) - F(\xi^*)| < \delta, \qquad \mathbf{x} \in R \qquad (39)$$

(the existence of R is a consequence of the continuity of $F(\mathbf{x})$), let $\varepsilon > 0$ be the number occurring in the definition of condition X, and let J be such that

$$\max_t |\psi_t(\xi_j)| < \varepsilon, \qquad j \geqslant J. \qquad (40)$$

Let the dependence on j of σ and θ be implied, so that ζ_j minimizes expression (4). Condition X is such that for $j \geqslant J$ we can let $\zeta_j \in R$ be a vector which satisfies the equations

$$\psi_t(\zeta_j) = \psi_t(\xi_j) \quad (t = 1, 2, ..., m) \qquad (41)$$

so that we obtain the identity

$$\phi(\xi_j) = F(\xi_j) + \sum_t \sigma_t [\theta_t + \psi_t(\xi_j)]^2$$

$$= F(\xi_j) + \sum_t \sigma_t [\theta_t + \psi_t(\zeta_j)]^2$$

$$= \phi(\zeta_j) + F(\xi_j) - F(\zeta_j). \qquad (42)$$

Now because the definition of ξ_j implies $\phi(\zeta_j) \geqslant \phi(\xi_j)$, we find that $F(\xi_j) \leqslant F(\zeta_j)$, whence from the relation (39) we derive

$$F(\xi_j) \leqslant F(\xi^*) + \delta. \qquad (43)$$

But $\delta > 0$ is arbitrary, so that if the numbers $F(\xi_j)$ converge the limit does not exceed $F(\xi^*)$, which proves the inequality (38), from which we deduce the correctness of the theorem.

Proof of Theorem 5. The statement of the theorem is independent of the choice of coordinate vectors in the space of the variables, therefore in order to simplify our algebra we let the first m coordinate vectors be the gradient vectors of the constraint functions $\psi_t(\mathbf{x})$ $(t = 1, 2, ..., m)$ calculated at $\mathbf{x} = \xi^*$; it is assumed that they are linearly independent.

To calculate the partial derivatives $\partial \xi_i(\sigma, \theta)/\partial \theta_q$, occurring in the definition (17), we note that the conditions

$$\left[\frac{\partial \phi(\mathbf{x}, \sigma, \theta)}{\partial x_i} \right]_{\mathbf{x} = \xi} = 0 \qquad (i = 1, 2, ..., n) \qquad (44)$$

relate ξ to θ. In particular if changing θ to $\theta+\eta$ causes ξ to change to $\xi+\delta$, we obtain the equation

$$\left[2\sum_{t=1}^{m}\sigma_t\,\eta_t\,\frac{\partial\psi_t\,(\mathbf{x})}{\partial x_i}+\sum_{j=1}^{n}\delta_j\,\frac{\partial^2\phi(\mathbf{x})}{\partial x_i\,\partial x_j}\right]_{\mathbf{x}=\xi}=O\left(\|\eta\|^2+\|\delta\|^2\right).\qquad(45)$$

Therefore if the Hessian is non-singular, and we let its inverse be H, then

$$\frac{\partial\xi_i\,(\sigma,\theta)}{\partial\theta_q}=-\sum_{j=1}^{n}H_{ij}\left[2\sigma_q\,\frac{\partial\psi_q\,(\mathbf{x})}{\partial x_j}\right]_{\mathbf{x}=\xi}.\qquad(46)$$

By substituting this result in Eqn (17), and by using our choice of coordinate vectors, we obtain the simple result

$$J^*_{tq}=-2\sigma_q\sum_{i=1}^{n}\sum_{j=1}^{n}H^*_{ij}\left[\frac{\partial\psi_t\,(\mathbf{x})}{\partial x_i}\quad\frac{\partial\psi_q\,(\mathbf{x})}{\partial x_j}\right]_{\mathbf{x}=\xi^*}$$

$$=-2\sigma_q\,H^*_{tq}.\qquad(47)$$

The theorem concerns the dependence of J^* on σ, so we find that the definition of the Hessian G^* is ($=H^{*-1}$)

$$G^*_{ij}=\left[\frac{\partial^2 F}{\partial x_i\,\partial x_j}+2\sum_{t=1}^{m}\sigma_t\left\{\theta_t\,\frac{\partial^2\psi_t\,(\mathbf{x})}{\partial x_i\,\partial x_j}+\delta_{ti}\,\delta_{tj}\right\}\right]_{\mathbf{x}=\xi}.\qquad(48)$$

Therefore, because of Theorem 2, the form of G^*, including the dependence on σ, is just

$$\begin{pmatrix} A_{11}+2\sigma_1 & A_{12} & \cdots & A_{1m} & A_{1m+1} & \cdots & A_{1n} \\ & & & & \vdots & & \\ A_{21} & A_{22}+2\sigma_2 & \cdots & A_{2m} & A_{2m+1} & \cdots & A_{2n} \\ \vdots & \vdots & \ddots & \vdots & \vdots & & \vdots \\ A_{m1} & A_{m2} & \cdots & A_{mm}+2\sigma_m & A_{mm+1} & \cdots & A_{mn} \\ A_{m+1\,1} & A_{m+12} & \cdots & A_{m+1m} & A_{m+1\,m+1} & \cdots & A_{m+1\,n} \\ \vdots & \vdots & & \vdots & \vdots & \ddots & \vdots \\ A_{n1} & A_{n2} & \cdots & A_{nm} & A_{n\,m+1} & \cdots & A_{nn} \end{pmatrix}\qquad(49)$$

and the Jacobian is obtained from the leading $m\times m$ submatrix of the inverse of G^*. The inequality (20) is derived from the following Lemma.

LEMMA *If the matrix* (49) *is positive definite when* $\sigma_t=\bar{\sigma}_t\geqslant 0$ $(t=1,2,...,m)$, *and if* $\sigma_t\geqslant\bar{\sigma}_t$, *then there exists a number* $b(A,\sigma)$ *such that for* $t,q=1,2,...,m$,

$$|H^*_{tq}-\delta_{tq}/2\sigma_t|<b/(\sigma_t-\bar{\sigma}_t)\,(\sigma_q-\bar{\sigma}_q)\qquad(50)$$

where H^* *is the inverse of the matrix* (49).

Proof. For $t = 1, 2, ..., m$ express the tth diagonal element of the matrix (49) as

$$(A_{tt} + 2\bar{\sigma}_t) + 2(\sigma_t - \bar{\sigma}_t).$$

Then express each number H_{tq}^* as a ratio of a cofactor to the determinant of the matrix, and expand the denominator and the various numerators in powers of the parameters $(\sigma_t - \bar{\sigma}_t)$. For instance the determinant of the matrix is of the form

$$\sum_{i_1=0}^{1} \sum_{i_2=0}^{1} \cdots \sum_{i_m=0}^{1} \beta_{i_1 i_2 \cdots i_m} (\sigma_1 - \bar{\sigma}_1)^{i_1} (\sigma_2 - \bar{\sigma}_2)^{i_2} \cdots (\sigma_m - \bar{\sigma}_m)^{i_m}.$$

Because the matrix is positive definite when $\boldsymbol{\sigma} = \boldsymbol{\sigma}$, all the numbers $\beta_{i_1 i_2 \cdots i_m}$ are strictly positive, therefore because the numerator of the expression for H_{tq}^* $(t \neq q)$ is independent of σ_t and σ_q it is apparent that

$$(\sigma_t - \bar{\sigma}_t)(\sigma_q - \bar{\sigma}_q) H_{tq}^*$$

is bounded. Moreover if $\{2(\sigma_t - \bar{\sigma}_t) H_{tt}^* - 1\}$ is expressed as a ratio, the numerator is independent of σ_t, so that

$$(\sigma_t - \bar{\sigma}_t)\{2(\sigma_t - \bar{\sigma}_t) H_{tt}^* - 1\}$$

is also bounded. By combining these results we obtain that there exists a number $\bar{b}(A, \bar{\boldsymbol{\sigma}})$ such that, for $t, q = 1, 2, ..., m$,

$$|H_{tq}^* - \delta_{tq}/2(\sigma_t - \bar{\sigma}_t)| < \bar{b}/(\sigma_t - \bar{\sigma}_t)(\sigma_q - \bar{\sigma}_q). \tag{51}$$

Because of the inequality

$$|H_{tt}^* - 1/2\sigma_t| \leqslant |H_{tt}^* - 1/2(\sigma_t - \bar{\sigma}_t)| + \bar{\sigma}_t/2\sigma_t (\sigma_t - \bar{\sigma}_t)$$
$$\leqslant |H_{tt}^* - 1/2(\sigma_t - \bar{\sigma}_t)| + \bar{\sigma}_t/2(\sigma_t - \bar{\sigma}_t)^2 \tag{52}$$

the lemma is a consequence of the inequality (51).

We now return to the proof of Theorem 5. Note that the set of vectors $\boldsymbol{\sigma}$ for which the matrix (49) is positive definite is open, so there exists a vector $\hat{\boldsymbol{\sigma}}$ in this set such that $\hat{\sigma}_t < \bar{\sigma}_t$ $(t = 1, 2, ..., m)$. We apply the lemma to $\hat{\boldsymbol{\sigma}}$ and we deduce the existence of a constant $\hat{b}$ such that

$$|H_{tq}^* - \delta_{tq}/2\sigma_t| < \hat{b}/(\sigma_t - \hat{\sigma}_t)(\sigma_q - \hat{\sigma}_q). \tag{53}$$

Therefore from Eqn (47) we obtain the inequality

$$|-J_{tq}^* - \delta_{tq}| = |2\sigma_q H_{tq}^* - \delta_{tq}|$$
$$< 2\sigma_q \hat{b}/(\sigma_t - \hat{\sigma}_t)(\sigma_q - \hat{\sigma}_q)$$
$$< \{2\bar{\sigma}_q \hat{b}/(\bar{\sigma}_q - \hat{\sigma}_q)\}/(\sigma_t - \bar{\sigma}_t) \tag{54}$$

so that a number γ exists for which the theorem holds.

4. Discussion

It should now be apparent that the method described in Section 2 will usually converge to the required solution, and if the functions $F(\mathbf{x})$ and $\psi_t(\mathbf{x})$ have continuous second derivatives then, because of the analogy with the Newton iteration, the ability to increase σ can achieve linear convergence at the rate of the inequality (9). Moreover, if the test (9) were replaced by a more stringent one, say $c_k \leqslant c_{k-1}/40$, then the new rate could also be achieved. The reason for preferring the slower rate is that faster convergence tends to make the separate unconstrained minimizations more difficult. However, we have not carried out a series of experiments to find out whether the factor 1/4 is a particularly good choice, but we would have done so if the numerical examples of Tables I and II had not yielded so easily. Probably it would be worthwhile to aim for faster convergence if more than six decimals accuracy were required, and then we would need high precision in the calculation of the various functions.

A very promising feature of the new method is the dependence of the Hessian on the parameters σ and θ, the Hessian being

$$G_{ij} = \frac{\partial^2 F}{\partial x_i \, \partial x_j} + 2 \sum_{t=1}^{m} \sigma_t \left[\theta_t \frac{\partial^2 \psi_t}{\partial x_i \, \partial x_j} + \psi_t \frac{\partial^2 \psi_t}{\partial x_i \, \partial x_j} + \frac{\partial \psi_t}{\partial x_i} \frac{\partial \psi_t}{\partial x_j} \right]. \quad (55)$$

The reason it is important is that it dominates the behaviour of $\phi(\mathbf{x})$ at the unconstrained minimum ξ, and the methods for unconstrained minimization which we have quoted are successful because they take account of the separate elements of the Hessian matrix. However, gaining this detailed information uses a substantial proportion of the total computing time, so we would like to be able to carry information forward from one unconstrained minimization to the next. Therefore we ask how the Hessian changes as σ and θ are adjusted. The change in θ given by Eqn (7) alters the Hessian by

$$2 \sum_{t=1}^{m} \sigma_t \psi_t \frac{\partial^2 \psi_t}{\partial x_i \, \partial x_j} \quad (56)$$

and this quantity is likely to be small because of the factors ψ_t; moreover it is zero if all the constraints happen to be linear. Therefore when only θ is altered it would seem to be well worthwhile to carry forward an approximation, like the estimate of H obtained by Davidon's method, to the next unconstrained minimization. On the other hand, if both σ and θ are adjusted so that $\sigma_t \theta_t$ remains fixed, then the change in the Hessian is the change in

$$2 \sum_{t=1}^{m} \sigma_t \left[\psi_t \frac{\partial^2 \psi_t}{\partial x_i \, \partial x_j} + \frac{\partial \psi_t}{\partial x_i} \frac{\partial \psi_t}{\partial x_j} \right]$$

and we note that the first part of this expression is just the quantity (56). Therefore usually the change is dominated by the alteration in

$$2 \sum_{t=1}^{m} \sigma_t \frac{\partial \psi_t}{\partial x_i} \frac{\partial \psi_t}{\partial x_j}$$

which can be quite large, supporting our reluctance to increase σ. But it may be feasible to take explicit account of this change because only first derivatives of the constraint functions are involved. Because of these remarks we are hopeful that very efficient algorithms can result from merging the method of this paper with the various unconstrained minimization techniques.

We are also very encouraged by the fact that it seems to be straightforward to extend the method to take account of nonlinear inequality constraints. The main extra difficulty which is introduced by inequalities is the problem of identifying which constraints actually limit the position of the final solution, for when these have been identified it is often sufficient to regard these constraints as equalities, and to ignore the other constraints. Therefore an iterative process might start with a guess of the important constraints, in which case it would have to be able to change the estimate. Deciding when to include extra constraints is easy, because direct calculation shows when a condition is violated, but to decide whether to drop a particular constraint from the current list of equalities requires either a prediction or a trial to find out which way the solution would move. The reason the method of this paper is particularly promising is that it provides a direct prediction through the sign of θ, (see Theorem 2). This observation may lead to a good algorithm for the general constrained minimization problem.

References

1. W. C. Davidon (1959). "Variable metric method for minimization", A.E.C. Research and Development Report, ANL-5990 (Rev.)

2. R. Fletcher and M. J. D. Powell (1963). A rapidly convergent descent method for minimization, *The Computer Journal*, **6**, 163.

3. M. J. D. Powell (1964). An efficient method for finding the minimum of a function of several variables without calculating derivatives, *The Computer Journal*, **7**, 155.

4. A. V. Fiacco and G. P. McCormick (1967). The slacked unconstrained minimization technique for convex programming, S.I.A.M. *J. Appl. Math.*, **15**, 505.

5. C. W. Carroll (1961). The created response surface technique for optimizing non-linear restrained systems, *Operations Research*, **9**, 169.

6. J. F. Traub (1964). "Iterative Methods for the Solution of Equations", Prentice-Hall Inc.

Discussion

SASSON. (Imperial College). Do you consider this method to be useful for the problem of solving sets of nonlinear equations? It has already been stated that when minimizing the sum of square one has no control over the individual terms, but by minimizing one equation subject to equality constraints on the others by using this method, then there would be some control on each individual equation.

POWELL. I think that in the case you suggest, the final values of the parameter θ_t would have to equal zero, so the main point of the method is lost.

20. A Modified Newton Method for Optimization with Equality Constraints

Y. Bard and J. L. Greenstadt

IBM, New York, U.S.A.

We wish to find the vector $\mathbf{x}$ at which the function $f(\mathbf{x})$ attains its maximum, subject to a set of equality constraints

$$\mathbf{g}(\mathbf{x}) = \mathbf{0}. \tag{1}$$

Let λ be a vector of Lagrange multipliers. We form the function

$$\Phi(\mathbf{x}, \lambda) \equiv f(\mathbf{x}) + \lambda^\tau \mathbf{g}(\mathbf{x}). \tag{2}$$

As we know, the constrained maximum of $f(\mathbf{x})$ will occur at a stationary point of Φ which is a saddle point in the $(\mathbf{x}, \lambda)$ space [1, p.231]. We now approximate $\Phi(\mathbf{x}, \lambda)$ by a quadratic function, and attempt to locate the saddle point of the quadratic approximation.

Assuming f and $\mathbf{g}$ to be twice differentiable, we can form the vector $\mathbf{b}$ and the matrices $\mathbf{G}$ and $\mathbf{A}$ defined by:

$$\mathbf{b}_i \equiv \frac{\partial \Phi}{\partial x_i} = \frac{\partial f}{\partial x_i} + \sum_j \lambda_j \frac{\partial g_j}{\partial x_i} \tag{3}$$

$$\mathbf{G}_{ij} \equiv \frac{\partial g_i}{\partial x_j} \tag{4}$$

$$\mathbf{A}_{ij} \equiv -\frac{\partial^2 \Phi}{\partial x_i \partial x_j} = -\frac{\partial^2 f}{\partial x_i \partial x_j} - \sum_k \lambda_k \frac{\partial^2 g_k}{\partial x_i \partial x_j}. \tag{5}$$

The Taylor series expansion around a point $\mathbf{x} = \mathbf{x}_0, \lambda = \lambda_0, \Phi = \Phi_0$ yields, up to terms of second order,

$$\Phi(\mathbf{x}_0 + \delta\mathbf{x}, \lambda_0 + \delta\lambda) = \Phi_0 + \mathbf{b}^\tau \delta\mathbf{x} + \mathbf{g}^\tau \delta\lambda - \tfrac{1}{2}\delta\mathbf{x}^\tau \mathbf{A}\delta\mathbf{x} + \delta\lambda^\tau \mathbf{G}\delta\mathbf{x}. \tag{6}$$

299

We wish now to transform the independent variables δx and $\delta \lambda$ so that the positive definite and negative definite parts of this expansion can be segregated. This transformation to new variables ϕ and ψ involves the following. Letting

$$\mathbf{B} \equiv \mathbf{GA}^{-1}\mathbf{G}^{\tau} \tag{7}$$

$$\phi \equiv \delta\mathbf{x} - \mathbf{A}^{-1}(\mathbf{G}^{\tau}\delta\lambda + \mathbf{b}) \tag{8}$$

$$\psi \equiv \delta\lambda + \mathbf{B}^{-1}(\mathbf{GA}^{-1}\mathbf{b} + \mathbf{g}) \tag{9}$$

we find that

$$\Phi = \Phi_0 - \tfrac{1}{2}\phi^{\tau}\mathbf{A}\phi + \tfrac{1}{2}\psi^{\tau}\mathbf{B}\psi - \tfrac{1}{2}(\mathbf{GA}^{-1}\mathbf{b} + \mathbf{g})^{\tau}\mathbf{B}^{-1}(\mathbf{GA}^{-1}\mathbf{b} + \mathbf{g}) + \tfrac{1}{2}\mathbf{b}^{\tau}\mathbf{A}^{-1}\mathbf{b} \tag{10}$$

which may be verified by substitution. The only part of Φ which depends on $\delta\mathbf{x}$ and $\delta\lambda$ is

$$\Phi^* \equiv -\tfrac{1}{2}\phi^{\tau}\mathbf{A}\phi + \tfrac{1}{2}\psi^{\tau}\mathbf{B}\psi. \tag{11}$$

We assume, for the time being, that $\mathbf{A}$ is positive definite. Then so is $\mathbf{B}$. Both Φ^* and Φ then have a saddle point at $\phi = 0$, $\psi = 0$. It is a maximum relative to variations in ϕ, and a minimum relative to variations ψ. Φ^* may be regarded as the generalization in many dimensions of a function of the form $f(x, y) = -ax^2 + by^2 (a, b > 0)$, which has a saddle point at $(0, 0)$. If we were at a point away from the origin, we would clearly want to make a step in the x direction so as to increase f, and a step in the y direction so as to decrease f. If these steps had the proper lengths, we would end up at the origin, which is the saddle point. Thus, in our present case, we take the following two-part step.

(1) We make $\phi = 0$ while leaving ψ unchanged. This implies

$$\phi = \delta\mathbf{x}_1 - \mathbf{A}^{-1}(\mathbf{G}^{\tau}\delta\lambda_1 + \mathbf{b}) = 0 \tag{12}$$

$$\psi = \delta\lambda_1 + \mathbf{B}^{-1}(\mathbf{GA}^{-1}\mathbf{b} + \mathbf{g}) = \psi_0 = \mathbf{B}^{-1}(\mathbf{GA}^{-1}\mathbf{b} + \mathbf{g}). \tag{13}$$

Solving for $\delta\mathbf{x}_1$ and $\delta\lambda_1$, we find

$$\delta\lambda_1 = 0 \tag{14}$$

$$\delta\mathbf{x}_1 = \mathbf{A}^{-1}\mathbf{b}. \tag{15}$$

Along this step, the function Φ^* (and hence Φ) must *increase*, as is evident from Eqn (11).

(2) We make $\psi = 0$ while leaving ϕ unchanged. This implies

$$\phi = \delta x_2 - A^{-1}(G^t \delta \lambda_2 + b) = \phi_0 = -A^{-1}b \qquad (16)$$

$$\psi = \delta \lambda_2 + B^{-1}(GA^{-1}b + g) = 0. \qquad (17)$$

Solving for δx_2 and $\delta \lambda_2$ we find

$$\delta \lambda_2 = -B^{-1}(GA^{-1}b + g) \qquad (18)$$

$$\delta x_2 = -A^{-1}G^t B^{-1}(GA^{-1}b + g). \qquad (19)$$

Along this step, the function Φ^* (and hence Φ) must *decrease*, as is evident from Eqn (11).

The steps given by Eqns (14), (15), (18) and (19) are valid if Φ is really a quadratic function of x and λ, that is if f is quadratic in x and the g are linear in x. In general, this will not be the case, so that it is necessary to use an interpolation procedure to make sure that progress is being made towards the solution.

Still assuming that A is positive definite, our procedure is as follows.

(1) We define $\alpha = 1$.

(2) Compute $\Phi_{1,\alpha} \equiv \Phi(x_0 + \alpha \delta x_1, \lambda_0 + \alpha \delta \lambda_1)$.

(3) If $\Phi_{1,\alpha} \leqslant \Phi_0$ we replace α by $\alpha/2$ and return to step (2). Unless $b = 0$ we are guaranteed that eventually $\Phi_{1,\alpha} > \Phi_0$, whereupon we proceed to step (4).

(4) Compute $\Phi_{2,\alpha} \equiv \Phi(x_0 + \alpha \delta x_1 + \alpha \delta x_2, \lambda_0 + \alpha \delta \lambda_1 + \alpha \delta \lambda_2)$.

(5) If $\Phi_{2,\alpha} \geqslant \Phi_{1,\alpha}$ we replace α by $\alpha/2$ and return to step (2). Eventually, we will have $\Phi_{2,\alpha} < \Phi_{1,\alpha}$, whereupon our iteration is completed. We replace x_0, λ_0 with $x_0 + \alpha \delta x_1 + \alpha \delta x_2, \lambda_0 + \alpha \delta \lambda_1 + \alpha \delta \lambda_2$ and return to step (1).

It is easily verified that an iteration with $\alpha = 1$ corresponds precisely to a single step of the Newton-Raphson method; for the gradient of Φ is given by

$$\delta \equiv \begin{bmatrix} \dfrac{\partial \Phi}{\partial x} \\[2mm] \dfrac{\partial \Phi}{\partial \lambda} \end{bmatrix} = \begin{bmatrix} b \\[2mm] g \end{bmatrix} \qquad (20)$$

and its Hessian by

$$
\mathbf{H} \equiv
\begin{bmatrix}
\dfrac{\partial^2 \Phi}{\partial \mathbf{x}\,\partial \mathbf{x}} & \dfrac{\partial^2 \Phi}{\partial \mathbf{x}\,\partial \lambda} \\[2ex]
\dfrac{\partial^2 \Phi}{\partial \lambda\,\partial \mathbf{x}} & \dfrac{\partial^2 \Phi}{\partial \lambda\,\partial \lambda}
\end{bmatrix}
=
\begin{bmatrix}
-\mathbf{A} & \mathbf{G}^\tau \\[2ex]
\mathbf{G} & \mathbf{O}
\end{bmatrix}
\tag{21}
$$

The Newton-Raphson step is given by

$$
\begin{bmatrix} \delta\mathbf{x} \\ \delta\lambda \end{bmatrix} = -\mathbf{H}^{-1}\boldsymbol{\delta}.
\tag{22}
$$

It may be verified by substitution that

$$
\mathbf{H}^{-1} = \mathbf{P} + \mathbf{Q}
\tag{23}
$$

with

$$
\mathbf{P} \equiv
\begin{bmatrix}
\mathbf{A}^{-1} & \mathbf{O} \\
\mathbf{O} & \mathbf{O}
\end{bmatrix}
=
\begin{bmatrix}
\mathbf{I} \\
\mathbf{O}
\end{bmatrix}
\mathbf{A}^{-1}\,[\mathbf{I}; \mathbf{O}]
\tag{24}
$$

$$
\mathbf{Q} \equiv -
\begin{bmatrix}
\mathbf{A}^{-1}\mathbf{G}^\tau\mathbf{B}^{-1}\mathbf{G}\mathbf{A}^{-1} & \mathbf{A}^{-1}\mathbf{G}^\tau\mathbf{B}^{-1} \\
\mathbf{B}^{-1}\mathbf{G}\mathbf{A}^{-1} & \mathbf{B}^{-1}
\end{bmatrix}
= -
\begin{bmatrix}
\mathbf{A}^{-1}\mathbf{G}^\tau \\
\mathbf{I}
\end{bmatrix}
\mathbf{B}^{-1}[\mathbf{G}\mathbf{A}^{-1}; \mathbf{I}]
\tag{25}
$$

It may also be verified that

$$
\begin{bmatrix} \delta\mathbf{x}_1 \\ \delta\lambda_1 \end{bmatrix} = \mathbf{P} \begin{bmatrix} \mathbf{b} \\ \mathbf{g} \end{bmatrix}
\tag{26}
$$

and

$$
\begin{bmatrix} \delta\mathbf{x}_2 \\ \delta\lambda_2 \end{bmatrix} = \mathbf{Q} \begin{bmatrix} \mathbf{b} \\ \mathbf{g} \end{bmatrix}
\tag{27}
$$

Thus, the total Newton-Raphson step (22) is the sum of the two steps (14-15) and (18-19). What we have gained is the following: when we take step (22) we have no way of telling whether f or Φ should increase or decrease. Hence, we have no guidance as to whether the length of our step needs to be reduced to ensure progress. On each partial step we do, however, have a guide: Φ must increase on the first step, and decrease on the second. This situation arises due to the fact that Φ has a saddle point at the solution.

When $\mathbf{A}$ is non-singular but not positive definite, we replace it by a matrix $\mathbf{A}^*$ which is positive definite, using Greenstadt's method [2, 3]. Let the spectral decomposition of $\mathbf{A}$ be given by

$$\mathbf{A} = \sum_i \mu_i \mathbf{v}_i \mathbf{v}_i^\tau \tag{28}$$

where the μ_i and $\mathbf{v}_i$ are the eigenvalues and associated eigenvectors of $\mathbf{A}$ (which is symmetric). Then we define

$$\mathbf{A}^* \equiv \sum_i |\mu_i| \mathbf{v}_i \mathbf{v}_i^\tau \tag{29}$$

and

$$(\mathbf{A}^*)^{-1} = \sum_i |\mu_i|^{-1} \mathbf{v}_i \mathbf{v}_i^\tau. \tag{30}$$

We simply replace $\mathbf{A}$ with $\mathbf{A}^*$ in Eqns (7), (15), (18) and (19), and proceed with steps (1)-(5) as above.

The situation becomes more complicated when $\mathbf{A}$ is singular. As we shall see later, this happens in many interesting problems. By proper partitioning of $\mathbf{A}$ it is possible, however, to decompose the total step into *three* successive parts in which Φ must alternately increase and decrease.

Let $\mathbf{V}$ denote the matrix whose columns are the eigenvectors of $\mathbf{A}$ associated with zero eigenvalues (that is, they span the null space of $\mathbf{A}$). Let μ_i and $\mathbf{u}_i$ denote non-zero eigenvalues and their associated eigenvectors. We define

$$\mathbf{A}^+ \equiv \sum_i |\mu_i|^{-1} \mathbf{u}_i \mathbf{u}_i^\tau \tag{31}$$

where $\mathbf{A}^+$ is the pseudo-inverse of $\mathbf{A}^*$. Furthermore, let

$$\mathbf{C} \equiv \mathbf{G}\mathbf{A}^+\mathbf{G}^\tau \tag{32}$$

and

$$\mathbf{D} \equiv \mathbf{V}^\tau\mathbf{G}^\tau\mathbf{C}^{-1}\mathbf{G}\mathbf{V}. \tag{33}$$

Then one may easily verify that

$$-\mathbf{H}^{-1} = \mathbf{R}_1 + \mathbf{R}_2 + \mathbf{R}_3 \tag{34}$$

with

$$\mathbf{R}_1 \equiv \begin{bmatrix} \mathbf{I} \\ \mathbf{O} \end{bmatrix} \mathbf{A}^+[\mathbf{I}; \mathbf{O}] \tag{35}$$

$$R_2 \equiv - \begin{bmatrix} A^+ G^\tau \\ I \end{bmatrix} C^{-1}[GA^+ ; I] \tag{36}$$

$$R_3 \equiv \begin{bmatrix} A^+ G^\tau C^{-1} G - I \\ C^{-1} G \end{bmatrix} V D^{-1} V^\tau [G^\tau C^{-1} G A^+ - I; G^\tau C^{-1}] \tag{37}$$

Clearly, R, is positive semi-definite (positive definite relative to variations in x outside the null space of A), R_2 is negative definite, and R_3 is positive definite. If we define

$$\begin{bmatrix} \delta x_i \\ \delta \lambda_i \end{bmatrix} \equiv R_i \begin{bmatrix} b \\ g \end{bmatrix} \qquad (i = 1, 2, 3) \tag{38}$$

then the steps corresponding to $i = 1, 2, 3$ should cause, respectively, an increase, decrease, and increase in Φ. To our algorithm we only have to add steps (6) and (7), in which $\alpha \delta x_3$ and $\alpha \delta \lambda_3$ are added, and an increase in Φ is expected.

For purposes of computation, the three steps are defined recursively as follows:

$$\begin{aligned} \delta \lambda_1 &= 0 & \delta x_1 &= A^+ b \\ \delta \lambda_2 &= -C^{-1}(G \delta x_1 + g) & \delta x_2 &= A^+ G^\tau \delta \lambda_2 \\ g &= -V D^{-1} V^\tau (G^\tau \delta \lambda_2 + b) & & \\ \delta \lambda_3 &= C^{-1} G g & \delta x_3 &= A^+ G^\tau \delta \lambda_3 - g \end{aligned} \right\} \tag{39}$$

Admittedly, the justification of this method is somewhat *ad hoc*, but its performance on practical problems has been very good. These problems were parameter estimations of the following kind.
Let

$$h(x, \theta) = 0 \tag{40}$$

represent a mathematical model relating certain variables x and some unknown parameters θ. We conduct n experiments in each of which we record the variables x, whose measured values for the μth experiment we designate x_μ. These values are subject to measurement errors e_μ, the true but unknown values of the variables at the μth experiment being given by

$$x_\mu{}^* = x_\mu + e_\mu. \tag{41}$$

We assume that the true values satisfy Eqn (29) exactly, that is

$$h(x_\mu{}^*, \theta) = 0 \qquad (\mu = 1, 2, ..., n) \tag{42}$$

We also assume that we know the joint probability density function of the errors e_μ, given by a function $p(e_1, e_2, ..., e_n) = p(x_1^* - x_1, x_2^* - x_2, ...)$. By the method of maximum likelihood we take as our estimates for θ and x_μ^* ($\mu = 1, 2, ..., n$) the values of these variables which maximize p, subject to Eqns (42) acting as constraints.

We see immediately that the objective function does not depend on θ. Hence, if some parameter θ_i appears as a constant term in Eqn (40), the corresponding rows and columns in the second derivative matrix A will be zero, and A will be singular. Hence, the importance of the singular case.

The problems which were solved by means of the algorithm presented above all involved two dimensional vectors x_μ, that is

$$\mathbf{x}_\mu = \begin{bmatrix} x_{\mu 1} \\ x_{\mu 2} \end{bmatrix} \tag{43}$$

Each $e_{\mu i}$ ($i = 1, 2$) was assumed independently and normally distributed with variance α_i^{-1}. Thus

$$p = (2\pi)^{-n}(\alpha_1 \alpha_2)^{n/2} \exp{-\tfrac{1}{2} \sum_{\mu=1}^{r} [\alpha_1 (x_{\mu 1}^* - x_{\mu 1})^2 + \alpha_2 (x_{\mu 2}^* - x_{\mu 2})^2]}. \tag{44}$$

With α_1 and α_2 given constants, maximizing p is equivalent to maximizing

$$f \equiv -\sum_{\mu=1}^{r} [\alpha_1 (x_{\mu 1}^* - x_{\mu 1})^2 + \alpha_2 (x_{\mu 2}^* - x_{\mu 2})^2]. \tag{45}$$

The following four models were used:

$$h \equiv x_{\mu 1}^* - \theta x_{\mu 2}^* = 0 \tag{46}$$

$$h \equiv x_{\mu 1}^* - \theta_1 - \theta_2 x_{\mu 2}^* = 0 \tag{47}$$

$$h \equiv x_{\mu 1}^* - \theta_1 - \theta_2 \exp \theta_3 x_{\mu 2}^* = 0 \tag{48}$$

$$h \equiv x_{\mu 1}^{*3} + \theta_1 x_{\mu 1}^{*2} x_{\mu 2}^* + \theta_2 x_{\mu 1}^* x_{\mu 2}^{*2} + \theta_3 x_{\mu 2}^{*3} + \theta_4 + z_\mu = 0 \tag{49}$$

In each case, appropriate data were supplied in the form of values of $x_{\mu 1}$ and $x_{\mu 2}$ ($\mu = 1, 2, ..., n$) (and for Eqn (49) values of z_μ). The value of n varied between five and eight. Various values of α_1 and α_2 were tried, as well as various starting guesses for θ and λ; the starting guesses for the x_μ^* were always taken equal to the x_μ. Convergence to the correct values were always obtained in seven to thirteen iterations.

The main computational drawback of this method for parameter estimation is the requirement of computing the second derivatives of the constraints, that is of the model equations. If we simply replace these derivatives with zeros wherever they are required, we are carrying out the analog of the Gauss method for conventional nonlinear estimation. When this was attempted, very efficient convergence was obtained in some cases, but no convergence at all in others. Further work is required in this direction.

In parameter estimation problems, the number of unknowns x_μ^* may be very large, due to the possibly large value of n. The separable nature of both the objective function and the constraints makes it possible to arrange the calculations in a very economical manner, both in time and storage space requirements.

A natural way of overcoming the need for calculating second derivatives is to generate an approximation to $\mathbf{A}^+$ by a Davidon-like algorithm. This is the next step we hope to undertake in the development of this method.

References

1. R. Courant and D. Hilbert (1953). "Methods of Mathematical Physics", Vol. 1. Interscience, New York.
2. H. Eisenpress and J. L. Greenstadt (1966). The estimation of non-linear econometric systems. *Econometrica*, **34**, 851-861.
3. J. L. Greenstadt (1967). On the relative efficiencies of gradient methods. *Math. Comput.* **21**, 360-367.

Discussion

BEALE. Do both the $\mathbf{x}$ and the λ change at each stage?

BARD. The x and λ are transformed to new variables ϕ and ψ, isolating the positive and negative definite parts of the expansion. Then one step of the process optimizes over ϕ and the other over ψ.

BEALE. There is some connection is there not between this method and the previous method described by Powell? Both seem to be trying to cope with the Lagrange multipliers at the same time as trying to solve the $\mathbf{x}$ problem.

BARD. There is also a connection between this method and a method of Arrow and co-workers. but I am not quite sure what it is.

21. Variable Metric Methods and Unconstrained Optimization

G. P. McCormick and J. D. Pearson

Research Analysis Corporation, McLean, Virginia, U.S.A.

1. Introduction

This paper is the outgrowth of computer investigations into first-order methods for solving the mathematical programming problem,

$$\text{minimize } f(x) \tag{1}$$

where $x = (x_1, ..., x_n)'$ is an n-dimensional column vector. First-order methods are those which require only up to first partial derivatives of the function $f(x)$. (We indicate the vector of first partial derivatives by $g(x)$ and use the notation g^k to denote $g(x^k)$.) The symbol G^k will indicate the $n \times n$ matrix whose i, jth element is $\partial^2 f(x^k)/\partial x_i \, \partial x_j$ (called the Hessian matrix of f).

Problems of the form (1) arise in many areas such as maximum likelihood estimation, nonlinear regression, and in recent years from attempts to solve the general nonlinear programming problem,

$$\text{minimize } \phi(x) \tag{2}$$

$$\text{subject to } h_i(x) \geqslant 0 \quad (i = 1, ..., m)$$

by converting this to a sequence of problems involving the unconstrained minima of an auxiliary function. Some of the test problems used in the computer investigation were problems of minimizing the auxiliary function

$$f(x) = \phi(x) - r \sum_{i=1}^{m} \ln h_i(x)$$

in the region $R^0 \equiv \{x | h_i(x) > 0, i = 1, ..., m\}$ for some value $r > 0$. Under appropriate assumptions, the unconstrained minima of this function are close to constrained solutions of (2) if r is small; thus methods to find unconstrained minima have application to the more general problem (2). Other references to this in [1–3].

Although several different methods and variations were tried on each problem, only a small part of the results will be discussed in this paper. (Full details will be published elsewhere; and some are contained in [4]). The main purpose of this paper is to present in some detail the results of two of the best known first-order algorithms: that of Davidon, later put in a conjugate gradient context by Fletcher and Powell, and one called the Projection method first proposed by Zoutendijk [5]. The statement of these two algorithms is given in Section 2; in Section 3 the problem of "convergence" of the algorithms to a solution of (1) is briefly considered; and Section 4 analyzes in detail the "rate of convergence" of the two algorithms. Conclusions are made in this paper regarding the relative efficiency of the algorithms and it is attempted to bring together the two points of view for first-order algorithms—"quasi–Newton" and "conjugate gradient".

2. Quasi-Newton and Conjugate Gradient Algorithms

Whilst acknowledging that the goal of minimizing techniques is to find the solution to problem (1), in practice one is usually satisfied to place less restrictive requirements on an algorithm. Generally speaking, algorithms are said to converge to a solution to problem (1) if they find a stationary point of f, that is a point x^* where

$$g^* \equiv \nabla f^* = 0. \tag{3}$$

This is not a guarantee that x^* will be a solution to (1) unless the function f happens to be a convex function, that is

$$f(y) \geqslant f(x) + (y-x)' g(x) \qquad \text{for all } x, y. \tag{4}$$

Convex functions have the nice property that local minima to problem (1) are global minima. Many problems are not convex programming problems and hence one must usually settle for finding local minima to problem (1). The following two lemmas state point conditions which can be used to test for local minima of a twice differentiable function.

LEMMA 1. *Necessary conditions that a point x^* be a local minimum to problem (1) are that (3) hold, and*

$$\nabla^2 f(x^*) = G^* \text{ be a positive semi-definite matrix.} \tag{5}$$

LEMMA 2. *Sufficient conditions that a point x^* be a local minimum to problem (1) are that (3) hold, and that*

$$G^* \text{ be a positive definite matrix.} \tag{6}$$

(A symmetric matrix is said to be positive definite if $z'Az>0$ for all $z \neq 0$; it is semi-definite if $z'Az\geqslant 0$ for all z.)

Henceforth we adopt the notion that an algorithm converges if a point satisfying the first-order necessary conditions for a local minimum is found by the algorithm. In Section 4 it will be seen that Lemma 2 plays an important role in the *rate* of convergence of algorithms.

The oldest algorithm for solving (1) was proposed in 1847 by Cauchy [6] and is generally called the method of Steepest Descent. Briefly stated the algorithm is as follows:

(1) Let x^0 be the starting point of the procedure.

(2) Let x^k denote the value of x at the kth iteration.
Let s^k (the "direction" vector at the kth iteration) be given by

$$s^k = -g. \tag{7}$$

(3) Let $x^{k+1} = x^k + t^k s^k$ where t^k is the smallest local minimum to the programming problem

$$\underset{t \geqslant 0}{\text{minimize}} f(x^k + ts^k).. \tag{8}$$

(4) Return to step 2.

Curry proved that every limit point of the sequence $x^0, x^1, ...,$ was a stationary point, that is satisfied (3).

The biggest dissatisfaction with the method of steepest descent is that although it converges, it often does so extremely slowly. This is in contrast to another method used to solve (1) which uses second partial derivatives and is called the Generalized Newton method.

The form of the Generalized Newton algorithm is the same as that given in the steps above with the single difference that the direction vector in (2) is now

$$s^k = -(G^k)^{-1}g^k. \tag{9}$$

In a neighborhood of a point x^* satisfying the sufficiency conditions of Lemma 2, the Generalized Newton method exhibits desirable properties regarding its rate of convergence to x^*. The drawbacks to the use of (9) were that until recently no satisfactory modification of the choice of the direction vector was formulated when G^* was an indefinite form (see [7] for a modification of this), and also that for many problems, computation of the matrix of second partial derivatives of f was impractical. What was desired then was a minimizing algorithm which used only first partial derivatives and which had the same rapid convergence properties in the neighborhood of an optimum as the Generalized Newton method.

Roughly speaking, there are two different kinds of first-order techniques which are used to speed convergence to an optimum. The first class we call quasi-Newton in that their rationale is in some way to gain enough information about the second derivatives of f to generate an $n \times n$ matrix H approximating G^{-1} and letting the direction vector be $-Hg$. The second class, called methods of conjugate gradients, uses an entirely different rationale for minimizing an unconstrained function.

2.1. *Conjugate Gradient Methods*

Conjugate Gradient methods are best explained by describing their behavior on problems which are a special case of (1). Consider the following quadratic programming problem

$$\text{minimize } f(x) = c + a'x + x'Gx/2 \qquad (10)$$

where G is a symmetric positive definite $n \times n$ matrix.

A set of vectors $z_1, \ldots, z_p$ is said to be *conjugate* with respect to the matrix G if

$$z_i' G z_j = 0 \qquad \text{for all } i \neq j. \qquad (11)$$

A conjugate gradient algorithm is one which, starting from an initial point x^0, generates a sequence of conjugate vectors, $\{\sigma^k\}$, where $\sigma^k = x^{k+1} - x^k$, all k, and where x^{k+1} is minimum from x^k in the direction σ^k (that is (8) is used at each iteration). There are many conjugate gradient algorithms and they differ in the manner in which the conjugate directions are generated. All of them however have the following important property.

LEMMA 3. *If f is a positive definite quadratic form as given by* (10), *and* $\{\sigma^k\}$ *is a sequence of n conjugate vectors given by a conjugate gradient algorithm, then x_n is the solution to problem* (10).

These algorithms accomplish the overall minimization by minimizing in n conjugate directions. For the same problem (10), the generalized Newton method would accomplish the overall minimization in just one iteration. The amount of "work" required to do this for the conjugate methods is n times the amount of work required to generate one conjugate direction plus the work required to do a one dimensional search as required by (8). An analysis of most conjugate direction algorithms shows that the effective ones require in the order of n^2 operations to generate one conjugate direction. Hence, excluding the work involved in the single variable unconstrained minimization (8), the amount of work required to minimize (10) using conjugate directions is of the order of n^3. For the generalized Newton method, computation of the vector given by (9) is approximately $n^3/6$ operations (multiplications and additions). Hence both methods vary in difficulty as the cube of the dimensionality of the problem.

Very little has been done with the theory of conjugate gradient algorithms when applied to general problems of type (1). This paper intends to explore two particular conjugate gradient algorithms, one of which can also be interpreted as a quasi-Newton method, and explore their convergence on several problems of the type (1).

2.2. *Projection Algorithm for Minimizing an Unconstrained Function*

The following conjugate direction algorithm has been proposed by Zoutendijk [5].

(It is convenient to establish the general pattern of an algorithm with cycles, each cycle having up to n iterations. The general notation for this is indicated by x_c^k, the kth point within the cth cycle. Also, the last point of the cth cycle is the same as the 0th point of the $(c+1)$th cycle. Hence the general pattern of points is

$$\{x_1^0, ..., (x_1^{n_1} \equiv x_2^0), x_2^1, ..., (x_2^{n_2} \equiv x_3^0), x_3^1, ...\}$$

where $n_1, n_2, ..., \leqslant n$. Wherever possible we will drop the subscript denoting the cycle number.)

For any cycle c, let y^k and Y^k be defined by

$$y^k \equiv g^{k+1} - g^k \qquad (k = 0, 1, ...) \tag{12}$$

$$Y^k \equiv [y^0, ..., y^{k-1}] \tag{13}$$

and define P^k

$$P^k \equiv [I - Y^k (Y^{k\prime} Y^k)^{-1} Y^{k\prime}] \tag{14}$$

as the projection matrix at iteration k into the subspace orthogonal to that spanned by $y^0, y^1, ..., y^{k-1}$.

The algorithm is as follows.

Iteration 0. The starting point x^0 is the last point of cycle $c-1$, let

$$P^0 \equiv I.$$

Iteration k. Compute P^k using Eqn (14) or the equivalent formula

$$P^k = P^{k-1} - P^{k-1} y^k (y^{k\prime} P^{k-1} y^k)^{-1} y^{k\prime} P_k^{k-1}. \tag{15}$$

If $P^k g^k \neq 0$, let the direction vector be

$$s^k \equiv -P^k g^k. \tag{16}$$

Then compute x^{k+1} using the single variable optimization of (8), Perform iteration $k+1$.

If $P^k g^k = 0$, and $g^k = 0$, x^k is a stationary point. No further steps need be taken. If $P^k g^k = 0$ and if $g^k \neq 0$, start cycle $c+1$.

There are many important properties of the Projection algorithm which we state here without proof.

LEMMA 4. *The algorithm is a "descent" algorithm and terminates after a finite number of steps if and only if $g^k = 0$.*

Proof. The matrix P^k is a projection matrix, that is $P^k \cdot P^k = P^k$ from which it follows that P^k is a positive semi-definite matrix

$$-g_{k'}P^k g^k < 0 \quad \text{if} \quad P^k g^k \neq 0, \ t^k > 0, \ x^{k+1} \neq x^k, \ f^{k+1} < f^k, \ g^{k+1'} s^k = 0. \tag{17}$$

Using Taylor's expansion

$$g^{k+1} = g^k + G(\eta^k) s^k t_k \tag{18}$$

where the j, ith element of $G(\eta^k)$ is given† by $\partial^2 f(\eta_j^k)/\partial x_i \partial x_j$; η_j^k being the point on the line between x^k and x^{k+1} for which Taylor's expansion is exact for g_j^k. Multiplying (18) by $s^{k'}$ and using (17) yields

$$t^k = -s^{k'} g^k / s^{k'} G(\eta^k) s^k > 0. \tag{19}$$

These relationships holding among quantities of the algorithm will be useful later.

The projection algorithm has the property of *left conjugacy*, that is successive direction vectors have the property that

$$s^{i'} G(\eta_j) s_j = 0, \qquad j < i \tag{20}$$

where η_j is the point used in (18).

Two other properties are given in the following lemmas.

LEMMA 5. *If $P^k g^k \neq 0$ and if $(Y^{k'}Y^k)^{-1}$ exists, then $[(Y^{k+1})' Y^{k+1}]^{-1}$ exists.*

LEMMA 6. *If $P^i g^i \neq 0$, $i = 0, ..., k$, then $\{s_0, ..., s_k\}$ is a linearly independent set.*

Lemma 5 guarantees that the algorithm makes sense at iteration $k+1$.

Note that no more than n iterations are possible in any one cycle since $P^n = 0$. Hence a new cycle must occur every n steps or fewer. Since the first direction of every cycle is the direction of steepest descent $(-P^0 g^0 = -g^0)$, the existence of a limiting stationary point of the algorithm is easy to prove. (See Section 3 for a discussion of this.)

Because for a quadratic programming problem of the form (10), left conjugacy is equivalent to conjugacy, the projection algorithm is a conjugate direction algorithm and has the attendant desirable convergence properties.

THEOREM 1. *The projection algorithm minimizes a positive definite quadratic form in n steps or fewer.*

The second major algorithm will now be described.

2.3. *The Variable Metric Method of Davidon*

Davidon [8] proposed a method with several variations for minimizing an unconstrained function. Fletcher and Powell [9] examined one of these,

---

† The authors point out that the use of G in this way is slightly inconsistent with its use as $\nabla^2 f$. However, the important property is that $G(\eta^k) \to G^*$ as $k \to \infty$, so even if $G(\eta^k)$ is not symmetric it is eventually positive definite. *ed.*

putting it into the framework of conjugate gradient algorithms, and proved that it minimized problem (10) in n or fewer steps.

Step 0. Let x^0 be the initial point, and H^0 be a positive definite matrix.

Step k. Form

$$H^k = H^{k-1} - H^{k-1}y^k\,(y^{k\prime}H^{k-1}y^k)^{-1}\,y^{k\prime}H^{k-1} + \sigma^k\,(y^{k\prime}\sigma^k)^{-1}\,\sigma^{k\prime} \qquad (21)$$

let

$$s^k = -H^kg^k \qquad (22)$$

and choose x^{k+1} as in (8).

They showed that:

LEMMA 7. *If H^{k-1} is a positive definite matrix, then so is H^k; the algorithm terminates at x^k if and only if $g^k = 0$; the algorithm is a descent method, $f^{k+1} < f^k$, and all the other properties of the projection algorithm summarized in Lemma 4 hold.*

There is one fundamental difference between the algorithms. After n steps the matrix P^n vanishes, having started out as the identity matrix. Hence there is no itention of trying to estimate $(G^*)^{-1}$. The Davidon method, on the other hand, was developed with the idea that as $k \to \infty$, $H^k \to (G^*)^{-1}$, and as such could be classified as a quasi-Newton method. The Fletcher–Powell development showed that it is also a method of conjugate gradients. In fact, for the quadratic case $g^n = 0$, and the exact Newton move which would occur on the next step is never made.

An obvious variation on the Davidon method is to allow for a cyclic method similar to the Projection algorithm. That is, after $n+1$ steps, the metric H^{n+1} is *reset* to some positive definite matrix. This variation was coded and a comparison with results for the other methods will be given in Section 5. (Note that one more move per cycle is made in their algorithm compared with the Projection method.)

3. Convergence of the Algorithms

At the end of the previous section we outlined three different methods (Projection, Davidon, Davidon with Reset) for minimizing an unconstrained function. The first question which must be answered is whether the methods converge, that is whether they are limit points of their minimizing sequences stationary points.

For the Projection algorithm, because at every n steps or fewer a steepest descent move is made, the answer to this question is yes with a slight modification.

LEMMA 8. *For the projection algorithm every point of accumulation of $\{x_c^0\}$ is a stationary point.*

Proof. The sequence $\{x_c^0\}$ is of course the sequence of x's which are first in each cycle. Since x_c^1 is obtained as $x_c^1 = x_c^0 + t_c^0(-g_c^0)$, and t_c^0 is chosen by (8), the usual proof [7], that $\{x_c^0\}$ tends to a stationary point, follows.

Let x^* be a point of accumulation of $\{x_c^0\}$. Then, because of the descent property, every point of accumulation of the entire sequence of the Projection algorithm must have the value $f(x^*)$. Thus although any other point of accumulation may not be a stationary point, its value is the same as the stationary point x^*, Since the stationarity of x^* is not sufficient to prove its optimality, the stationarity of any other point of accumulation loses its interest.

LEMMA 9. *If $f(x)$ is a convex function, then all points of accumulation of the Projection algorithm are stationary points and hence global solutions of problem* (1). *(Proof omitted.)*

We note briefly that for the third algorithm, Davidon with Reset, if the set of initial matrices used to map the gradient at the start of each cycle $\{H_c^0\}\, c = 1, ...,$ have the property that they come from a compact set of positive definite matrices (roughly that the eigenvalues of any limiting matrix of $\{H_c^0\}$ are bounded below by some number greater than zero, and above by some finite value) then the two convergence lemmas just stated for the Projection algorithm also apply to the Davidon with Reset method.

An interesting question about the original Davidon method is whether its points of accumulation are necessarily stationary points. From convergence theory it is possible to show that a sufficient condition that any point of accumulation of the infinite sequence $\{x^k\}$ be a stationary point is that (assuming $\{x^k\}$ represents also any converging subsequence)

$$\lim_{x^k \to x^*} -\frac{g^{k\prime}s^k}{\|s^k\|} > 0.$$

For the Davidon algorithm this means

$$\lim_{x^k \to x^*} +g^{k\prime}H^k g^k/\|H^k g^k\| > 0.$$

No one has yet been able to prove that this is the case. Thus the possibility that the Davidon method does not converge to a stationary point must be kept open.

Some of the computational results of the experiments indicated that even if the convergence could be proved, it was possible under certain circumstances to get "hung up" using the regular Davidon algorithm. Hence the regular method without reset was dangerous.

All the following are penalty function unconstrained minimizations of the form

$$\text{minimize } \phi(x) - r \sum_{i=1}^{m} \ln h_i(x)$$

the value of r, $\phi(x)$, $\{h_i(x)\}$ to be specified for each problem.

Problem 1. This problem had 15 variables, 25 constraints. The objective function was a nonconvex cubic, the constraints were quadratic forms. The resulting penalty function to be minimized was not convex in the region of minimization ($R^0 = \{x|h_i(x) > 0, \text{ all } i\}$). For this problem $r = 1 \cdot 0$ [4].

Problem 2. This problem had 9 variables and 13 inequality constraints. The penalty function was nonconvex. For this problem $r = 10^{-4}$. The results of these two problems are summarized in Table I [4].

TABLE I

Number of Iterations for Three Minimization Alogrithms

	Problem 1 iterations	Problem 2 iterations
Projection	120	37
Davidon Reset	97	24
Davidon Regular	> 406	> 140

The "greater than" symbols indicate that Regular Davidon method had not converged after that number of iterations had been made.

On the same two problems all the non-resetting quasi-Newton type algorithms failed to converge. However, nonconvergence was not observed on any convex penalty function problems tried.

In summary, the resetting concept for algorithms of the quasi-Newton type may be more than a theoretical nicety allowing convergence proofs. At least for some problems, Table I provides evidence that a non-resetting algorithm can get "stuck" or "hung up" and thus fail to converge to a stationary point.

4. Rate of Convergence

If convergence of the minimizing algorithms to a stationary point were our only concern, then Cauchy's steepest descent method would be just as satisfactory a method as any of the others. From a computational point of view, however, it is important to use algorithms which require the fewest number of iterations. Attempting to formalize the concept of "fast" algorithms leads to interesting mathematical problems gathered roughly under the title the "rate of convergence of algorithms."

From the computer experiments there emerged patterns relating to the rate of convergence of the methods under discussion. This pattern will be analyzed with respect to a somewhat artificial unconstrained minimization problem.

Problem 3.

Minimize $$100(x_2 - x_1^2)^2 + (1 - x_1)^2 + 90(x_4 - x_3)^2$$
$$+ (1 - x_3)^2 + 10 \cdot 1[(x_2 - 1)^2 + (x_4 - 1)^2] + 19 \cdot 8(x_2 - 1)(x_4 - 1).$$

The solution to this unconstrained problem is at $x^* = (1, 1, 1, 1)$ where $f(x^*) = 0$. The problem is designed to trap algorithms at a nonoptimal stationary point.

The results are summarized in Figs 1, 2 and 3, where

$$\|x - y\|^{\frac{1}{4}} = \left[\sum_{i=1}^{n} (x_i - y_i)^2 \right]^{\frac{1}{4}}.$$

The important thing to notice about the Projection algorithm and the Davidon–Reset algorithm is the cyclic way convergence is achieved. For the Projection algorithm (as the solution is approached) no significant progress is made within a cycle until the nth or last move in that cycle. At that time the difference $\|x_c^n - x^*\|$ becomes much smaller than $\|x_c^{n-1} - x^*\|$. Then a new cycle begins and no new significant progress is made until x_{c+1}.

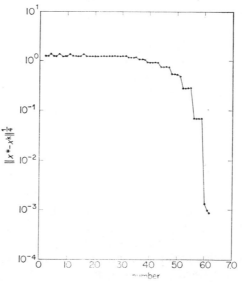

Fig. 1.   Plot of $\|x^* - x^k\|^{\frac{1}{4}}$ for projection algorithm.

The same pattern emerges in the results of the Davidon–Reset algorithm with $H_c{}^0 = I$, after n moves in the cycle, the difference between the solution and $x_c{}^n$ decreases dramatically. However this cycle contains one more move

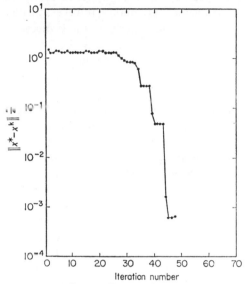

FIG. 2. Plot of $\|x^* - x^k\|^{\frac{1}{4}}$ for reset Davidon algorithm.

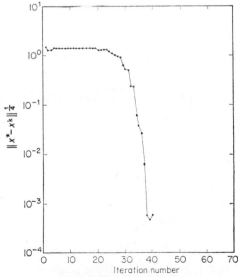

FIG. 3. Plot of $\|x^* - x^k\|^{\frac{1}{4}}$ for regular Davidon algorithm.

than the projection method; this move corresponds to a Newton move since only now has enough information been gained to construct a good approximation to the inverse of the Hessian matrix. On this move the algorithm achieves a significant decrease in the difference from the optimum. This decrease is not of an order of magnitude. However, this extra step gives the Davidon–Reset algorithm an advantage over the Projection method.

The main point here is that in both cases the efficiency of convergence is effected by moving in n almost-conjugate directions, not in approximating the Hessian inverse. The approximation in the case of the Davidon–Reset method is a bonus.

This cyclic pattern of convergence is not as obvious in the regular Davidon method. As the minimum is approached, and the "almost-conjugacy" conditions hold, every point exhibits a marked improvement toward the solution. Up until a certain takeoff point, however, the regular algorithm, the Davidon–Reset algorithm and the Projection algorithm behave almost identically.

We outline here a proof of this cyclic pattern.

Proof of rate of convergence of conjugate gradient algorithms. Without specifying the particular conjugate gradient algorithm used, we assume that one exists which generates directions having the following properties.

A sequence of sequences $\{s_c^0, ..., s_c^{n-1}\}$ $(c = 1, 2, ...)$ corresponding to a sequence of a sequence of points $\{x_c^0, ..., x_c^{n-1}\}$ $(c = 1, 2, ...)$, where

$$x_{c+1}^0 = x_c^{n-1} + t_c^{n-1} s_c^{n-1} \tag{23}$$

$$x_c^{k+1} = x_c^k + t_c^k s_c^k \qquad (k = 0, ..., n-2) \tag{24}$$

where t_c^k is the smallest (in magnitude) solution to

$$\min_t f(x_c^k + t s_c^k). \tag{25}$$

We further assume that each s_c^k is of the same order of magnitude of $-g_c^k$, and points in roughly the same direction, that is for c large there exists three values $\gamma_1, \gamma_2, \gamma_3 > 0$ such that

$$0 < \gamma_3 \leqslant -g_c^{k\prime} s_c^k / \|s_c^k\| \cdot \|g_c^k\| \tag{26}$$

$$\gamma_1 \|s_c^k\|^2 \leqslant -g_c^{k\prime} s_c^k \leqslant \gamma_2 \|s_c^k\|^2 \qquad \text{all } k, c. \tag{27}$$

(These properties obviously hold for the Davidon–Reset algorithm.)

We also assume the algorithm generates directions such that

$$t_c^k = 0 \quad \text{if and only if} \quad g_c^k = 0. \tag{28}$$

Hence the algorithm converges after a finite number of steps if and only if the gradient vanishes at some point.

We assume also that the sequence of points has a single point of accumulation x^* where

$$g^* = 0 \tag{29}$$

and $\qquad\qquad$ G^* is a positive definite matrix. $\qquad\qquad$ (30)

(Recall that these two conditions are sufficient to ensure that x^* is an isolated local minimum.)

Finally we assume the sequences form almost conjugate directions in that

$$s_c^{i\prime} G^* s_c^{\,j} = \|s_c^{\,i}\| \cdot \|s_c^{\,j}\| \varepsilon_c^{i,j} \qquad \text{all } i \neq j \text{ within a cycle} \tag{31}$$

where $\qquad\qquad$ $\lim_{c \to \infty} \varepsilon_c^{i,j} = 0 \qquad \text{all } i, j$ $\qquad\qquad$ (32)

For any particular conjugate gradient algorithm, we must of course prove that the above properties hold—and we assume here that they do.

THEOREM 2. (Rate of Convergence of Conjugate Gradient Algorithms). *Assume $f(x)$ is a twice continuously differentiable function. If a conjugate gradient algorithm has generated a sequence of directions and points having the above properties, then*

$$\lim_{c \to \infty} \frac{\|x_{c+1}^{\,0} - x^*\|}{\|x_c^{\,0} - x^*\|} = 0.$$

Proof. We will here simply outline the steps of the proof.

(1) For c large enough, the almost conjugacy property (31) ensures that the tet $\{s_c^{\,0}, ..., s_c^{\,n-1}\}$ is independent.

(2) For c large the $t_c^{\,k}$'s are uniformly bounded above and below. This follows by using Taylor's series,

$$g_c^{k+1} = g_c^{\,k} + G(\eta_c^{\,k}) s_c^{\,k} t_c^{\,k}$$

she fact that from (17), $s_c^{\,k\prime} g_c^{k+1} = 0$, yielding

$$t_c^{\,k} = -s_c^{\,k\prime} g_c^{\,k} / s_c^{\,k\prime} G(\eta_c^{\,k}) s_c^{\,k}. \tag{34}$$

Using (30) then (27) provides the last step.

(3) For any infinite sequences $\{y_c\}, \{z_c\}$ where $y_c \to x^*, z_c \to x^*, f(y_c) > f(z_c)$, for all large c there exists a $\delta_1 > 0$ such that

$$\|y_c - x^*\| > \|z_c - x^*\| \delta_1.$$

Using Taylors series, (29),

$$f(y_c) = f(x^*) + \|x^* - y_c\|^2 \lambda_c$$
$$f(z_c) = f(x^*) + \|x^* - z_c\|^2 \gamma_c \tag{35}$$

where for c large, using (30),

$$0 < \Pi_{\min} \leqslant \lambda_c, \gamma_c \leqslant \Pi_{\max}. \tag{36}$$

Rearranging,

$$\frac{\|y_c - x^*\|^2}{\|z_c - x^*\|^2} \geqslant \frac{\Pi_{\min}\,(f(y_c) - f(x^*))}{\Pi_{\max}\,(f(z_c) - f(x^*))} > \frac{\Pi_{\min}}{\Pi_{\max}}. \tag{37}$$

This proves the result. Geometrically this result is quite clear.

(4) There exist $\alpha_1, \alpha_2, \beta_1, \beta_2$ all greater than zero such that for all k, c

$$\alpha_1 \|s_c^{\,k}\| \leqslant \|g_c^{\,k}\| \leqslant \alpha_2 \|s_c^{\,k}\| \tag{38}$$

$$\beta_1 \|g_c^{\,k}\| \leqslant \|s_c^{\,k}\| \leqslant \beta_2 \|g_c^{\,k}\|. \tag{39}$$

These relationships follow in a straightforward manner from (26) and (27).

(5) There exists $\alpha_3 > 0$ such that for y, z sufficiently close to x^*,

$$\frac{\|x^* - y\|}{\|x^* - z\|} \leqslant \alpha_3 \frac{\|g(y)\|}{\|g(z)\|}. \tag{40}$$

Using Taylor's expansion,

$$g(y) = g(x^*) + G(\eta)\,(y - x^*).$$

Using (29) then (30), (40) follows.

Let C denote the sequence of integers $c = 1, 2, \ldots,$ defined by the cycles of the algorithm.

It is now necessary to divide the infinite sequence of $c \in C$ into two sequences, C^0 and C^1, defined in the following way. Consider

$$\min_{i > j} \{\liminf_{c \to \infty} \|s_c^{\,i}\|/\|s_c^{\,j}\|\}.$$

This quantity is well defined: let C^0 and C^1 be any collection of indices where $C^0 \cup C^1 = C$, when for at least one pair $(I, J), I > J$

$$\left\{ \lim_{\substack{c \to \infty \\ c \in C^0}} \|s^I{}_c\|/\|s_c^{\,J}\| \right\} = 0 \tag{41}$$

and

$$\min_{i > j} \left\{ \liminf_{\substack{c \to \infty \\ c \in C^1}} \|s_c^{\,i}\|/\|s_c^{\,j}\| \right\} = \delta > 0. \tag{42}$$

(It is possible that either of these sets is empty, but not both.)

(6) For $c \in C^0$, (33) holds. Let I, J be one set of indices, $I > J$, for which (41) holds. Then, since

$$f(x_{c+1}^0) < f(x_c^I), f(x_c^J) < f(x_c^0)$$

(3) can be applied twice yielding

$$\frac{\|x^* - x_{c+1}^0\|}{\|x^* - x_c^0\|} \leqslant \omega_1 \frac{\|x^* - x_c^I\|}{\|x^* - x_c^J\|}, \quad \omega_1 > 0.$$

Using (5), for c large,

$$\frac{\|x^* - x_c^I\|}{\|x^* - x_c^J\|} \leqslant \omega_2 \frac{\|g(x_c^I)\|}{\|g(x_c^J)\|}, \quad \omega_2 > 0.$$

Using (4),

$$\frac{\|g(x_c^I)\|}{\|g(x_c^J)\|} < \omega_3 \frac{\|s(x_c^I)\|}{\|s(x_c^J)\|}, \quad \omega_3 > 0.$$

Combining all these inequalities yields,

$$\frac{\|x^* - x_{c+1}^0\|}{\|x^* - x_c^0\|} < \omega_4 \frac{\|s(x_c^I)\|}{\|s(x_c^J)\|}, \quad \omega_4 > 0. \tag{43}$$

For $c \in C^0$, the limit on the right in (42) is, from (41), equal to zero; hence (33) follows. This case of convergence corresponds to the possibility in a quadratic programming problem that the gradient is zero before n conjugate moves are made.

(7) For $c \in C^1$,

$$t_c^k = -s_c^{k\prime} g_c^0 / s_c^{k\prime} G(\eta_c^k) s_c^k + \varepsilon_c^k$$

where $\|\varepsilon_c^k\| \to 0$ as $c \to \infty$.

Using Taylor's theorem,

$$g_c^k = g_c^0 + G(\eta) \left(\sum_{i=0}^{k=1} s_c^i t_c^i \right).$$

Multiplying by $(s_c^k)'$, using (32), and the almost conjugacy property (31) which, because $x_c^k \to x^*$ applies to any $G(\eta)$, yields

$$t_c^k = \frac{-s_c^{k\prime} g_c^0 - \sum_{i=0}^{k=1} \|s_c^i\| \cdot \|s_c^k\| \cdot \varepsilon_c^{i,k} t_c^i}{s_c^{k\prime} G(\eta_c^k) s_c^k}. \tag{44}$$

Because of (30), for c large the denominator of (44) is bounded by $\Pi_{\min}\|s_c^k\|^2$, $\Pi_{\max}\|s_c^k\|^2$.

For large $c \in C^1$ and $i < k$, $\|s_c^i\|/\|s_c^k\| < \omega_5$ by (42).

Since by (2) the t_c^i's are bounded and $\varepsilon_c^{i,\,j} \to 0$ by (31), (7) follows.

(8) For $c \in C^1$, and large, there exists an $\omega_6 > 0$ such that

$$\|s_c^i\| < \omega_6 \|g_c^0\| \qquad (i = 1, \ldots, n-1)$$

This follows from applying the inequalities in (6), those in (4) and (37).

(9) For $c \in C^1$,

$$\sum_{i=0}^{n-1} s_c^i t_c^i = -(G^*)^{-1} g_c^0 + \varepsilon^c \qquad (45)$$

where $\|\varepsilon^c\|/\|g_c^0\| \to 0$ as $c \to \infty$.

Now, using (1), (7), (8)

$$\sum_{i=0}^{n-1} s_c^i t_c^i = -S_c \Delta^{-1} S_c' g_c^0 + \varepsilon_1^c \qquad (46)$$

where $S_c = [s_c^0, \ldots, s_c^{n-1}]$, $\Delta = \mathrm{diag}[s_c^{j\prime} G^* s_c^j]$ and $\|\varepsilon_1^c\|/\|g_c^0\| \to 0$ as $c \to \infty$.

Similarly, using (1) and (31),

$$\begin{aligned}
-(G^*)^{-1} g_c^0 &\equiv -S_c (S_c^{-1})' (G^*)^{-1} (S_c^{-1})' g_c^0 \\
&\equiv -S_c (S_c' G^* S_c)^{-1} S_c' g_c^0 \qquad (47) \\
&= -S_c \Delta^{-1} S_c' g_c^0 + \varepsilon_2^c
\end{aligned}$$

where $\|\varepsilon_2^c\|/\|g_c^0\| \to 0$ as $c \to \infty$. Combining Eqns (47) and (45) gives the result.

(10) For $c \in C^1$, (33) follows. Using Taylor's expansion, and (29)

$$g_c^0 = 0 + G(\xi_c^0)(x_c^0 - x^*)$$

or

$$[G(\xi_c^0)]^{-1} g_c^0 = x_c^0 - x^*. \qquad (48)$$

Using (9),

$$x_{c+1}^0 = x_c^0 - (G^*)^{-1} g_c + \varepsilon^c \qquad (49)$$

where $\|\varepsilon^c\|/\|g_c^0\| \to 0$ as $c \to \infty$. Subtracting these expressions yields

$$x_{c+1}^0 - x^* = \{[G(\xi_c^0)]^{-1} - (G^*)^{-1}\} g_c^0 + \varepsilon^c. \qquad (50)$$

Taking norms, dividing (50) by (48), using (30) yields the result.

(11) Since (10) provides (33) for $c \in C^1$, it is true for all C; hence the theorem.

If an additional Lipschitz property is available on the second derivatives of f, and if the particular conjugate gradient algorithm generates $\varepsilon_c^{i,\,j}$'s in

(31) with a similar property, it is possible to show that the convergence established is "quadratic". Lacking these hypotheses, the rate of convergence is still significant as stated in (33).

5. Conclusions

Three conclusions can be drawn form these results.

(1) Use of metric methods without a reset option can result in convergence to a non-stationary point, or at least tremendously slow down its convergence in the early iterations.

(2) The method of Davidon, while it can be regarded as a quasi-Newton method, derives much of its important rate of convergence properties from the fact that it is a conjugate gradient method.

(3) Probably the most efficient metric algorithm of all would be one which used resetting options in its early stages and switched to a non-resetting method when it was close enough to the solution that its almost-conjugate properties held.

References

1. A. V. Fiacco and G. P. McCormick (1967). The slacked unconstrained minimization technique for convex programming. *J. Soc. Ind. Appl. Math.* **15** (3), 505-515.
2. A. V. Fiacco and G. P. McCormick (1963). "Programming under Nonlinear Constraints by Unconstrained Minimization: A Primal-Dual Method", Research Analysis Corp., McLean, Virginia, RAC-TP-96.
3. W. I. Zangwill (1967). Nonlinear programming via penalty functions. *Mgmt. Sci.* **13**, (5), 344-58.
4. J. D. Pearson (1968). "On Variable Metric Methods of Minimization," RAC-TP-302, Research Analysis Corporation, McLean, Virginia.
5. G. Zoutendijk (1960). "Methods of Feasible Directions", Elsevier Publishing Company, Amsterdam and New York.
6. A. Cauchy (1847). Méthode Généralè pour la resolution Des systèmes d'équations simultanées. *C.r.hebd. Seanc. Acad. Sci. Paris* **25**, 536-538.
7. G. P. McCormick and W. I. Zangwill. "A Technique for Calculating Second Order Optima". (To be published).
8. W. C. Davidon (1959). "Variable Metric Method for Minimization," Research and Development Report ANL-5990 (Rev.) U.S. Atomic Energy Commission, Argonne National Laboratories, 1959.
9. R. Fletcher and M. J. D. Powell (1963). A rapidly convergent descent method for minimization. *Comput. J.* **6**, 163-168.

Discussion

SARGENT. Are you saying that all methods which have quadratic termination also generate conjugate directions? If so I would dispute this, because if in the methods which I described earlier in the conference an arbitrary point is taken rather than the minimum in each search direction, then the methods have the former property but not the latter.

PEARSON. I agree: the same is true of the Projection methods where the fact that you have minimized down the line is irrelevant. The only reason for choosing conjugate directions is that the quadratic termination property then follows, but not *vice versa*. In fact one can generate all kinds of methods with $1, 2, \ldots n, n+1,$ $\ldots, 2n, \ldots$ step convergence although these have not been mentioned. It depends what flexibility one allows in the choice of directions. For instance, one can easily estimate the inverse Hessian by taking n probes, followed by a Newton step; this method would have $n+1$ step convergence. However your point is well taken.

BARD. I think some of the examples you present show what is wrong with always searching for the minimum along a line.

PEARSON. That is quite true: the trouble with going to the minimum in our opinion is that away from the solution you can waste an awful lot of time doing it. This is obvious: however it is not obvious how one should organize an algorithm which is fail-safe without it. There are some things you can do, like going a fixed part of the way, but these are heuristics which we try to avoid.

FLETCHER. I said in my paper that if you take methods with property Q but without the pseudo-Newton property (such as the Projection methods), and compare them against methods with both properties (such as Davidon's), then the former do not work as well in practice as the latter.

PEARSON. Yes I agree, the Projection method has to go through nearly the whole cycle before it shows any improvement. Its advantage however is that it is extremely simple, probably the simplest method after Fletcher and Reeves.

FLETCHER. Do you not think then that, because both of them give exactly the same set of directions in the quadratic case, you cannot argue that property Q is the only major factor which determines the rate of convergence of these methods away from the minimum, and that the pseudo-Newton factor might be another?

PEARSON. Near the minimum the conjugate gradient aspect is important. As we showed in the paper this can be demonstrated by considering a nearly quadratic region about the minimum. Further away from the minimum this does not hold. We find that in this region it doesn't much matter which method is used, and no methods really "take-off" until near the minimum.

ANON. Your comparisons are given in terms of number of steps: would it not be a more valid to compare numbers of function evaluations, because these are what take up the time?

PEARSON. This is a good point. All the methods used the same search in which the minimum was found extremely accurately but at great expense in function evaluations. In fact I daren't even put these figures on the board. We hope to do something about accelerating this part of the program: more than 50 per cent of the total time is taken up in the searches. This is more critical to us than the choice of directions, which we feel we understand. However I don't think we understand the search down the line nearly as well and I thought the relevant parts of the paper of Fletcher and McCann were most interesting.

ANON. When solving constrained problems using penalty functions are there any criteria on how accurate each line minimum should be, and how accurate each r-minimum should be to get a certain accuracy overall? A simple rule would suffice.

BARD. As I mentioned earlier in discussion, there is a trade-off between how many function evaluations are required per search and the total number of linear searches to solve the problem. Using the concept of "total equivalent function evaluations" as Box did in his paper (counting each function evaluation as 1 and each derivative as 1) I found that there was a fairly flat minimum when total function evaluations were plotted againt the accuracy required in the linear search. The optimum accuracy was in the range 1:300 to 1:3000, so that 1:1000 would be a satisfactory accuracy to required in the step length for a variety of problems.

FLETCHER. With regard to how accurately each surface needs minimizing when solving a constrained problem using CRST, we have taken the attitude that if one has an approximate minimum for any surface, then to obtain the exact minimum will require only a few more evaluations because of property Q in the minimization method. However, the determination of exact minima allows any acceleration techniques to work more reliably and this can be of much greater importance in determining the overall rate of convergence. Hence we minimized each surface as accurately as possible.

PEARSON. I think a most valuable feature would be an error analysis showing what the effect of errors is at all stages in the process so that we know exactly what action to take.

22. Generalized Lagrangian Functions and Mathematical Programming

J. D. Roode

Atomic Energy Board, Pelindaba, Pretoria, S. Africa

1. Generalized Lagrangian Functions

For the mathematical programming problem

$$\text{minimize } f(x) \tag{1}$$

$$\text{subject to } f_i(x) \leqslant 0 \quad (i = 1, \dots, m)$$

where f and f_i $(i = 1, \dots, m)$ are real-valued functions defined on n-dimensional euclidean space E^n, the associated Lagrangian function is given by

$$L(x, y) := f(x) + \sum_{i=1}^{m} y_i f_i(x)$$

defined for $(x, y) \in E^n \times E_+^m$, where E_+^m is the non-negative orthant of m-dimensional euclidean space.

As is well known, the saddle points of $L(x, y)$, if they exist, are important for problem (1), since we have:

THEOREM 1. *If $(\bar{x}, \bar{y}) \in E^n \times E_+^m$ is a saddle point of $L(x, y)$, then $\bar{x}$ solves problem* (1).

Proof.
$$L(\bar{x}, y) \leqslant L(\bar{x}, \bar{y}) \leqslant L(x, \bar{y}).$$

Since the left-hand inequality holds for all $y \in E_+^m$, we must have

$$f_i(\bar{x}) \leqslant 0 \quad (i = 1, \dots, m)$$

For $y = 0$ we have

$$f(\bar{x}) \leqslant f(\bar{x}) + \sum_{i=1}^{m} \bar{y}_i f_i(\bar{x})$$

so that

$$\sum_{i=1}^{m} \bar{y}_i f_i(\bar{x}) = 0$$

must hold. The right-hand inequality then reads

$$f(\bar{x}) \leqslant f(x) + \sum_{i=1}^{m} \bar{y}_i f_i(x).$$

Hence for all feasible $x \in E^n$,

$$f(\bar{x}) \leqslant f(x)$$

and $\bar{x}$ solves problem (1).

As Kuhn and Tucker [1] showed, if f and f_i ($i = 1, ..., m$) are convex functions, then, provided a regularity condition is satisfied, if $\bar{x}$ solves (1), there exists a $\bar{y} \in E_+^m$ such that $(\bar{x}, \bar{y})$ is a saddle point of the Lagrangian function.

These results, together with local optimality conditions obtainable when we specialize to differentiable functions, answered the question regarding the optimality of a given feasible point and gave a valuable impetus to the development of the theory of mathematical programming. For convex programs, Arrow and co-workers [2] developed a Lagrangian differential gradient method based on the Kuhn–Tucker theory, and in quadratic programming a number of methods were proposed which directly utilized the optimality conditions. However, efficient algorithms rarely make computational use of the Lagrangian function of the given problem.

In 1963, Everett [3] pointed out that the Lagrangian may be used to convert a constrained optimization problem into a sequence of unconstrained problems. This follows from:

THEOREM 2. *Suppose for $\bar{y} \in E_+^m$ the Lagrangian function $L(\cdot, \bar{y})$ is minimized over E^n at $\bar{x}$. Then $\bar{x}$ is a solution of the following problem*

$$\min \{f(x) \mid x \in E^n, \ f_1(x) \leqslant f_i(\bar{x}), \quad i = 1, ..., m\}.$$

Proof.
$$L(\bar{x}, \bar{y}) \leqslant L(x, \bar{y}) \qquad \text{for all } x \in E^n$$

$$f(\bar{x}) \leqslant L(x, \bar{y}) - \sum_{i=1}^{m} \bar{y}_i f_i(\bar{x})$$

$$= f(x) + \sum_{i=1}^{m} \bar{y}_i (f_i(x) - f_i(\bar{x})).$$

Since $\bar{y} \in E_+^m$,

$$f(\bar{x}) \leqslant f(x) \qquad \text{for all } x \in E^n$$

such that
$$f_i(x) \leqslant f_i(\bar{x}) \qquad (i = 1, ..., m)$$

This theorem shows that in principle we may try to solve problem (1) by iterating in E_+^m, the space of Lagrange multipliers, together with corresponding unconstrained minimizations of the Lagrange function in E^n. However, even in simple examples, there may be no $y \in E_+^m$ for which the Lagrangian function assumes its minimum at a finite point in E^n, and even if it is possible to specify a set $D \subset E_+{}^m$ such that for each $\bar{y} \in D$, $L(x, \bar{y})$ assumes its minimum over E^n, two questions remain unanswered:

(1) Does the set D contain an "optimal" $\hat{y}$, that is a $\hat{y}$ such that if $\hat{x}$ minimizes $L(\cdot, \hat{y})$, $\hat{x}$ also solves problem (1)?

(2) If such a $\hat{y} \in D$ exists, how must the iterations in E_+^m be carried out in order to arrive at $\hat{y}$? Surely, a trial and error method would be most unsatisfactory?

In 1965, Falk and co-workers [4, 5] treated these questions and established interesting results. We shall give a brief description of Falk's approach by stating his main results.

Let $D := \{y \mid y \in E_+^m,\ L(\cdot, y)$ attains its minimum over $E^n\}$

On D we define a function $g: D \to R$ by

$$g(y) := \min_{x \in E^n} L(x, y)$$

and consider the following problem, to be called the auxiliary problem of (1):

$$\max_{y \in D} g(y) \qquad (2)$$

It is easily seen that g is a concave function over convex subsets of D.

LEMMA 1. *If x and y are feasible for problems* (1) *and* (2) *respectively, then*

$$g(y) \leqslant f(x).$$

Proof.

$$g(y) = \min_{\xi \in E^n} L(\xi, y) \leqslant L(x, y) \leqslant f(x)$$

for any feasible $x \in E^n$.

Suppose now that (1) is a convex program and satisfies the regularity condition

$$\exists\, x^0 \in E^n \text{ such that } f_i(x^0) < 0 \qquad (i = 1, \ldots, m).$$

If problem (1) has a solution $\bar{x}$, we know that there exists a $\bar{y} \in E_+^m$ such that $(\bar{x}, \bar{y})$ is a saddle point of the Lagrangian function:

$$L(\bar{x}, y) \leqslant L(\bar{x}, \bar{y}) \leqslant L(x, \bar{y})$$

for all $(x, y) \in E^n \times E_+^m$. Hence $\bar{y} \in D$ and

$$g(\bar{y}) = L(\bar{x}, \bar{y}) = f(\bar{x})$$

(see proof of Theorem 1). By Lemma 1, it then follows that $\bar{y}$ solves (2). Hence we have proved

THEOREM 3. *If (1) is a convex program satisfying the regularity condition and has a solution $\bar{x}$, the auxiliary problem (2) is feasible and has a solution $\bar{y}$. The optimal values are equal*

$$f(\bar{x}) = g(\bar{y}).$$

To show that every solution of (2) provides a solution of (1), it is necessary to require additionally that f be strictly convex. (The regularity condition need not be imposed.)

Under these conditions it can be shown that

(1) D is a convex set, relatively open in E_+^m, and

(2) g is differentiable with continuous partial derivatives in the interior of D, and if $\bar{y} \in D$ and $\bar{y}_i = 0$, the right-hand partial derivative exists at $\bar{y}$:

$$\bar{y} \in D \Rightarrow \begin{cases} \left. \dfrac{\partial g}{\partial y_i^+} \right|_{\bar{y}} = f_i(\bar{x}) \\[2em] \left. \dfrac{\partial g}{\partial y_i^-} \right|_{\bar{y}} = f_i(\bar{x}) \end{cases}$$

where $\bar{x}$ is the unique point at which $L(\cdot, \bar{y})$ is minimized.

THEOREM 4. *Suppose g attains its maximum over D at $\bar{y}$. Then $\bar{x}$, defined by $L(\bar{x}, \bar{y}) = \min\limits_{x \in E^n} L(x, \bar{y})$, solves (1) and $g(\bar{y}) = f(\bar{x})$.*

Proof. g is maximized at $\bar{y} \in D$, so that

$$\left. \frac{\partial g}{\partial y_i} \right|_{\bar{y}} = 0 \quad \text{if} \quad \bar{y}_i > 0$$

and

$$\left. \frac{\partial g}{\partial y_i^+} \right|_{\bar{y}} \leqslant 0 \quad \text{if} \quad \bar{y}_i = 0.$$

This implies that $\bar{x}$ is feasible and

$$\sum_{i=1}^{m} \bar{y}_i f_i(\bar{x}) = 0.$$

Hence $g(\bar{y}) = L(\bar{x}, \bar{y}) = f(\bar{x})$ and by Lemma 1, $\bar{x}$ solves (1).

The problem of solving (1) has therefore been reduced to that of maximizing a concave differentiable function over an open convex set with non-negativity conditions, and the gradient of the objective function of the auxiliary problem can be calculated by solving an unconstrained problem.

For a restricted class of problems this approach therefore offers a computational method which directly utilizes the Lagrangian function of (1). Falk actually described a more general approach where the inequalities in (1) are partioned and showed that for separable programming useful results can be obtained.

It seems that the capability of the classical Lagrangian function to provide computational methods for the solution of (1) is limited to problems having a special structure. In what follows we shall generalize the concept of a Lagrangian function and show that this enables us to develop a generalized duality theory and useful computational methods for more general classes of problems.

With the program (1) we associate a function ψ, defined as follows:

$$\psi: E^n \times E_+^m \to R$$

and

(1) $\psi(x, y) \leqslant f(x)$ for all feasible $x \in E^n$ and all $y \in E_+^m$.

(2) For any $x \in E^n$,

$$\psi(x, 0) = f(x).$$

(3) If $x \in E^n$ is such that there exists at least one index i_0 such that

$$f_{i_0}(x) > 0$$

then there exists a sequence $\{y^k(x)\}$, $\|y^k(x)\| \to \infty$ such that

$$\limsup_{k \to \infty} \psi(x, y^k(x)) = \infty.$$

For the function ψ thus defined, we consider the following problem:

$$\text{minimize } \psi(x, y) \tag{3}$$

subject to

$$(x, y) \in S: = \{(x, y) \mid x \in E^n, y \in E^m_+, \psi(x, y) = \sup_{\eta \in E^m_+} \psi(x, \eta)\}.$$

We shall establish the equivalence of problems (1) and (3), in the sense that a solution of one problem provides a solution of the other, and the optimal values are equal.

To show this, we define the set S^0 by

$$S^0: = \{(x, y) \mid x \in E^n, y \in E^m_+, f_i(x) \leqslant 0, i = 1, ..., m \text{ and } \psi(x, y) = f(x)\}$$

and first prove:

LEMMA 2. *If $S \neq 0$ and $S^0 \neq 0$, then $S = S^0$.*

Proof. Let

$$(x^0, y^0) \in S^0.$$

Then

$$\psi(x^0 y^0) = f(x^0) \geqslant \psi(x^0, y) \quad \text{for all } y \in E^m_+$$

so that

$$\psi(x^0, y^0) = \sup_{y \in E^m_+} \psi(x^0, y),$$

hence

$$(x^0, y^0) \in S \quad \text{and} \quad S^0 \subset S.$$

Conversely, if $(\bar{x}, \bar{y}) \in S$, then

$$\psi(\bar{x}, \bar{y}) \geqslant \psi(\bar{x}, y) \text{ for all } y \in E^m_+$$

Suppose there is an index i_0 such that $f_{i_0}(\bar{x}) > 0$. Then there exists a sequence

$$\{y^k(\bar{x})\}, \qquad y^k(\bar{x}) \in E^m_+ \text{ for all } k, \qquad \|y^k(\bar{x})\| \to \infty$$

and

$$\limsup_{k \to \infty} \psi(\bar{x}, y^k(\bar{x})) = \infty$$

contradicting the fact that

$$\psi(\bar{x}, \bar{y}) \geqslant \psi(\bar{x}, y^k(\bar{x})) \quad \text{for all } y^k(\bar{x}).$$

Hence

$$f_i(\bar{x}) \leqslant 0 \quad \text{for } i = 1, ..., m.$$

Furthermore,

$$f(\bar{x}) \geqslant \psi(\bar{x}, \bar{y}) \geqslant \psi(\bar{x}, y) \quad \text{for all } y \in E^m_+$$

hence also for

$$y = 0 \in E^m_+$$

so that $\qquad\qquad f(\bar{x}) \geqslant \psi(\bar{x}, \bar{y}) \geqslant \psi(\bar{x}, 0) = f(\bar{x}).$

Hence $\qquad\qquad \psi(\bar{x}, \bar{y}) = f(\bar{x})$ and $(\bar{x}, \bar{y}) \in S^0$

implying $S \subset S^0$. We therefore have

$$S = S^0.$$

THEOREM 5. *If $(\bar{x}, \bar{y})$ solves problem (3), then $\bar{x}$ is an optimal solution of problem (1), and if $\bar{x}$ is an optimal solution of (1), then $(\bar{x}, \bar{y})$ solves (3) for any $y \in L_+^m$ such that $\psi(\bar{x}, \bar{y}) = f(x)$.*

Proof. Suppose $(\bar{x}, \bar{y})$ solves (3). Then $(\bar{x}, \bar{y}) \in S$ and by Lemma 2, $(\bar{x}, \bar{y}) \in S^0$. If $\bar{x}$ is not optimal for (1), so that there exists an $\hat{x} \in E^n$ satisfying

$$f(\hat{x}) < f(\bar{x}), \quad f_i(\hat{x}) \leqslant 0 \quad (i = 1, ..., m)$$

then $\qquad\qquad \psi(\bar{x}, \bar{y}) = f(\bar{x}) > f(\hat{x}) = \psi(\hat{x}, 0)$

where $(\hat{x}, 0) \in S^0$ would imply

$$\psi(\hat{x}, 0) < \psi(\bar{x}, \bar{y})$$

contradicting the fact that

$$\psi(\bar{x}, \bar{y}) = \inf_S \psi(x, y) = \inf_{S^0} \psi(x, y).$$

Hence $\bar{x}$ is optimal for (1).

Conversely, suppose $\bar{x}$ is an optimal solution of (1). Let $\bar{y} \in E_+^m$ be such that $\psi(\bar{x}, \bar{y}) = f(\bar{x})$. Then

$$\psi(\bar{x}, \bar{y}) = f(\bar{x}) \leqslant f(x) = \psi(x, y) \quad \text{for all } (x, y) \in S^0 = S.$$

Hence $\qquad\qquad \psi(\bar{x}, \bar{y}) = \inf_S \psi(x, y)$

and $(\bar{x}, \bar{y})$ solves (3).

COROLLARY. *Suppose $(\bar{x}, \bar{y})$ is a saddle point of $\psi(x, y)$. Then $\bar{x}$ solves problem (1).*

Theorem 5 establishes the equivalence of the problems (1) and (3). Instead of (1), we may therefore consider problem (3).

It is clear that the classical Lagrangian function of (1) satisfies all the properties which we have required of the function ψ, so that in view of the above results, we shall refer to any function $\psi(x, y)$ satisfying properties

(1)–(3), as a Generalized Lagrangian function for problem (1). We now define the *dual* of (3) as follows:

$$\text{maximize } \psi(x, y) \tag{4}$$

subject to

$$(x, y) \in T := \left\{ (x, y) | x \in E^n, y \in E^m_+, \psi(x, y) = \inf_{\xi \in E^n} \psi(\xi, y) \right\}.$$

This introduces a generalized duality concept, since instead of considering problem (1) and its dual induced by the classical Lagrangian function, the dual problems now are

$$\inf_{x \in E^n} \sup_{y \in E^m_+} \psi(x, y)$$

and

$$\sup_{y \in E^m} \inf_{x \in E^n} \psi(x, y).$$

The generality offered by the introduction of the generalized Lagrangian function can be exploited in the following ways.

(1) A general theory of duality in nonlinear programming has been developed by Stoer [6], Mangasarian and Ponstein [7] and, most recently, Karamardian [8]. They studied minimax problems for general functions ϕ defined on $E^n \times E^m$, satisfying certain conditions. To obtain duality results in nonlinear programming, the function ϕ is specialized to the classical Lagrangian function. Using instead a generalized Lagrangian function, we can obtain new and general duality results for nonlinear programs.

(2) We can apply Falk's approach of an auxiliary problem based on the classical Lagrangian function to the generalized Lagrangian function to obtain computational methods and duality results.

(3) We can require more properties of the generalized Lagrangian function ψ such that a saddle point exists *a priori*, and then study the computational consequences of this.

We shall in this paper only point out how the possibility (3) can be realized. It will be shown that a dual method applied to such a restricted generalized Lagrangian function reduces in special cases to certain well-known indirect methods for the solution of mathematical programming problems.

2. A Computationally Convenient Generalized Lagrangian Function

We require the following additional property of $\psi(x, y)$:

(4) For fixed $x \in E^n$, ψ is upper semi-continuous and non-decreasing in $y \in E_+^m$.†

If $x \in E^n$ is feasible for problem (1), we have that for all $y \in E_+^m$

$$f(x) = \psi(x, 0) \leqslant \psi(x, y)$$

and

$$\psi(x, y) \leqslant f(x)$$

hence

$$\psi(x, y) = f(x).$$

We note that

$$\psi(x, y) = f(x)$$

also if

$$\|y\| = \infty \qquad x \in E^n \text{ feasible}$$

LEMMA 3. *For any non-feasible* $x \in E^n$, *let* $y^k(x) \in E_+^m$ *be such that* $y^k(x) \to \bar{y}$, $\|\bar{y}\| = \infty$ *and*

$$\limsup_{k \to \infty} \psi(x, y^k(x)) = \infty.$$

Then

$$\psi(x, \bar{y}) = \infty.$$

Proof. Since ψ is upper semi-continuous in y, we have

$$\infty = \limsup_{k \to \infty} \psi(x, y^k(x)) \leqslant \psi(x, \bar{y}),$$

hence

$$\psi(x, \bar{y}) = \infty.$$

COROLLARY. *If* $\bar{y} = (\infty)$, *where* (∞) *is the vector with all components equal to* $+\infty$, *then* $\psi(x, \bar{y}) = \infty$ *for all non-feasible* $x \in E^n$.

LEMMA 4. *Let* $\bar{y} = (\infty)$. *Then*

$$\inf_{x \in E^n} \psi(x, \bar{y}) = \inf \{f(x) | f_i(x) \leqslant 0, \ i = 1, \dots, m\}.$$

Proof. For any feasible x we have $\psi(x, \bar{y}) = f(x)$, while

$$\psi(x, \bar{y}) = \infty \qquad \text{for any non-feasible } x.$$

THEOREM 6. *Let* $\bar{y} = (\infty)$ *and* $\bar{x} \in E^n$ *be such that*

$$\inf \{f(x) | x \in E^n, f_i(x) \leqslant 0, \ i = 1, \dots, m\} = f(\bar{x}).$$

Then $(\bar{x}, \bar{y})$ *is a saddle point of the function* $\psi(x, y)$.

Proof. $\psi(x, \bar{y}) = f(x) \geqslant f(\bar{x}) = \psi(\bar{x}, y)$ for any feasible $x \in E^n$, and

$$\psi(x, \bar{y}) = \infty \qquad \text{for any non-feasible } x \in E^n.$$

---

† In this section E_+^m is assumed to be the product $\overbrace{\hat{R}_+ \times \hat{R}_+ \ .. \ \times \hat{R}_+}^{m}$, where $\hat{R}_+$ is the augmented set of non-negative real numbers (that is R_+ together with the point $+\infty$), and accordingly ψ is assumed to be defined satisfying (1)–(4) for all $(x, y) \in E^n \times \hat{E}_+^m$, $\psi: E^n \times \hat{E}_+^m \to \hat{R}$.

Also $\psi(\bar{x}, y) = f(\bar{x}) = \psi(\bar{x}, \bar{y})$ for all $y \in E_+^m$.

Hence $\psi(\bar{x}, y) = \psi(\bar{x}, \bar{y}) \leqslant \psi(x, \bar{y})$ for all $(x, y) \in E^n \times E_+^m$.

THEOREM 7. *Let $y^k \in E_+^m$ be such that the sequence $\{y^k\}$ is increasing and*

$$\lim_{k \to \infty} y^k = \bar{y} = (\infty).$$

Suppose there is an index k_0, such that for $k \geqslant k_0$, an $x^k \in E^n$ exists such that

$$\inf_{x \in E^n} \psi(x, y^k) = \psi(x^k, y^k).$$

Then $\{\psi(x^k, y^k\}$ converges to the value $\psi(\bar{x}, \bar{y})$, where

$$\psi(\bar{x}, \bar{y}) = \inf \{f(x) | x \in E^n, f_i(x) \leqslant 0, \ i = 1, ..., m\}.$$

Proof. Suppose $k \geqslant k_0$. Then

$$\psi(x^k, y^k) \leqslant \psi(\bar{x}, y^k) = \psi(\bar{x}, \bar{y})$$

and $\psi(x^k, y^k) \leqslant \psi(x^{k+1}, y^k) \leqslant \psi(x^{k+1}, y^{k+1})$ for all $k \geqslant k_0$.

The sequence $\{\psi(x^k, y^k\}$ therefore converges since it is monotonically increasing and bounded above, and we have

$$\lim_{k \to \infty} \psi(x^k, y^k) = \sup_{y \in E_+^m} \inf_{x \in E^n} \psi(x, y).$$

By Theorem 6,

$$\sup_{y \in E_+^m} \inf_{x \in E^n} \psi(x, y) = \inf_{x \in E^n} \sup_{y \in E_+^m} \psi(x, y) = \psi(\bar{x}, \bar{y}),$$

so that

$$\lim_{k \to \infty} \psi(x^k, y^k) = \psi(\bar{x}, \bar{y}) = \inf \{f(x) | x \in E^n, f_i(x) \leqslant 0, \ i = 1, ..., m\}.$$

This theorem shows that the method described in it solves problem (1) through a sequential unconstrained exterior point technique (we have $f(x^k) = \psi(x^k, 0) \leqslant \psi(x^k, y^k) \leqslant f(\bar{x})$). It is a dual method, since the points (x^k, y^k) are feasible for the (generalized) dual problem (4).

To obtain results concerning the convergence of subsequences of $\{x^k\}$ to an optimal solution of (1) more structure has to be required of problem (1).

It is easily seen that the penalty functions described by Zangwill [9] in his treatment of sequential unconstrained exterior point methods belong to the class of generalized Lagrangian functions described above.

As is well known, there is a close relationship between the barrier function of Fiacco and McCormick's SUMT method [10, 11] and the classical Lagrangian function. In a previous paper [12] the author described a general theory of interior point methods for the solution of mathematical programming problems. It can be shown that an analogous relationship exists between classes of interior point methods and generalized Lagrangian functions.

Acknowledgement

The author wishes to acknowledge the financial support of the Atomic Energy Board of South Africa who made this research possible. The research was done while the author was at the Centraal Reken-Instituut, University of Leiden, The Netherlands.

References

1. H. W. Kuhn and A. W. Tucker (1951). Non-linear programming. *In* "Proceedings of the Second Berkeley Symposium on Mathematical Statistics and Probability," (J. Neyman ed.) University of California Press, California.
2. K. J. Arrow, L. Hurwicz and H. Uzawa (1958). "Studies in Linear and Non-linear Programming". Stanford University Press, Stanford, California.
3. H. Everett (1963). Generalized Lagrange multiplier method for solving problems of optimal allocation of resources. *Ops Res.* **11**, 399-417.
4. J. E. Falk and R. M. Thrall, (1965). A Constrained Lagrangian Approach to Nonlinear Programming. The University of Michigan: Technical Report, Department of Mathematics.
5. J. E. Falk (1967). Lagrange multipliers and nonlinear programming. *J. Math. anal. Appl.* **19**, 141-159.
6. J. Stoer (1963). Duality in nonlinear programming and the minimax theorem. *Num. Math.* **5**, 371-379.
7. O. L. Mangasarian and J. Ponstein (1965). Minimax and duality in nonlinear programming. *J. Math. anal. Appl.* **11**, 504-518.
8. S. Karamardian (1967). Strictly quasi-convex (concave) functions and duality in mathematical programming. *J. Math. anal. Appl.* **20**, 344-358.
9. W. I. Zangwill (1967). Non-linear programming via penalty functions. *Mgmt. Sci.* **13**, 344-358.
10. A. V. Fiacco and G. P. McCormick (1964). The sequential unconstrained minimization technique for non-linear programming: a primal-dual method. *Mgmt. Sci.* **10**, 360-364.
11. A. V. Fiacco (1967). Sequential Unconstrained Minimization Methods for Nonlinear Programming, Doctoral Dissertation, Northwestern University, Evanston, Ill., June 1967.
12. J. D. Roode (1968). Interior Point Methods for the Solution of Mathematical Programming Problems. Paper presented at the Sixth International Symposium on Mathematical Programming, Princeton, N.J., August, 1967.

Discussion

ANON. Dr Roode makes an interesting point that there are a whole class of Lagrangian functions. I wonder if he could give any indication of what might be the appropriate choice or appropriate grounds for choice of Lagrangian function in any particular case? Particularly one still feels that the best Lagrangian function is one where ψ is linear in y, and that it has a special role because it clarifies the situation.

ROODE. It is certainly possible to choose $\psi(x, y)$ as linear in y, but to be honest I haven't started thinking about the grounds for choice.

BEALE. It is interesting that Dr Roode has reminded us that conventional Lagrangian techniques do fall apart when we are using non-convex functions. Powell's contribution earlier showed that you need not have a minimum value of one of your parameters to make the method work for non-convex problems. This shows in a quantitative way how the conventional Lagrangian technique does not work when the problem is non-convex, and also shows what to do about it. The method is related to Lagrangian methods but not restricted to convex problems.

POWELL. In fact in my contribution I overcame a lot of the convexity problems by having a square term coming into the function. The proofs of convergence do depend upon slight restrictions on the functions and I think this square term in the generalized Lagrangian can help quite a lot.

23. Panel Discussion

Chairman

    P. WOLFE   IBM Ltd, Yorktown Heights

Panel Members

J. ABADIE	Université de Paris
E. M. L. BEALE	Scientific Control Systems Ltd.
C. G. BROYDEN	University of Essex
D. DAVIES	ICI Ltd.
J. L. GREENSTADT	IBM Ltd.
M. J. D. POWELL	AERE Harwell

1. What are the most useful books or papers that people in this area must read?†

"Mathematical Programming in Practice" by E. M. L. Beale, published by Pitmans 1968.

BEALE. The book is written primarily for people in operational research groups in industry or those about to join such organizations. It is divided into three parts: the first is called "Elementary Linear Programming", the second is "Organization of L.P. calculations", and is concerned with practical difficulties with formulation problems for computers and with what to do with the results. Finally there is a section on "Special Procedures" which does go into nonlinear problems to some extent, but specifically covers those problems which can and have been tackled by extensions of linear programming. It does not say anything about hill-climbing techniques.

"Nonlinear Programming" by J. Abadie, published by Nord-Hall 1967.

WOLFE. A collection of papers by well-known authors.

"Nonlinear Optimization Techniques" I.C.I. monograph No. 5 by M. J. Box, D. Davies and W. H. Swann, published by Oliver and Boyd 1969.

DAVIES. This is written for chemical engineers and other people who want to know what methods will solve their problems. It gives brief descriptions of methods and our opinions as to which methods are the better ones.

"Finding Minima of Functions of Several Variables" by M. J. D. Powell, a paper in the book "Numerical Analysis—an Introduction" by J. Walsh, published by Academic Press 1966.

WOLFE. An excellent review paper for the area.

---

† Topics for discussion were supplied beforehand in written form by conference delegates and collated by the chairman.

"A Survey of Nonlinear Programming" by W. Dorn, published in the journal "Management Science" 1964.

WOLFE Possibly a little out of date now.

"A Comparison of Several Current Optimization Methods and the Use of Transformations in Constrained Problems" by M. J. Box, published in the "Computer Journal" 1966.

BEALE. A review of alternative methods for optimization which in turn gives many useful references.

"Constructive Real Analysis" by A. A. Goldstine, published by Harper and Rowe 1967.

BROYDEN. Good background reading for the subject.

2. What is the availability of computer routines for the various methods?

National Computer Centre. Index of Computer Programs. (N.C.C., Quay Street Manchester.)

ANON. The NCC will provide the name and address of authors of working programs to members of organizations who have paid their subscriptions to the NCC.

C.E.R.N. (Mr. Kopanyi, C.E.R.N., Geneva.)

ANON. This organization has FORTRAN card decks and listings for many well-known methods such as Rosenbrock, Fletcher–Powell–Davidon, Powell (1964) and Powell (sum of squares).

I.B.M. New York. (Either A. R. Colville or Y. Bard, I.B.M. Scientific Center, 410 E 62nd Street, New York.

BARD. Programs are available from the IBM program library and also descriptions of the Colville nonlinear programming study.

I.C.I. (J. Payne, ICI Ltd. Management Services, Fulshaw Hall, Wilmslow, Cheshire.)

DAVIES. Many programs including Rosenbrock, DSC, Davidon, Davidon plus constraints, Complex, and so on, which are now available in Kautocode but will become available in PL1. There is a charge for industrial organizations.

A.C.M. and B.C.S. ALGOL 60 Algorithms supplements.

PEARSON. Many well-known procedures including Davidon–Fletcher–Powell.

Other useful sources. Simplex method, Rosen's method, and so on (ALGOL 60) —Mr Ribiere, Inst. Blaise Pascal, 23 Rue du Maroc, Paris. Reduced Gradient Method and so on (FORTRAN)—J. Abadie, 17 Ave. de la Liberation, 92 Clamart, Paris.

3. In practical minimization algorithms of descent type, what stopping rules would the panel recommend? ("*Safety versus extravagence*".)

BEALE. When solving nonlinear regression problems we carry out five to ten iterations of our Levenberg/Marquardt type scheme and then look at the answers. We do of course arrange for the iteration to be restarted readily but in practice it is often necessary to change the problem at this stage.

POWELL. If a cheap unsafe criterion is required when using a method with rapid ultimate convergence then it is usually sufficient to stop when the changes in the parameters are sufficiently small. But for a safe criterion, an excellent procedure is to make a perturbation to a point which satisfies the previous criterion and continue the iteration from that perturbed point. This is not too expensive because the various approximations which have been accumulated will still apply. If the point ultimately reached from the perturbed point is substantially different from that obtained originally, then it is wise to treat any results with suspicion.

GREENSTADT. The problem of stopping is very complicated and it depends on what results one is interested in. If you are interested in the value of the objective function, and if this is very flat about the optimum, then it is not necessary to obtain the parameters to high accuracy. On the other hand, if it is required to obtain the parameters accurately, and if the function is very flat, then it will be very difficult to achieve your objective. One of the ways round this perhaps is to examine the solution and see if it is more or less what you want, although I would not recommend this in general.

WOLFE. I suppose for unconstrained problems we think of getting the gradient below a certain tolerance. Is this good enough?

GREENSTADT. You might have a low value of the gradient over a certain region such that the tolerance was satisfied and yet the solution was not optimum.

BROYDEN. I disagree slightly with that last remark. There is a mapping about the optimum of the gradient on to the error, and I would have thought that with some knowledge of bounds on the inverse Hessian, one could easily have obtained a reasonable bound on the error in the solution, given a bound on the gradient.

McCANN. When minimizing penalty functions it is very difficult to reduce the gradient to small values and yet it is possible to obtain the parameters very accurately. When the value of the parameter r becomes sufficiently small, then the penalty function is almost discontinuous and it is almost impossible to achieve small gradients. In this case if a criterion of convergence were based on the gradient, one would have to vary this criterion with the parameter r.

BROYDEN. This problem is well-conditioned because the minimum can be located very accurately; because the curvature is large, the gradient maps on to a very small error. This is the converse of the case which Dr Greenstadt mentioned in which there is a very flat optimum; then a very small gradient maps on to a large error. Hence for an adequate criterion you must have bounds on the inverse Hessian, and in many methods you have this sort of information in the iteration matrix H.

BEALE. The context of the problem is something which one should also bear in mind. Going back to this nonlinear regression problem, as soon as the changes in the value of the objective function in an iteration are small compared with the residual mean square, then it is very probable that the parameters at this point are as good from the point of view of the fit as any that you can obtain, although the iteration might well continue for some time.

DAVIES. We have built a criterion into our programs very similar to that described by Powell.

POWELL. I think we must bear in mind that it is a fact of life that these methods do tend to get "stuck" away from the solution, and then start again. It is a situation fraught with difficulty.

4. Could the panel say which descent algorithms are invariant with respect to: (i) Change of scale of the variables, (ii) General linear transformations of the variable space?

ABADIE. Davidon's method satisfies both (i) and (ii); in fact any method which uses the Hessian is invariant under these transformations. I think this is a significant point in favour of Davidon's method.

BARD. For this to be true a change must be made to the Hessian matrix when the scaling of the variables is changed. It is not true if you stick to an initial unit matrix all the time.

DAVIDON. If the matrix is transformed appropriately, then the method is invariant under all affine transformations on the variables.

WOLFE. How about the vector methods, for instance Fletcher and Reeves?

ABADIE. No. Possibly Powell's method without derivatives?

POWELL. Yes, except that you start going along coordinate directions. One would have to start going along other directions, after which this property would apply.

FLETCHER. I believe that the projection matrix methods of which Pearson was talking would be invariant if the appropriate transformations were made to the Projection matrix.

POWELL. I think that in a certain sense, algorithms that have this property are extremely bad methods. Why I say this is that when it appears to one of these algorithms that you have got bad scaling, so that you can move a long way in one direction with a slight change in the function and only a short way in the other then the conclusion that the method makes is that it should move along a direction which is nearly orthogonal to the steepest descent direction. If the conclusion is correct and there is this stretching out behaviour due to poor scaling then that is fine. What often happens is that these matrices tend to be singular by chance, and in these situations the behaviour of the methods in going almost at right angles to the steepest descent direction is quite the wrong behaviour. If the user could provide sensible scaling beforehand I would think that the person designing algorithms could produce even better methods than we have at present.

BROYDEN. This sounds like a confession of defeat for the algorist. I can see that this can happen of course but I feel that we must provide algorithms that require the minimum of intervention from the user.

GREENSTADT. I take exactly the opposite view on this. When you seem to have long ridges, then you do not want to go to the top of the ridge, but to proceed parallel to the top of the ridge toward the ultimate optimum. When you have proceeded far enough, so that the component of the gradient in that direction gets small enough, you will then turn around and go to the top of the ridge.

POWELL. But it is a fact that when these sort of algorithms are used on least squares problems where you are trying to fit a curve by parameters one of which is a constant, these algorithms will not even shift the approximation up or down so that it passes approximately through the curve corresponding to adjusting this constant term. This is because they are looking at a bad scaling matrix and they tend to change even this constant parameter very slowly. Things like this look very bad to the computer user, but they happen.

BEALE. From the point of view specifically of sums of square problems, most of my experience has been with pseudo-Newton methods which are scale invariant. If you then put in Levenberg's idea of adding something to the diagonal, this is no longer true. When we first started working on this, we simply added the Levenberg parameter into the diagonal. In practice this works fairly badly, and it works better if instead of adding the λ in, you multiply the diagonal elements by $1+\lambda$. This means that we have a method which is invariant under change of scale of the variables but is not so under general linear transformations. I don't want to make any general philosophical remarks about the relative metits of these, I only say that the method works fairly well.

5. Is any comparison available for one-dimensional interpolation methods (for example, cubic)?

DAVIES. Such a comparison appears in I.C.I. Monograph No. 5 mentioned earlier. We have also an internal report in which we evaluated several of these methods. We did not try cubic interpolation, but we concluded that the best method was when we tried quadratic interpolation techniques after doubling up along the line to bracket the minimum. We found this better than methods which extrapolate with a quadratic.

BEALE. You must put in a stopping rule which sometimes gets overlooked. It is possible to get better and better estimates which tend asymptotically to a finite limit at infinity, and which is not the optimum.

6. What suggestions have the panel for searching for global optima? To what extent is Bender's procedure successful and are there any other methods?

GREENSTADT. Some work was done by Spielberg at the IBM Scientific Center but he has now moved to Paris, I think to the Paris Scientific Center. He has tried different variants of Benders' procedure, although I do not know whether his results were favourable or not.

WOLFE. I take it this is for integer/linear programming type problems? What about the nonlinear problem, even without constraints.

BEALE. If your objective function can be reduced to separable form, that is to nonlinear functions of single arguments, and if you are prepared to make piecewise linear approximations to these functions, then the problem is reduced to one in integer linear programming. We have solved some large problems of this type, much larger than Spielberg has been writing about, using branch and bound techniques. In order to find the global optimum of a non-convex problem in general you have to know such a lot about the structure of the problem. If the problem is not separable I think it will be very difficult to get a handle on what the problem is, before making any guaranteed statements.

ABADIE. I would like to know how we recognize a problem as convex or not. In my experience we receive problems with very complicated functions—maybe they are convex, maybe not; we do not even know.

GREENSTADT. I think people have been spoiled by linear programming. In linear programming you know the problem is convex, but when you go further you know nothing. People are expecting the same neatness and niceness in nonlinear programming as in the linear case.

FLETCHER. We had a try at Leeds at a procedure for getting from local optimum to a global optimum. When solving n nonlinear equations in n unknowns you can recognize a minimum as being local by looking at the sum of squares of residuals in relation to the gradient. At such a local minimum one might look at the third derivatives of the function at the minimum, say along the eigenvectors of the Hessian, and pick a direction in which the second derivative is reducing most rapidly. Then one would search along that direction hoping to go first to a maximum of the function and then downhill again. At this point the optimization routine would be entered again with hope that convergence would take place to a global minimum, and hence to a solution of the equations. It is certainly not guaranteed to work, although it did in the one case in which it was tried.

ANON. One other obvious point is to choose various starting points at random and select the best solution of those obtained from each starting point.

BEALE. There are some cutting plane methods which have been described briefly in the literature, but I do not know to what extent they have been used in practice. There is one published in Russia for maximizing a convex function subject to linear constraints. (Hoang Tui, "Concave Programming under Linear Constraints" *Dokl. Acad. Nauk. SSSR*. A.M.S. Translation, 5, 1437, 1964). All local optima are at vertices and additional linear inequalities can be imposed, having found a local optimum; therefore it is known that there cannot be a better local optimum unless you go to the other side of the constraint which cuts off your current trial solution. Candler and Townsley have done the same thing (*Management Science*, 10, 515, 1964) with a quadratic objective function and linear constraints. I do not know to what extent anyone has used this.

GREENSTADT. Could these problems not be formulated as integer linear programming problems?

BEALE. Cutting planes are a method of tackling integer programming problems, and so far experience suggests that these are not so successful as other types of method.

ABADIE. When maximizing a convex function subject to linear inequalities do you not have to examine all the vertices of the polyhedron?

BEALE. No. You can see this from a geometrical argument which extends to many variables.

MITRA. I have tried the method mentioned here on problems of two variables by hand, and a small five or six variable problem. The method tends to converge but the only thing is that the cutting planes tend to grow.

BEALE. The paper does define a method for finding a global solution. It is a cutting plane method but I don't think that there is any finiteness proof: you might have to generate an infinite number of cuts to find the solution. This is the trouble with cutting plane methods.

7. How does dynamic programming fit into the framework of nonlinear optimization techniques? If it suffers from the handicap of too large a number of function evaluations, is it not fast because it does away with repetitive matrix algebra? Are decomposition techniques related to dynamic programming?

WOLFE. I can answer the third part—No. Perhaps we should also ask to what type of problem it is best applicable. Are they problems of the type we are discussing?

GREENSTADT. In a general way, dynamic programming can be applied to control problems, and insofar as people are trying to do control problems by the methods we have discussed here, then there is some connection.

BEALE. I have been involved in using dynamic programming to some extent. There was one particular problem which we set out to solve using our mixed integer programming package, when the client said that there was an economy drive on and he could only do work on his own computer, so could I suggest some way which could be solved on his computer. We reformulated the problem as one of dynamic programming and it gives answers in a reasonably short space of time. The trouble basically is that dynamic programming is such a fragile technique. If you can formulate the problem in such a way that you have not more than four state variables, then if other things are favourable, you can get global optima for non-convex problems. However, the addition of one state variable increases the size of the problem by enormously more than the addition of one integer variable (or even half a dozen) to an integer programming problem (although this would be bad enough).

8. To what extent should an algorithm be tested before publication?

WOLFE. I asked Davies after his presentation what he meant by "reliable" and he pointed out then that an algorithm ought to solve the problem most of the time. I very often want to know what the author of an algorithm means by "satisfactory". I have had a couple of experiences in refereeing papers, one of which illustrates this. I was presented with an algorithm; I didn't think it was a very good algorithm and it didn't seem to teach me anything. Surely, it gave solutions to the problem, but I pointed out that one could do this by programming the computer to try all possible values of all the variables. Thus I asked the author for some statement as to why he should present this algorithm rather than some of its competitors, and to say something about its performance. The revised version of the paper said that he had programmed the algorithm and it had performed "very satisfactorily". I rejected the paper.

I think it may be too demanding in an area as new as this to expect a complete comparison with all other methods. I feel personally that the work is incomplete if an author has not been able to make any evaluation except to feel that it does not do too badly. Although the principle of *caveat emptor* does apply to some extent, the problem is parallel in many ways to that of how much a drug should be tested before it is put on the market. We are offering these algorithms as solutions to problems. We have certainly had experience of algorithms which do not solve problems and I personally would like to see things run more tightly where possible.

GREENSTADT. You may have put things a bit strongly here. I know of no case where misuse of a program has actually killed somebody. There may have been many wishes in that direction though!

WOLFE. I always remember a statement of Householder's in which he said that he was very nervous whenever he boarded a plane because he knew it had been designed with floating-point arithmetic.

BEALE. I think the answer to this does depend on the algorithm, and some algorithms give insight into some aspects of a problem even if you never use them. For example, take the algorithm that Powell presented; even if that had not been tested at all, I would have thought this was well-worth presenting if the time had come for the conference and the programs were not working. It is a fairly simple idea and it does really throw some light on how penalty functions work. This is in contrast to an algorithm which is just a long series of steps which may not tell you much, so that the only justification of this algorithm would be that it worked. Some algorithms can be interesting whether or not they work.

BROYDEN. I think a lot depends upon the spirit in which the algorithm is presented. I agree with Beale's point that an algorithm can be a fine constructive proof, and even if no work has been done on it then it should be published. I think too that where the problem has not been solved satisfactorily by any other algorithm, publication of a new algorithm might be justified without much experimental work provided that this fact is stressed, and provided that the user is made aware that the algorithm might break down. The argument here is that it might help someone, sometime, in some place, and as such might be useful. I think however that if we are going to issue an algorithm as a standard one, we ought to provide convergence proofs, even if only for a limited class of functions, and to suggest under what circumstances the algorithm might be expected to break down.

DAVIES. Speaking as someone in industry, we get many papers which we have to look through to decide which methods we should be using. One of the first criteria which I adopt is that if they haven't shown me that it is going to work on any problems, then I am afraid it is just forgotten about. I think it is very well worth-while publishing some results.

BEASLEY. (English Electric Computers). The effort required to test algorithms is extremely large and it can be extremely boring. It may not teach you much and there may be a lot to be said for having a standard library of test problems which could be run off on any particular computer and which would be a very valuable asset.

WOLFE. Yes, it is a lot of work, but if a person is not in a position to do it then he probably does not have contact with a machine, and I think that if a person is not in contact with a computer he should not be inventing algorithms. (Loud cries of "hear! hear!" from audience. *ed.*) Research is a lot of work.

BEASLEY. But there is a difference in testing a method out on three problems and on two hundred.

BEALE. I think there is a lot to be said for testing at least some methods on only three problems of fifty or seventy variables, rather than on two hundred problems with no more than twelve variables. Using a standard library you are likely to be involved only in fairly small problems.

GREENSTADT. These small problems seem capable of breaking many methods: small problems are not easy to solve. A possible compromise might be to use these well-known problems, both for the constrained and unconstrained cases, as a standard battery of tests. If the algorithm breaks down on 30 per cent of these cases then it ought to be obvious even to the inventor that it is not very good.

BEALE. I am not altogether sure about that. An algorithm that deals with a large easy problem may be quite useful even if it doesn't deal with a small difficult one.

ANON. I think we may have overlooked here that the numerical comparison of algorithms might depend critically on how the algorithm is programmed. We have only to look at the question of stopping criterion mentioned earlier and the choice of starting point, and possibly quite trivial details of how you compare two numbers. A slight change in the algorithm might make a two to one difference in the number of function evaluations. When one reads comparisons of methods *A, B* and *C* one should excercise considerable caution in interpreting the results. Programming details are extremely important.

9. Would the panel speculate on the developments which might take place in optimization techniques over the next year or so? In particular what are the chances of a do-it-yourself set of packages becoming available? Will there ever be a nonlinear programming code in the sense that there are perfectly general and reliable LP codes?

BEALE. I am sure the answer to the last question is No—because nonlinear programming is too wide a field to be covered effectively by a single code. Nearly linear problems want to be solved by a method based on linear programming ideas, other problems which do not have this special structure will obviously want quite different methods.

POWELL. Since you ask for an opinion, I would like to make two forecasts as to the development of algorithms. I think that people are going to stop searching for minima along lines, and I think they are going to pay more attention to steepest descent methods.

BROYDEN. I will speculate that the variable variance methods of the Davidon type are going to come into far greater prominence than they are even now.

GREENSTADT. I agree with everyone else. One comment about this nonlinear programming code in which I agree with Dr Beale is that to refer to a perfectly general LP code is like referring to a perfectly general set of numbers from one to ten. The linear case is so special as compared with the infinite complexity and generality of nonlinear cases, and anything which is done in the linear case just doesn't apply in the nonlinear.

DAVIES. I am not quite sure what is meant by a do-it-yourself set of packages. If this is something for a user just to plug into, then this is what we have already got. I think an algorithm should be able to handle linear constraints as such, in particular those constraints which are simple bounds. I am in agreement with Beale that we are not likely to get a perfectly general code, certainly not in the next few years.

ABADIE. A battery of codes would be good: we take a problem and shoot at it with each method in succession until at least two have given the same result.

BEALE. I would hope that more effort will be devoted to the specific problem of incorporating simple upper and lower bounds on the values of variables in otherwise unconstrained problems. These are very important in themselves and we can use the ideas of Wolfe's reduced gradient method to reduce other constraints to that form as they arrive. I would like to see effort devoted to this special case of linear constraints, rather than more general linear constraints.

WOLFE. I want to thank the panel for their contributions and also all those that have spoken up from the floor. Speaking for myself may I say what great pleasure I have obtained from attending this conference which has presented so much exciting new material.

Author Index

The numbers in italics refer to the Bibliography pages where the references are listed in full.

A

Abadie, J., 22 (7), *33*, 46 (12), *47*, 66 (1), *83*, 187 (1), *200*
Akeroyd, A. J., 32 (20), *34*
Ang, A., 127 (19), *150*
Aris, R., 167 (14), *170*
Arrow, K. J., 190 (5), *200*, 328 (2), *337*

B

Barnes, J. G. P., 9 (20), *11*, 219 (5), *246*
Beale, E. M. L., 22 (5, 6), 29 (6), 30 (5), 31 (5, 14, 15, 17), 33 (5), *33*, *34*, 106 (24), *113*
Bellman, R., 151 (2), *169*
Berge, C., 109 (31), *114*
Best, G. C., 127 (31), *150*
Boltyanskii, V. G., 151 (1), *169*
Box, G. E. P., 21 (2), *33*
Box, M. J., 1 (1), 2 (3), *11*, 28 (10), 32 (10), *34*, 188 (2, 3), 193 (7), 196 (3), *200*, 259 (3), 266 (3), 267 (3), *268*
Breakwell, J. V., 153 (5, 6), 154 (5), 168 (5), *169*
Broisc, P., 87 (3), *97*
Brooks, S. H., 106 (25), *113*
Brown, D. M., 127 (19), *150*
Broyden, C. G., 9 (21), *11*, 15 (2), 20 (2), *20*, 220 (6), 234 (6), *246*, 251 (6), *258*
Bryson, A. E., 153 (5), 154 (5, 8), 155 (12), 168 (5, 8), *169*, *170*
Buehler, R. J., 5 (7), *11*, 104 (21), *113*

C

Carathéodory, C., 223 (8), *246*
Carpentier, J., 46 (12), *47*, 187 (1), *200*

Carroll

Carroll, C. W., 196 (14), *201*, 203 (1), *210*, 204 (5), *207*
Cauchy, A., 13 (1), *20*, 309 (6) *323*
Cavallaro, L., 116 (1), *149*
Chernoff, H., 127 (23), *150*, 167 (15), 168 (15), *170*
Colville, A. R., 32 (19), *34*, 46 (11), *47*, 206 (6), 208 (6), *213*
Cooper, L., 195 (13), *201*
Cottle, R. W., 271 (3), 272 (5), *272*
Courant, R., 299 (1), *306*
Crockett, J. B., 127 (23), *150*, 167 (15), 168 (15), *170*

D

Dantzig, G. B., 21 (1, 3), 24 (1), 26 (9), 29 (3, 12), *33*, *34*, 69 (3), *83*, 87 (1), *97*, 99 (1), 100 (6, 7), 103 (17), 112 (17), *112*, *113*, 271 (3), *272*
Davidon, W. C., 5 (9), *11*, 16 (3, 4), 17 (4), 19 (4), *20*, 105 (23), 109 (23), *113*, 168 (16), *170*, 215 (3), *246*, 273 (1), 278 (4), *280*, 283 (1), *297*, 312 (8), *323*
Davies, D., 1 (1), *11*, 188 (3), 196 (3), 199 (22), *200*, *201*
Davies, O. L., 264 (6), *268*
de Angelis, V., 31 (16), *34*, 66 (2), 74 (4), 79 (4), *83*
Deist, F. H., 10 (26), *12*
Denham, W. F., 155 (12), *170*
Denn, M. M., 167 (14), *170*
de Silva, B. M. E., 118 (3), 129 (32, 33), *149*, *150*
d'Esopo, D. A., 45 (4), *47*
Dorn, W. S., 252 (7), *258*
Dreyfus, S. E., 151 (2), *169*
D'Sylva, E., 122 (13), *149*

349

Subject Index

(p indicates that the subject is referred to in the Preface.)

A

Alternating directions method, 7
Approximation programming (MAP), p, 22, 31, 35, 100

B

Barrier functions, 247
Behaviour variables, 115 ff
Black box, 116, 124
Boundary region, 192
Branch and bound, 343

C

Complex method, 340
Conjugate directions, 3, 4, 7,
Conjugate gradients method, p, 5, 46, 47, 103, 170, 200, 308
Control variable, 151
Covariance, 259
Cutting plane method, p, 35, 101, 105, 127, 255, 344

D

Davidon–Fletcher–Powell method, 12, 168, 215, 273, 278, 283, 296, 308, 312, 340
Decomposition, p, 29, 85 ff., 345
Design variables, 115 ff.
Difference approximations, 8, 10, 98
Direct search methods, 7, 259
DSC method, 7, 197, 340
Duality, p, 37, 43
Dynamic programming, 344

E

Epsilon-centre, 51
Evolutionary operation, 259

G

Generalized (pseudo) inverse, 9, 205, 303
Generalized least squares method, 9,10, 260
Generalized reduced gradient method, 37 ff.
Generalized upper bounds, 99
Geometrical constraint, 115 ff.
Global optimum, 29, 47, 67, 103, 276, 308, 343
Gradient projection method, p, 127, 140, 215

H

Hill climbing, p, 21, 171

I

Integer programming, 30, 35, 343

K

Knapsack problem, 35
Kuhn–Tucker conditions, 190, 230, 328

L

Lagrange multipliers, p, 152, 204, 230, 299, 329
Lagrangian function, 127, 190, 247, 327 ff.
Linear programming, p. 21 ff., 45, 63, 344, 347
Linear (univariate-) search, 1, 9, 12, 17, 210 ff., 253, 324, 343
Local minimum, 1, 29, 67, 126, 237, 308

DATE DUE